Thirteen Papers in Algebra, Functional Analysis, Topology, and Probability, Translated from the Russian

American Mathematical Society

TRANSLATIONS

Series 2 • Volume 149

Thirteen Papers in Algebra, Functional Analysis, Topology, and Probability, Translated from the Russian

by

V. S. Afraĭmovich
Ya. G. Berkovich
V. Z. Grines
L. G. Khanin
I. Ya. Novikov
B. M. Pogrebinskiĭ
V. V. Sharko
V. P. Shchedrik
L. P. Shil′nikov
P. Skachkova
V. A. Tolokonnikov
N. A. Vavilov
B. I. Zil′ber

American Mathematical Society
Providence, Rhode Island

Translation edited by SIMEON IVANOV

1991 *Mathematics Subject Classification.* Primary 15A, 20B, 20D, 20E, 30C, 30H, 46J, 57R, 58F; Secondary 18G, 19B, 57Q, 46A, 46B.

Library of Congress Cataloging-in-Publication Data
Thirteen papers in algebra, functional analysis, topology, and probability, translated from the Russian/by V. S. Afraĭmovich...
[et al.; translation edited by Simeon Ivanov].
p. cm. – (American Mathematical Society translations, ISSN 0065-9290; ser. 2, v. 149)
Includes bibliographical references.
ISBN 0-8218-3137-2 (alk. paper)
1. Mathematics. I. Afraĭmovich, V. S. (Valentin Senderovich) II. Ivanov, Simeon. III. Title: 13 papers in algebra, functional analysis, topology, and probability, translated from the Russian. IV. Series.
QA3.A572 ser. 2, vol. 149
[QA7]
510 s–dc20 91-24552
[510] CIP

The paper used in this book is acid-free and falls within the guidelines
established to ensure permanence and durability. ♾

This publication was typeset using AMS-TEX,
the American Mathematical Society's TEX macro system.

10 9 8 7 6 5 4 3 2 1 96 95 94 93 92 91

Contents

Russian Contents*

Б. И. Зильбер, Группы и кольца, теория которых категорична, Fund. Math. 95 (1977), № 3, 173–188

В. П. Щедрик, К выделению действительного линейного унитального множителя из матричного многочлена, Мат. методы и физ.-мех. поля 1986, № 24, 7–10

Я. Г. Беркович, Подгруппы симметрических и знакопеременных групп, Изв. Сев.-Кавказ. науч. центра. высш. школы. Естеств. н. 1981, № 1, 5–9

Я. Г. Беркович, Обобщение второй теоремы Бернсайда о $[p,q]$-группах, Изв. Сев.-Кавказ. науч. центра высш. школы. Естеств. н. 1985, № 4, 18–22

Я. Г. Беркович и Б. М. Погребинский, Конечные группы большой степени, Изв. Сев.-Кавказ. науч. центра высш. школы. Естеств. н. 1986, № 1, 21–26

Н. А. Вавилов, Максимальные подгруппы гпупп Шевалле, содержашие максимальный расщепимый тор, Кольца и модули. Предел. Теоремы теории вероятностей. вып. 1, Ленинград, 1986, 67–75

В. А. Толоконников, Теорема о короне в алгебрах ограниченных аналитических функций, Ин-т аналит. приборостр. научн.-тех. об-ния АН СССР, Ленинград, 1983

Л. Г. Ханин, Структура замкнутых идеалов в некоторых алгебрах гладких функций, Исслед. по теории функций мног. веществ. переменных. Ярославль, 1986, 84–92

*The American Mathematical Society scheme for transliteration of Cyrillic may be found at the end of index issues of *Mathematical Reviews*.

В. В. Шарко, K-теория и теория Морса. I, Ин-т мат. АН УССР. Препр., 1986, № 39

В. В. Шарко, K-теория и теория Морса. II, Ин-т мат. АН УССР. Препр., 1986, № 40

В. З. Гринес, Диффеоморфизмы двумерных многообразий с транзитивными слоениями, Методы качеств. теории дифференц. уравнений, Горкий, 1982, 62–72

В. С. Афраймович и Л. П. Шильников, Инвариантные двумерные торы, их разрушение и стохастичность, Методы качеств. теории дифференц. уравнений, Горкий, 1983, 3–20

И. Я. Новиков и О. П. Скачкова, Устойчивые случайные величины и вылуклость банаховых функциональных пространств, Исслед. по теории функций мног. веществ. переменных. Ярославль, 1986, 84–92

Amer. Math. Soc. Transl.
(2) Vol. **149**, 1991

Groups and Rings Whose Theory is Categorical

B. I. ZIL'BER

Abstract. We investigate groups that are interpretable in models of $\aleph_1$-categorical theories. As an application of the results obtained, we answer a question of Macintyre: every skew field whose theory is $\aleph_1$-categorical is a commutative field.

Bibliography: 17 titles.

§0. Introduction

Let A be a model of an $\aleph_1$-categorical theory T and let G be a structure that is first-order interpretable in the model A. In this situation we say that G is weakly categorical. The aim of this paper is to investigate weakly categorical groups, and to apply the results to weakly categorical rings. In particular we prove the following, which answers a question of A. Macintyre [13].

THEOREM 6.1. *A weakly categorical skew field is commutative, i.e. is a field.*

It is easy to see that any algebraic group over an algebraically closed field is weakly categorical, since the theory of an algebraically closed field is $\aleph_1$-categorical. This example points to an analogy between algebraic and weakly categorical groups. It was by exploring this analogy that we obtained the basic (technical) theorems given in this paper, on weakly categorical groups. We make essential use of the model-theoretic technique of (Morley) rank, as developed by the author in [4], [5]. It is interesting to note that the rank of a definable subset is a good analogue of the dimension of an algebraic variety in algebraic geometry.

Among other results we note:

THEOREM 5.2. *The theory of a simple algebraic group over an algebraically closed field is $\aleph_1$-categorical.*

1991 *Mathematics Subject Classification*. Primary 30C35; Secondary 30C60.
Translation of Fund. Math. **95** (1977), 173–188; MR **56** #118.

We note that this result is stated in the spirit of traditional model theory, but it seems to be impossible to obtain it by classical methods.

S. A. Chikhachëv [8] shows that a question of M. A. Taĭtslin on the classification of $\aleph_1$-categorical universal theories of semigroups reduces to the question that we answer in the following theorem.

THEOREM 4.2. *If G is a group whose universal theory is $\aleph_1$-categorical, then G is abelian.*

We conclude by stating a number of questions connected with weakly categorical groups.

The basic results of this paper were obtained by the author in 1972, and a brief report on these results was published in [6].

The author is happy to express his gratitude to M. A. Taĭtslin for showing him such an interesting subject for investigation, and to O. V. Belegradek for a number of valuable pieces of advice, and for his active interest in the present work.

We fix the following notation. T is a complete theory of countable signature Ω. A (possibly with subscripts) is a model of the theory T. $\mathrm{St}(A)$ is the Stone space over A, and $\mathrm{St}^\alpha(A)$ is the subspace of points of rank $\geq \alpha$. If the type p in $\mathrm{St}(A)$ has rank α (i.e., if $p \in \mathrm{St}^\alpha(A)\backslash\mathrm{St}^{\alpha+1}(A)$) then we write $R(p) = \alpha$.

We do not distinguish between elements of A and the constant symbols which denote them. $\Omega(A)$ is the signature we get by adding constant symbols for elements of A to the signature Ω. $F_n(A)$ is the set of formulas of signature $\Omega(A)$ with n free variables. We do not distinguish formulas in $F_1(A)$ from the subsets of A that they define, and thus $F_1(A)$ can be thought of as the Boolean algebra of those subsets of the model A (or of extensions of A) that are definable in $\Omega(A)$.

The rank of a definable subset Φ in $F_1(A)$ is defined in the natural way from the ranks of types which contain Φ, and is written $R(\Phi)$; we put $R(\Phi) = \max\{R(p) | p \in \mathrm{St}(A), \Phi \in p\}$. We always use the letter v (possibly with subscripts) to represent a free variable. Sometimes we shall write $R(\Phi(v))$ to indicate the variable that is considered free in the case in question. In the text we use logical connectives $\rightarrow$, $\&$, $\vee$, together with $\subseteq$, $\cap$, $\cup$ as notation for the set-theoretical operations.

Our group-theoretical notation is standard. We use the letter e to denote the identity of a group; Φ^g means the conjugate of the set Φ by the element g.

§1. Preliminary results. Ranks of definable subsets

We assume the reader is familiar with the theory of $\aleph_1$-categoricity [11], [14]. We introduce without proof some results on the rank of definable subsets, which are contained in the author's papers [4] and [5].

The following proposition is contained in [16] and was independently noted by the author in [5].

PROPOSITION P.1. *If* A *is* $\aleph_0$*-saturated, then* $\mathrm{St}^{\alpha+1}(A)$ *coincides with the set of limit points of* $\mathrm{St}^{\alpha}(A)$.

Baldwin [9] and a little later the author [4], [5] proved:

THEOREM P.2. *Suppose* T *is an* $\aleph_1$*-categorical theory. Then* $R(T)$ *is finite; in other words,* $R(\Phi)$ *is finite for any* $\Phi \in F_1(A)$.

The formula $\Psi(v_0, v_1)$ in $F_2(A)$ is called a *layering* of the formula $\Phi(v_1)$ along the formula $\Delta(v_0)$ if:

(a) $A \vDash \Psi(v_0, v_1) \rightarrow \Delta(v_0)$,
(b) $A \vDash \exists v_0 \Psi(v_0, v_1) \leftrightarrow \Phi(v_1)$.

If moreover $R(\Psi(a, v_1)) < \alpha$ for any $a \in \Delta(A)$ then we say Ψ is a *layering of rank* $< \alpha$ (or likewise $\leq \alpha$ or $= \alpha$).

LEMMA P.3 ([4], [5]). *Suppose* T *is an* $\aleph_1$*-categorical theory and* A *is a model of* T; $\Phi, \Delta \in F_1(A)$; $R(\Phi) = \alpha > 0$ *and* Δ *is a strongly minimal subset of* A. *Then there exists a layering* $\Psi(v_0, v_1)$ *of* rank $< \alpha$ *of the definable subset* $\Phi(v_1)$ *along* $\Delta(v_0)$.

(In fact the converse is also true, i.e., the existence of a layering of rank $< R(\Phi)$ for all formulas Φ with $R(\Phi) > 0$ implies $\aleph_1$-categoricity for a totally transcendental theory T.)

The two theorems stated below are our basic technical tool.

THEOREM P.4 ([4], [5]). *Let* T *be an* $\aleph_1$*-categorical theory and* A *a model of* T. *Let* $\Phi, \Delta \in F_1(A)$. *If* $\Psi(v_0, v_1)$ *is a layering of rank* $\leq n$ *of the definable subset* Φ *along* Δ, *then*

$$R(\Phi) \leq n + R(\Delta).$$

THEOREM P.5 ([4], [5]). *Let* T *be an* $\aleph_1$*-categorical theory,* $\Psi(v_0, v_1, \ldots, v_k)$ *a formula of signature* Ω *and* m *a natural number. Then there exists a formula* $\Theta \in F_k(\varnothing)$ *such that for any model* A *of* T *and any choice of* $a_1, \ldots, a_k \in A$,

$$R(\Psi(v_0, a_1, \ldots, a_k)) \leq m \Leftrightarrow A \vDash \Theta(a_1, \ldots, a_k).$$

REMARK. We note that many of the theorems quoted above are analogous to the theorems of Shelah in [17] for ϕ-rank.

§2. Weakly categorical structures. Groups

None of the signatures to be considered will contain function symbols.

We shall say that a structure K of signature Ω_K is *interpreted in a theory* T (or in a model A of the theory T) *of signature* Ω if

1. The signature Ω of the theory T contains all the symbols of the signature Ω_K.

2. There exists a one-variable formula of the signature Ω, which we denote by the same symbol as the interpreted structure, i.e. K, such that the substructure $K(A)$ of A (understood as a structure of signature Ω_K) is isomorphic to the structure K.

We remark that in the notation we have just used, i.e. $K = K(A)$, the symbol on the left-hand side denotes a structure and on the right-hand side it denotes a formula which marks off a substructure. In what follows, except where we explicitly say otherwise, it will be convenient not to distinguish between these two uses of the letter K (or respectively G, Q, etc.). If the theory of an infinite structure K of signature Ω_K is $\aleph_1$-categorical, we shall express this briefly as follows: K is a *categorical structure.*

If a structure K is interpreted in a model A of an $\aleph_1$-categorical theory T, then we shall say that K is a *weakly categorical structure.*

A categorical structure is always understood to be infinite, but at the same time we shall allow a weakly categorical structure to be finite.

Let K be a structure which is interpreted in a model A of a theory T. Let K_0 be a one-variable formula in the signature $\Omega(A)$, such that $K_0(A)$ is a substructure (with respect to Ω_K) of the structure K. We shall say that $K_0(A)$ is a *definable substructure of* K (respectively, *definable subgroup, definable subring*, etc.).

Clearly if K is a weakly categorical structure and K_0 is a definable substructure of K, then K_0 is also weakly categorical.

We remark that if A_1 and A_2 are two models of the theory T, then $K(A_1)$ is elementarily equivalent to $K(A_2)$ in the signature Ω_K, since T is a complete theory. We shall write T_K for the theory of the structure $K(A)$ in the signature Ω_K.

REMARK 1. If A is an $\aleph_0$-saturated model of T, then $K(A)$ in the signature Ω_K is an $\aleph_0$-saturated model of T_K. In fact, to verify that $K(A)$ is $\aleph_0$-saturated it suffices to check that $K(A)$ realizes every set of formulas $\{\Phi_i(v_0) | i \in I,\ \Phi_i \in F_1(C)\}$ which is finitely satisfied in $K(A)$, where C is any finite subset of $K(A)$. But $\{\Phi_i(v_0) | i \in I\}$ is satisfied in $K(A)$ since A is $\aleph_0$-saturated and $K(A)$ is a definable subset of A.

If K is a weakly categorical structure, then T_K has many of the properties of an $\aleph_1$-categorical theory, in particular:

LEMMA 2. *Let T be an $\aleph_1$-categorical theory in which a structure K is interpreted, and let T_K be the theory of K in the signature Ω_K. Then*

(a) *T_K is a totally transcendental theory*;
(b) *T_K does not have the finite cover property (f.c.p.).*

PROOF. (a) We check that if A is a countable $\aleph_0$-saturated model of T then $\mathrm{St}(K)$ is countable, where $K = K(A)$ and $\mathrm{St}(K)$ is the Stone space of the model K in signature Ω_K. By Remark 1, K is $\aleph_0$-saturated and hence it is an $\aleph_0$-universal model of T_K, and therefore a proof that $\mathrm{St}(K)$

is countable will suffice to show that T_K is totally transcendental.

For a type $p \in \mathrm{St}(K)$ we write p^K for the type (not necessarily complete) of signature $\Omega(A)$, which we get by relativizing all formulas of p to K. It is clear that p_1^K agrees with p_2^K in A if and only if $p_1 = p_2$ (that is, $p_1^K = p_2^K$). Since T is totally transcendental, the set $\{p^K | p \in \mathrm{St}(K)\}$ is countable, and hence $\mathrm{St}(K)$ is countable. This proves (a).

(b) Assume the contrary, namely that the theory T_K, and consequently some formula $\Phi(v_0, v_1, \dots, v_n)$ of signature Ω_K, has the f.c.p. Then the relativization of Φ to K has the f.c.p. in T. This contradicts the $\aleph_1$-categoricity of T. The lemma is proved.

We emphasize that when a weakly categorical structure K is given, this presupposes that we are given T, A, and a formula $K(v)$. Then $R(K)$ denotes $R(K(v))$ in a model A of the theory T. In this situation, if A is a model of T we write

$$\mathrm{St}_K(A) = \{p \in \mathrm{St}(A) | K \in p\},$$
$$F_{1,K}(A) = \{\Phi \& K | \Phi \in F_1(A)\}.$$

REMARK 3. It is easy to see that $\mathrm{St}_K(A)$ is homeomorphic to the Stone space of the Boolean algebra $F_{1,K}(A)$. The homeomorphism takes each formula Φ in a type $p \in \mathrm{St}_K(A)$ to the formula $\Phi \& K$ in $F_{1,K}(A)$.

In particular, every automorphism of the Boolean algebra $F_{1,K}(A)$ induces a homeomorphism from $\mathrm{St}_K(A)$ to $\mathrm{St}_K(A)$.

We introduce some important examples of weakly categorical groups.

The first example is not explicit. It shows why we have to consider weakly categorical groups at the same time as weakly categorical rings.

EXAMPLE 1. Let K be a categorical (or weakly categorical) ring. Then (a) the additive group of the ring K is weakly categorical; (b) the multiplicative group of units of the ring K is weakly categorical.

EXAMPLE 2 (THEOREM OF MACINTYRE [12]). An abelian group G is categorical if and only if $G \simeq H(p^n) \dotplus B$ or

$$G \simeq B \dotplus \sum_{i \in I} C_{p_i}^{m_i} \dotplus Q$$

where $H(p^n)$ is a direct sum of cyclic groups of order p^n, B is a finite abelian group, C_p is a quasicyclic group of type p^∞, Q is 0 or a divisible torsion-free abelian group, p and p_i are primes, p_i is the ith prime, I is a subset of ω, and m_i are natural numbers.

EXAMPLE 3. Arbitrary algebraic groups [1] over an algebraically closed field are weakly categorical. In what follows we shall describe a way of interpreting a linear algebraic group over an algebraically closed field in a certain $\aleph_1$-categorical theory. The case of an abstract algebraic group requires a more cumbersome construction, which is a routine generalization of the linear case.

Thus, let polynomials $P_1(v_1, \dots, v_n), \dots, P_k(v_1, \dots, v_n)$ define an affine variety over an algebraically closed field K, and suppose that a rational correspondence

$$\phi(v_1, \dots, v_n, w_1, \dots, w_n): K^{2n} \to K^n$$

makes this variety into an algebraic group.

The signature Ω of the theory T in which we shall interpret the given algebraic group consists of two 1-place predicates K and G; three 3-place predicates S, Q, and O; constant symbols $a_1, \dots, a_m$ for all the coefficients of ϕ, $P_1, \dots, P_k$; and an $(n+1)$-place predicate f. We shall describe the theory T by means of its model A constructed as follows:

(a) $K(A) \cup G(A) = A$; $K(A) \cap G(A) = \varnothing$.

(b) $K(A)$, together with S and Q, forms a field isomorphic to K.

Here $S(x_1, x_2, x_3)$ is understood as $x_1 + x_2 = x_3$ and $Q(x_1, x_2, x_3)$ as $x_1 \cdot x_2 = x_3$. The constants $a_1, \dots, a_m$ are interpreted as the coefficients of ϕ, $P_1, \dots, P_k$. Thus the original variety is defined by the polynomials $P_1, \dots, P_k$ and the correspondence ϕ is definable in $K(A)$.

(c) The predicates $S(x_1, x_2, x_3)$ and $Q(x_1, x_2, x_3)$ are false if any of x_1, x_2, x_3 lie in $G(A)$. The predicate $O(x_1, x_2, x_3)$ is false if any of x_1, x_2, x_3 lie in $K(A)$.

(d) The predicate $O(v_1, v_2, v_3)$ defines a group operation on $G(A)$, and $O(x_1, x_2, x_3)$ is read as $x_1 \circ x_2 = x_3$.

The predicate f represents the graph of a partial function which we also denote by f and which has the following properties:

(1) the domain of f is the set of common zeros of the polynomials $P_1, \dots, P_k$;

(2) the codomain of f is $G(A)$;

(3) $f(x_1, \dots, x_n) = f(y_1, \dots, y_n) \Leftrightarrow x_1 = y_1, \dots, x_n = y_n$;

(4) $f(x_1, \dots, x_n) \circ f(y_1, \dots, y_n) = f(\phi(x_1, \dots, x_n, y_1, \dots, y_n))$.

Thus f "transports" to G the structure of an algebraic group.

Clearly the isomorphism type of any model A of the theory T is determined by the isomorphism type of the field $K(A)$ (together with the interpretation of the constants $a_1, \dots, a_m$). Since the theory of an algebraically closed field with distinguished constants is $\aleph_1$-categorical, the theory T is also $\aleph_1$-categorical and $\langle G, \circ \rangle$ is a weakly categorical group.

Among algebraic groups there are groups that are categorical in the original "group" structure. In particular, every simple algebraic group over an algebraically closed field is categorical (cf. Theorem 5.2).

On the other hand, there are algebraic groups that are not categorical in the signature of groups. The algebraic group $\mathrm{PSL}(n, K) \times \mathrm{PSL}(n, K)$ (where K is the field of complex numbers) is elementarily equivalent in the signature of groups to the group $\mathrm{PSL}(n, K) \times \mathrm{PSL}(n, K_0)$ (where K_0 is the field of algebraic numbers of characteristic 0). Both these groups have the cardinality of the continuum, but they are not isomorphic, since in the second case there is a countable normal subgroup, whereas in the first case every normal subgroup has the cardinality of the continuum.

§3. Definable subgroups

An element g of a group G can act on a definable subset Φ of G by left (or right) multiplication. We write $g\Phi$ for the result of this action, i.e.,

$$g\Phi = \{a \in G | \exists b\ (\Phi(b) \& a = gb)\}.$$

It is clear that left (or right) action by an element $g \in G$ defines an automorphism of the Boolean algebra $F_{1,G}(A)$. On the other hand, this automorphism induces a homeomorphism of the Stone space $\mathrm{St}_G(A)$, which is defined by the action of the element g on the types $p \in \mathrm{St}_G(A)$. We write gp for the image of p under this action of g, i.e., $gp = \{g\Phi | \Phi \in p\}$.

REMARK 4. Since the action of g in $G(A)$ gives a homeomorphism of $\mathrm{St}_G(A')$ for $A' \succeq A$, it follows as in Remark 3 and Proposition P.1 that $R(gp') = R(p')$ and $R(g\Phi') = R(\Phi')$ for any $p' \in \mathrm{St}_G(A')$, $\Phi' \in F_{1,G}(A')$, $g \in G(A)$, and hence $R(gp) = R(p)$, $R(g\Phi) = R(\Phi)$ for $p \in \mathrm{St}_G(A)$ and $\Phi \in F_{1,G}(A)$.

When p is a type in $\mathrm{St}_G(A)$ we put

$$\mathrm{Stab}(p) = \{g \in G(A) | gp = p\}.$$

Clearly $\mathrm{Stab}(p)$ is a subgroup of G. We call it the *stabilizer subgroup* of the type p.

THEOREM 3.1. *Suppose G is a weakly categorical group and $p \in \mathrm{St}_G(A)$. Then there is a formula $\Theta \in F_{1,G}(A)$ such that $\Phi(A) = \mathrm{Stab}(p)$.*

PROOF. Let $R(p) = m$. There exists $\Phi \in F_{1,G}(A)$ such that $\Phi \in p$ and p is the unique type of rank $\geq m$ that contains Φ. In particular, this implies that $R(\Phi) = m$ and if $gp \neq p$ then $g\Phi \in gp$ and $g\Phi \notin p$, since $g\Phi$ is also contained in a unique type of rank m. On the other hand $g\Phi \& \Phi \in p$ and $R(g\Phi \& \Phi) = m$ if $gp = p$. Thus

$$g \in \mathrm{Stab}(p) \Leftrightarrow R(g\Phi \& \Phi) = m.$$

Now we obtain the required formula Φ by Theorem P.5.

The idea of the following lemma is due to Macintyre [12], but we use a variant of the proof due to Yu. L. Ershov.

LEMMA 5. *Let $G_0 \geq G_1 \geq \cdots \geq G_n \geq \cdots$ be a chain of definable subgroups of a weakly categorical group G (where $G = G_0$). Then there exists n_0 such that $G_{n_0} = G_n$ for all $n \geq n_0$.*

PROOF. Suppose the opposite, i.e., that the sequence G_n is strictly decreasing in the model A of the theory T. Then for every n there exists $a_n \in A$ such that $a_n \in G_n(A) \backslash G_{n+1}(A)$. Now let η be any sequence of zeros and ones, and let $a^{\eta(i)}$ be a if $\eta(i) = 1$ and e if $\eta(i) = 0$.

Let the type p^η be $\{\Phi_n^\eta | n \in \omega\}$ where

$$\Phi_n^\eta = \left(\prod_{i\le n} a_i^{\eta(i)}\right) \cdot G_{n+1}.$$

It is easy to see that $\Phi_n^\eta \supseteq \Phi_{n+1}^\eta$, Φ_n^η is a coset of G_{n+1}, and $\Phi_n^{\eta_1}$ meets $\Phi_n^{\eta_2}$ only if $\eta_1(i) = \eta_2(i)$ for $i \in 0, 1, \dots, n$. Thus by varying η we obtain a continuum of distinct types p^η, contradicting the fact that T is totally transcendental.

COROLLARY 6. *Let $\{G_\alpha | \alpha \in I\}$ be a family of definable subgroups of the group G. Then there is a finite set of indices $\{\alpha_1, \dots, \alpha_k\} \subseteq I$ such that $\bigcap_{\alpha\in I} G_\alpha = G_{\alpha_1} \cap \dots \cap G_{\alpha_k}$.*

By the *closure* of a subgroup Q of G we mean the least definable subgroup $\overline{Q}$ such that $G \ge \overline{Q} \ge Q$. By Corollary 6 every subgroup Q has a closure. Let $C(Q)$ be $\{g \in G: \forall a \in Q, ga = ag\}$.

LEMMA 7. *If Q is an abelian subgroup of G then its closure $\overline{Q}$ is also abelian.*

PROOF. First, note that $C(Q) \in F_{1,G}(A)$. In fact, it is clear that $C(a) \in F_{1,G}(A)$ for any $a \in G(A)$. But the infinite intersection $\bigcap_{a\in Q} C(a) = C(Q)$ is equal to some finite intersection (by Corollary 6), i.e., $C(Q) = C(a_1) \cap \dots \cap C(a_k)$. Moreover, $C(Q) \supseteq Q$ and so $C(Q) \supseteq \overline{Q}$.

Now, let $x, y \in \overline{Q}$. Then $x \in C(Q)$, which means $C(x) \supseteq Q$, i.e. $C(x) \supseteq \overline{Q}$, and this implies $y \in C(x)$. The lemma is proved.

REMARK. It is not hard to show that the closure of a nilpotent (respectively solvable) group will be a nilpotent (respectively solvable) group of the same class.

For any definable subgroup Q we shall say that the definable subset Φ of the group G is *Q-decomposable* if $\Phi \subseteq a_1Q \cup \dots \cup a_kQ$ for some $a_1, \dots, a_k \in G$ and $k > 1$, but $\Phi \not\subseteq a_iQ$ for all i. If Φ is not Q-decomposable for any Q, we shall say that Φ is *indecomposable.* In the case where Φ is a subgroup we shall say *connected* instead of indecomposable.

It is clear that a weakly categorical group G is connected if and only if G does not contain a proper definable subgroup of finite index.

REMARK 8. Since the intersection of a finite number of subgroups of finite index is again a subgroup of finite index, any weakly categorical group contains a normal connected definable subgroup G^0 of finite index, which is defined by $G^0 = \cap Q^g$, where Q runs through all definable subgroups of G of finite index and g runs through G (cf. Corollary 6). We call G^0 the *connected component* of G. Clearly $R(G^0) = R(G)$.

LEMMA 9. *Let Φ be in $F_{1,G}(A)$ and suppose Φ is a normal subset of G (i.e. $\Phi^g = \Phi$ for all $g \in G$). If Φ is P-indecomposable for all normal subgroups P, then Φ is indecomposable.*

PROOF. Suppose to the contrary that Φ is Q-decomposable. Then for any $g \in G$, Φ is Q^g-decomposable since $\Phi^g = \Phi$. Suppose Φ splits into m cosets with respect to the subgroup Q (and Q^g). Then it is easy to check that for any $g_1, \dots, g_n \in G$, Φ splits into k cosets with respect to the subgroup $Q^{g_1} \cap \dots \cap Q^{g_n}$, with $m \le k \le m^n$. By Corollary 6 the normal subgroup $P = \bigcap_{g \in G} Q^g$ is definable and $P = Q^{g_1} \cap \dots \cap Q^{g_n}$ for some $g_1, \dots, g_n \in G$. As we remarked, P splits Φ into k pieces (where $m^n \ge k \ge m$). But we assumed that $m > 1$, and so Φ is P-decomposable for the normal subgroup P. The lemma is proved.

We say that a formula $\Phi \in F_1(A)$ is *irreducible* if for any $\Phi_1, \Phi_2 \in F_1(A)$, if $R(\Phi_1) = R(\Phi_2) = R(\Phi)$ and $\Phi = \Phi_1 \cup \Phi_2$ then $\Phi_1 \cap \Phi_2 \ne \varnothing$.

To state and prove the next lemma it will be convenient to introduce the notation $\Phi| \sim \chi$ (to be read as "χ is almost contained in Φ"). For formulas $\Phi, \chi \in F_1(A)$, $\Phi| \sim \chi$ means that $R(\chi \backslash \Phi) < R(\chi)$, i.e., the part of the definable subset χ that lies outside Φ has smaller rank than χ.

LEMMA 10. *Let $p \in \mathrm{St}_G(A)$, let $\Phi, \chi \in F_{1,G}(A)$, $\Phi \in p$, and $\chi \subseteq \mathrm{Stab}(p)$. Then there exists $g \in \Phi(A)$ such that $\Phi| \sim \chi g$. In particular, $\Phi| \sim \mathrm{Stab}(p) \cdot g$.*

PROOF. By induction on $R(\chi)$. Let $R(\chi) = 0$ and $\chi \subseteq \mathrm{Stab}(p)$, so $\chi = \{a_1, \dots, a_k\}$. Clearly $\Phi \& a_1\Phi \& \cdots \& a_k\Phi \in p$. Since p is a consistent type, there exists $g \in \Phi(A) \cap a_1\Phi(A) \cap \dots \cap a_k\Phi(A)$. But this means $g = a_1^{-1} g_1 = \dots = a_k^{-1} g_k$ for some $g_1, \dots, g_k \in \Phi(A)$. It is now clear that $\chi \cdot g = \{g_1, \dots, g_k\} \subseteq \Phi$.

Suppose we have proved the lemma for all $\chi' \subseteq \mathrm{Stab}(p)$ with $R(\chi') \le m$, and suppose $R(\chi) = m + 1$.

We fix some strongly minimal formula $\Delta \in F_1(A)$. We consider a layering $\Psi(v_0, v_1)$ of rank $\le m$ of the formula $\chi(v_1)$ along $\Delta(v_0)$. For each g in $G(A)$ the set

$$D_g(A) = \{a \in \Delta(A) | R(\Psi(a, A) \cdot g \backslash \Phi(A)) < m\}$$

is definable in the signature $\Omega(A)$ (Theorem P.5). On the other hand, for any $a_1, \dots, a_n \in \Delta(A)$, by induction hypothesis there exists $g_n \in \Phi(A)$ such that $\Phi| \sim (\bigcup_{i \le n} \Psi(a_i, A)) \cdot g_n$, i.e., for each n, $D_{g_n}(A)$ contains at least n elements. By the compactness theorem there are some $A' \succeq A$ and $g \in \Phi(A')$ such that $D_g(A')$ is infinite. Since Δ is strongly minimal, $\Delta(A') \backslash D_g(A')$ is finite. This last condition is first order, and so there is $g_0 \in \Phi(A)$ such that $\Delta(A) \backslash D_{g_0}(A)$ is finite. Using Theorem P.4 we infer $R(\chi \cdot g_0 \backslash \Phi) < m$, i.e., $\Phi| \sim \chi \cdot g_0$.

THEOREM 3.2. *A weakly categorical group is connected if and only if it is irreducible.*

PROOF. Let G be irreducible and G^0 the connected component of G.

Then $R(G) = R(G^0) = R(gG^0)$ for any $g \in G$. Since G is irreducible, $R(G\backslash G^0) < R(G)$. This can happen only if $G = G^0$, i.e., G is connected.

Now let G be a connected group and

$$P = \{p \in \mathrm{St}_G(A) | R(p) = R(G)\}.$$

Since P is always finite, we can write $P = \{p_0, \dots, p_k\}$.

Consider $\mathrm{Stab}(p_0)$. If $g \in G$ then $R(gp_0) = R(p_0)$ and therefore $gp_0 = p_i$ for some $i \in \{0, \dots, k\}$. Hence the index of $\mathrm{Stab}(p_0)$ in G is at most $k+1$, i.e., $\mathrm{Stab}(p_0)$ is a definable subgroup of G of finite index. Since G is connected, $G = \mathrm{Stab}(p_0)$.

Now, if $\Phi \in p_0$ and p_0 is the unique type of rank $R(G)$ containing Φ, then by Lemma 10, $\Phi| \sim G$, i.e., p_0 is the unique type of rank $R(G)$ that contains G. In other words, $P = \{p_0\}$. This means that G is irreducible. The theorem is proved.

COROLLARY 11. *If* $\mathrm{Stab}(p) \in p$ *for some* $p \in \mathrm{St}_G(A)$, *then* $\mathrm{Stab}(p)$ *is connected.*

PROOF. By Lemma 10, p is the unique type of rank $R(p)$ that contains $\mathrm{Stab}(p)$. From this it follows that $R(\mathrm{Stab}(p)) = R(p)$ and $\mathrm{Stab}(p)$ is irreducible and thus connected.

Let $\{S_i | i \in I\}$ be a family of subsets of the group G. Then $\mathrm{gp}\{S_i | i \in I\}$ denotes the subgroup generated by $\bigcup\{S_i | i \in I\}$. We shall say that $\mathrm{gp}\{S_i | i \in I\}$ has *finite breadth* with respect to $\{S_i | i \in I\}$ if there are $j_1, \dots, j_k \in I$ (where $k \in \omega$) such that for any $x \in \mathrm{gp}\{S_i | i \in I\}$ there are $z_1 \in (S_{j_1} \cup S_{j_1}^{-1}), \dots, z_k \in (S_{j_k} \cup S_{j_k}^{-1})$ with $x = z_1 \cdot \dots \cdot z_k$.

THEOREM 3.3. *Let* $\{S_i | i \in I, S_i \in F_{1,G}(A)\}$ *be a family of indecomposable definable subsets of the weakly categorical group* G, *all containing the identity* e *of* G. *Then* $\mathrm{gp}\{S_i | i \in I\}$ *has finite breadth with respect to* $\{S_i | i \in I\}$ *and is a connected definable subgroup of the group* G.

PROOF. By Theorem P.2, there is a product $S_{i_1} \cdot \dots \cdot S_{i_k} = S$ (with $k \in \omega$) such that $R(S) \geq R(S_{j_1} \cdot \dots \cdot S_{j_n})$ for any $n \in \omega$ and $j_1, \dots, j_n \in I$. Let $p \in \mathrm{St}_G(A)$, $S \in p$, and $R(p) = R(S) = m$.

Consider an arbitrary S_j $(j \in I)$. Since S_j is $\mathrm{Stab}(p)$-indecomposable and $e \in S_j$, two cases arise. The first is that $S_j \subseteq \mathrm{Stab}(p)$, and the second is that S_j meets an infinite number of cosets of $\mathrm{Stab}(p)$, i.e., $S_j \cap g_n \cdot \mathrm{Stab}(p) \neq \varnothing$ for $g_1, \dots, g_n, \dots \in G$ with $g_n \mathrm{Stab}(p) \neq g_k \mathrm{Stab}(p)$ if $n \neq k$. In the second case we can suppose that $g_n \in S_j$ and therefore $S_j S \supseteq g_n S$, i.e., $S_j S \in g_n p$ $(n = 1, 2, \dots)$ and $g_n p \neq g_k p$ for $n \neq k$. It follows that $R(S_j S) > R(p) = R(S)$ contradicting the maximality of $R(S)$. This implies that for any $j \in I$, $S_j \subseteq \mathrm{Stab}(p)$. Now, if $g \in \mathrm{Stab}(p)$ then $gS \& S \in p$ and so $gS \cap S \neq \varnothing$, that is, $g \in SS^{-1}$. This proves that

$\mathrm{gp}\{S_i | i \in I\} = S \cdot S^{-1} = \mathrm{Stab}(p)$ and moreover $\mathrm{Stab}(p) \in p$. Bearing in mind Corollary 11, we obtain the assertion of the theorem.

§4. Abelian subgroups

LEMMA 12. *If the infinite weakly categorical group G has no proper infinite definable subgroup then G is abelian.*

PROOF. It suffices to find an infinite abelian (not necessarily definable) subgroup Q of the group G, since then its closure $\overline{Q}$ is a definable abelian subgroup by Lemma 7, and $\overline{Q} = G$ by assumption.

If the group G contains an element a of infinite order than $\mathrm{gp}\{a\}$ is an abelian group, and so we can suppose at once that G is a periodic group.

Let C be the center of G. We suppose that $C \neq G$, and so by the hypothesis of the lemma, C is a finite subgroup. For any a in G which is not in the center C, we consider the map $\phi_a: G \to G$ which is definable from a as follows: $\phi_a(x) = x^{-1}ax$. It is easy to see that

$$\phi_a(x_1) = \phi_a(x_2) \Leftrightarrow x_1 x_2^{-1} \in C(a).$$

Now $a \notin C$ and therefore $C(a)$ is a proper definable subgroup, whence $C(a)$ is finite. So $\phi_a^{-1}(y)$ is finite for all $y \in \phi_a(G)$. We define $\Psi_a(v_0, v_1)$ to be the formula $\phi_a(v_1) = v_0$. From the above, it follows that $\Psi_a(v_0, v_1)$ is a layering of rank ≤ 0 of G along $\phi_a(G)$. By Theorem P.4, $R(\phi_a(G)) = R(G)$.

We note that the hypothesis of the lemma implies that G is connected, since $G = G^0$ (where G^0 is the connected component of G). So by Theorem 3.2, G is irreducible. Now if $a, b \in G/C$ then $\phi_a(G) \cap \phi_b(G) \neq \varnothing$ since G is irreducible. It is easy to see that this last condition implies that $\phi_a(G) = \phi_b(G)$, i.e., a and b are conjugate in the group G.

In fact we have proved that all elements of G/C (except e) are conjugate. As G/C is periodic, it follows that there exists a prime p with $g^p = e$ for all $g \in G/C$. Now $p \neq 2$, since otherwise G/C is abelian which is impossible by the condition of conjugacy. Let $a \in G/C$, $a \neq e$. Then by hypothesis there exists $g \in G/C$ such that $a^2 = g^{-1}ag$ (as $a^2 \neq e$). Hence $a^{2^n} = g^{-n}ag^n$ for any natural number n. In particular, $a^{2^p} = (g^p)^{-1}ag^p = a$, so $2^p \equiv 1 (\mathrm{mod}\ p)$, which is impossible as $p > 2$ and $2^{p-1} \equiv 1 (\mathrm{mod}\ p)$.

The assumption $C \neq G$ gives a contradiction and so $C = G$. The lemma is proved.

REMARK. In the group-theoretical part of our original proof of Lemma 12 we used strong results from the theory of groups. We simplified the proof after becoming acquainted with the work of Reineke [15].

COROLLARY 13. *Every infinite weakly categorical group G has an infinite abelian definable subgroup.*

PROOF. By Lemma 5 we can find in G a definable subgroup H which has no proper infinite definable subgroup. By Lemma 12, H is abelian.

LEMMA 14. *If H is a definable subgroup of the weakly categorical group G, and H has infinite index in G, then $R(G) > R(H)$.*

PROOF. Let $p \in \mathrm{St}_H(A)$ with $R(p) = R(H) = m$. By assumption, the number of cosets of H in G is infinite. Let S be a set of coset representatives. By Remark 4, for any g in S, $R(gp) = R(p) = m$. By choice of S, $g_1 p \neq g_2 p$ if $g_1 \neq g_2$ with $g_1, g_2 \in S$. Moreover, $gp \in \mathrm{St}_G(A)$ for all $g \in G$. Since S is infinite it follows from a remark above that $\mathrm{St}^m_G(A)$ is infinite, and hence $R(G) > m$. The lemma is proved.

THEOREM 4.1. *If G is a weakly categorical group and $R(G) = 1$, then G has a normal definable abelian subgroup G^0 of finite index, i.e., G is almost abelian.*

PROOF. It follows from Lemma 14 that any infinite definable subgroup of G has finite index in G. Let G^0 be the connected component of G. By Remark 8, G^0 has no proper definable subgroups of finite index, and $R(G^0) = 1$, so G^0 is infinite. Let G' be an infinite abelian definable subgroup of G (Corollary 13). Since G' has finite index in G, $G^0 \cap G'$ has finite index in G^0, so $G^0 \cap G' = G^0$ and G^0 is abelian. The theorem is proved.

As was proved by Chikhachëv [8], the classification of universal $\aleph_1$-categorical theories of semigroups reduces to the question that we answer in the following theorem.

THEOREM 4.2. *Let T be a universally axiomatizable $\aleph_1$-categorical theory of groups, and let G be an infinite model of T. Then G is abelian.*

PROOF. Clearly G is a categorical group. By Corollary 13, G has an infinite definable abelian subgroup G'. Since T is universal, G' is a model of T. Since the theory of infinite models of T is complete, G is also abelian. The theorem is proved.

§5. Simple weakly categorical groups

THEOREM 5.1. *Let G be an infinite nonabelian group which has no proper normal definable subgroups. Then G is a simple group.*

PROOF. Since G has no proper definable normal subgroups, its connected component G^0 coincides with G, i.e., G is a connected group.

We write $g^G = \{x^{-1}gx | x \in G\} \cup \{e\}$ for $g \in G$, $g \neq e$. Clearly g^G is a normal subset of G and $g^G \in F_{1,G}(A)$. Moreover g^G is infinite, since otherwise $C(g)$ has finite index in G, i.e., since G is connected, $G = C(g)$ and $g \in C$, contradicting the hypothesis of the theorem.

Now we can assert that g^G is indecomposable. In fact, it is enough to verify that it is Q-indecomposable for normal subgroups Q, i.e., $Q = G$ or

$Q = \{e\}$. The first case falls out at once, and the $\{e\}$-indecomposability of g^G follows from the fact that it is infinite.

By Theorem 3.3, $\mathrm{gp}\{g^G\}$ is a definable subgroup of G; since g^G is normal, $\mathrm{gp}\{g^G\}$ is normal too, i.e. $\mathrm{gp}\{g^G\} = G$, and this means that the normal closure of an element $g \in G$ coincides with the whole group G. Since g can be chosen at random, the theorem is proved.

We say that a totally transcendental theory T is *almost strongly minimal* if: (a) there is a strongly minimal formula $\Delta \in F_1(A_0)$, where A_0 is the prime model of T; (b) for all (for some) models A of the theory T there is a formula $\Psi(v_0, v_1, \dots, v_n) \in F(A)$ and a natural number k such that

$$A \models \forall v_1, \dots, v_n(\Delta(v_1)\& \cdots \&\Delta(v_n) \to \exists v_0^{\leq k}\Psi(v_0, v_1, \dots, v_n))$$
$$\&\forall v_0 \exists v_1, \dots, v_n(\Delta(v_1)\& \cdots \&\Delta(v_n)\&\Psi(v_0, v_1, \dots, v_n)).$$

Baldwin noted that

THEOREM (Baldwin [10]). *An almost strongly minimal theory is $\aleph_1$-categorical.*

For the proof of the next theorem we need:

LEMMA (Baldwin, Lachlan [11]). *If a totally transcendental theory T does not have the f.c.p., then there is a strongly minimal formula $\Delta \in F_1(A_0)$, where A_0 is the prime model of the theory T.*

This lemma is an easy generalization of Lemma 9 of [11], when one notes that for the proof of Lemma 9 it suffices to use the f.c.p.

THEOREM 5.2. *Let G be a simple weakly categorical group. Then the theory of G in the signature of groups is $\aleph_1$-categorical, i.e., G is a categorical group.*

PROOF. Let T_G be the theory of the group G in the signature of groups, let T be the original $\aleph_1$-categorical theory in which the group G was interpreted, and let A be a model of T. As usual we shall write $G = G(A)$.

By Lemma 2, T_G is totally transcendental. By the Baldwin-Lachlan lemma there is a strongly minimal formula Δ, $\Delta \in F_1(G_0)$, where G_0 is the prime model of the theory T_G. Since we can suppose that $G_0 \preceq G$, the formula Δ is in $F_{1,G}(A)$.

By Lemma 5 there is a minimal (in the sense of inclusion of subgroups) definable subgroup Q of the group G, for which the intersection $\Delta(A) \cap aQ$ is infinite, for some $a \in G$. It is clear that $\Delta(A) \cap aQ$ is indecomposable, since otherwise Q would not be minimal.

Let $g \in \Delta(A) \cap aQ$, and write $\Phi = g^{-1}(\Delta \cap aQ)$. It is clear that Φ is also indecomposable, and that it contains e. This same condition is satisfied by Φ^{g_1} for any $g_1 \in G$. By Theorem 3.3, $\mathrm{gp}\{\Phi^{g_1} | g_1 \in G\}$ is a definable subgroup of G. This subgroup is normal, and since G is simple, $G = \mathrm{gp}\{\Phi^{g_1} | g_1 \in G\}$. On the other hand, by Theorem 3.3, $\mathrm{gp}\{\Phi^{g_1} | g_1 \in G\}$ has finite width with respect to $\{\Phi^{g_1} | g_1 \in G\}$. In particular, this means that there is a group word $W(v_1, \dots, v_m, g, g_1, \dots, g_n)$ (where $g, g_1, \dots, g_n \in G$) such that for any $c \in G$, there are $b_1, \dots, b_m \in \delta(A)$ satisfying the identity $W(b_1, \dots, b_m, g, g_1, \dots, g_n) = c$. Clearly the formula

$$\Psi(v_0, v_1, \dots, v_m) \rightleftharpoons (W(v_1, \dots, v_m, g, g_1, \dots, g_n) = v_0)$$

satisfies the definition of an almost strongly minimal theory. Consequently T_G is almost strongly minimal, and by Baldwin's theorem T_G is an $\aleph_1$-categorical theory. The theorem is proved.

COROLLARY. *In particular, the theory of any simple algebraic group over an algebraically closed field is $\aleph_1$-categorical.*

§6. Skew fields

Macintyre proved that

THEOREM (Macintyre [13]). *An infinite field whose theory is totally transcendental is algebraically closed.*

Macintyre also posed the problem of describing rings, and in particular skew fields, which are $\aleph_1$-categorical. The following theorem gives an answer for skew fields. (It was announced in [6], and we have now obtained a complete solution of Macintyre's problem [7].)

THEOREM 6.1. *The infinite weakly categorical skew field K is an algebraically closed field.*

PROOF. We shall split the proof into paragraphs. We remark that in paragraph (c) we prove an inequality which follows from Galois theory for skew fields [2]. For the partial case which we need, we shall give a complete and self-contained proof (but following the pattern of the general case). Thus:

(a) In any weakly categorical skew field K there exists an infinite definable sub-skew field which is a field. In fact, by Corollary 13 the multiplicative group K^* has an infinite abelian definable subgroup H. Let N be a minimal definable sub-skew field of K (by Lemma 5) containing H and the center C of the skew field K. Since N is minimal, $N \subseteq C(H)$ (where $C(H)$ is the centralizer of H), and hence $(C \cup H) \subseteq (C(N) \cap N) \subseteq N$, that is $N = C(N) \cap N$. This last means that N is a field. Moreover, by Macintyre's theorem and Lemma 2, this field is algebraically closed.

(b) We can regard K as a left vector space over the field N. Let $V_1 \subseteq V_2 \subseteq \cdots \subseteq V_k \subseteq \cdots$ be a sequence of finite-dimensional vector subspaces (over N) of K such that $\dim_N(V_n) = n$. It is clear that V_n can be considered as a definable subgroup of the additive group of K. By Lemma 14, $R(V_{n+1}) > R(V_n)$. Since $R(K)$ is finite (by Theorem P.2), it follows that $K = V_m$ for some natural number m. We have proved that K is an m-dimensional vector space over N. (In symbols, $(K:N) = m$.)

(c) By construction $N \supseteq C$. We shall show $(N:C) \leq m$. We begin by noting the following. Let $\mathcal{N}$ be the set of all N-linear transformations from the vector space K to itself. We consider the linear transformation $r_a: K \to K$ which is associated with the element $a \in K$ by the rule $r_a(x) = xa$. Since $(K:N) = m$ it follows that $\mathcal{N}$ is an m-dimensional left vector space over K, or more precisely over $\{r_a | a \in K\}$. The product $r_a \cdot f$ $(f \in \mathcal{N})$ is read as follows: $r_a \cdot f(x) = f(x) \cdot a$.

Now we suppose $(N:C) > m$. Then there exist $b_1, \ldots, b_{m+1} \in N$ linearly independent over C. Clearly the transformation $l_b: K \to K$ given by $l_b(x) = bx$ lies in $\mathcal{N}$ if $b \in N$. Since the dimension of $\mathcal{N}$ is less than $m+1$, $l_{b_1}, \ldots, l_{b_{m+1}}$ are linearly dependent over K, that is, there exist $a_1, \ldots, a_{m+1} \in K$ not all zero, such that $r_{a_1} \cdot l_{b_1} + \cdots + r_{a_{m+1}} \cdot l_{b_{m+1}} = 0$. This last equation means that

$$b_1 x a_1 + \cdots + b_{m+1} x a_{m+1} = 0 \qquad (*)$$

for any $x \in K$. We can suppose $a_1 = 1$ and that $a_1, \ldots, a_{m+1}$ are chosen so that in $(*)$ the number of nonzero terms is as small as possible. By the choice of the b_i's, there must exist a_i such that $a_i \notin C$, that is $g a_i g^{-1} \neq a_i$ for some $g \in K$. The following identities hold: $b_1 x g + \cdots + b_i x g a_i + \cdots + g_{m+1} x g a_{m+1} = 0$, $b_1 x + \cdots + b_i x g a_i g^{-1} + \cdots + g_{m+1} x g a_{m+1} g^{-1} = 0$. By combining this last with the identity $(*)$ with $a_1 = 1$ and $g a_i g^{-1} \neq a_i$, we obtain a shorter linear dependence, which contradicts the choice of $a_1, \ldots, a_{m+1}$. This proves that $(N:C) \leq m$. But this means that $N = C$ since C is infinite and hence algebraically closed.

(d) Summing up, we have shown that K is a skew field of finite dimension over its algebraically closed center. If $x \in K$, the field generated by $\{C, x\}$ is a finite extension field of C. But C is algebraically closed and so $x \in C$ and $K = C$. The theorem is proved.

REMARK. Theorem 6.1 can be rephrased as follows: A weakly categorical skew field is a field. This formulation includes Wedderburn's theorem on finite fields [2].

Finally we state some questions about weakly categorical groups:

(1) Does there exist a nonabelian connected weakly categorical group that is not isomorphic to an algebraic group over an algebraically closed field?

There are abelian groups that are not algebraic, as one can easily deduce from Example 2.

(2) Do there exist any simple categorical groups apart from algebraic groups over algebraically closed fields?

The following question also interests us:

(3) Can one interpret the group $G = H(p) + H(q)$ in an $\aleph_1$-categorical theory, i.e., make G weakly categorical? Here $H(p)$ and $H(q)$ are infinite abelian groups of prime exponent p and q respectively.

Actually the question is related to pure model theory and is connected with the question:

(4) For any two $\aleph_1$-categorical theories T_1 and T_2, does there exist an $\aleph_1$-categorical theory T in which both T_1 and T_2 can be interpreted?

REFERENCES

1. A. Borel, *Linear Algebraic Groups*, Moscow, 1972.

2. N. Jacobson, *The Structure of Fields*, 1961.

3. M. A. Taĭtslin, *Model Theory*, Novosibirsk, 1970.

4. B. I. Zil'ber, *On the rank of formulas in an* $\aleph_1$*-categorical theory*, Thesis, Moscow, 1972, pp. 18–19.

5. ——, *On the transcendence rank of formulas in an* $\aleph_1$*-categorical theory*, Mat. Zametki **15** (2) (1974), 321–329.

6. ——, *Groups with categorical theories*, Fourth All-Union Symposium on the Theory of Groups, Novosibirsk, 1973, pp. 63–68.

7. ——, *Rings whose theories are* $\aleph_1$*-categorical*, Algebra i Logika **13** (2) (1974), 168–187.

8. S. A. Chikhachëv, $\aleph_1$*-categorical universal theories of semigroups*, Third All-Union Conference on Mathematical Logic, Novosibirsk, 1974, pp. 223–224.

9. J. T. Baldwin, α_T *is finite for* $\aleph_1$*-categorical* T, Trans. Amer. Math. Soc. **181** (1973), 37–51.

10. ——, *Almost strongly minimal theories*, J. Symbolic Logic **37** (1972), 487–493.

11. J. T. Baldwin and A. H. Lachlan, *On strongly minimal sets*, J. Symbolic Logic **36** (1971), 79–96.

12. A. Macintyre, *On* ω_1*-categorical theories of abelian groups*, Fund. Math. **70** (1970), 253–270.

13. ——, *On* ω_1*-categorical theories of fields*, Fund. Math. **71** (1971), 1–25.

14. M. D. Morley, *Categoricity in power*, Trans. Amer. Math. Soc. **114** (1965), 514–538.

15. J. Reineke, *Minimale Gruppen*, Z. Math. Logik Grundlag. Math. **21** (1975), 357–359.

16. G. E. Sacks, *Saturated Model Theory*, W. A. Benjamin, Reading, Mass., 1972.

17. S. Shelah, *Stability, the f.c.p. and superstability*, Ann. Math. Logic **3** (1971), 271–362.

Kemerova University

Translated by D. EVANS AND W. HODGES

Amer. Math. Soc. Transl.
(2) Vol. **149**, 1991

On the Separation of a Real Linear Monic Factor from a Matrix Polynomial

UDC 512.8

V. P. SHCHEDRIK

Abstract. The author presents a proof of a theorem on a necessary and sufficient condition for the existence of real linear unital factors of real matrix polynomials.

Bibliography: 2 titles.

Let $A(x) = A_0 x^s + A_1 x^{s-1} + \cdots + A_{s-1}x + A_s$, $|A(x)| \not\equiv 0$, be a matrix polynomial whose coefficients are $n \times n$ matrices over the field $\mathbb{R}$ of real numbers. The main result of the paper is the following.

THEOREM 1. *If the elementary divisors of the matrix $A(x)$ are relatively prime and from $A(x)$ one separates a linear factor $B(x)$, then $B(x)$ is real if and only if its characteristic polynomial $\varphi(x) = |B(x)|$ is real, i.e., $\varphi(x) \in \mathbb{R}[x]$.*

In order to prove the theorem we establish a series of facts which present interest in their own right.

LEMMA 1. *Assume that a complex $n \times n$ matrix D has the form*

$$D = \left\| \begin{matrix} a_1 \\ \overline{a}_1 \\ \vdots \\ a_s \\ \overline{a}_s \\ L \end{matrix} \right\| ,$$

1991 *Mathematics Subject Classification*. Primary 15A23.

Translation of Mat Metody i Fiz.-Mekh. Polya No. 24 (1986), 7–10; MR 88c:15014.

where

$$a_j = \|a_{j1} + b_{j1}i \quad \cdots \quad a_{jn} + b_{jn}i\|,$$
$$\overline{a}_j = \|a_{j1} - b_{j1}i \quad \cdots \quad a_{jn} - b_{jn}i\|,$$

$j = 1, \ldots, s$, *and the submatrix* L *is real. Then* $|D| = di^s$, *where* $d \in \mathbb{R}$.

PROOF. After elementary transformations on the rows of the matrix D, we obtain

$$|D| = \begin{vmatrix} b_{11}i & \cdots & b_{1n}i \\ a_{11} & \cdots & a_{1n} \\ \cdots & \cdots & \cdots \\ b_{s1}i & \cdots & b_{sn}i \\ a_{s1} & \cdots & a_{sn} \\ & L & \end{vmatrix} = di^s,$$

where $d \in \mathbb{R}$.

LEMMA 2. *Let* $p(x)$, $\psi(x)$, $\varphi_i(x)$, $i = 1, 2$, *be arbitrary polynomials in* $\mathbb{R}[x]$, $\psi(x)$ *being an irreducible monic polynomial in* $\mathbb{R}[x]$ *and* $(\varphi_i(x), \psi(x)) = 1$, $i = 1, 2$. *The equality*

$$\varphi_1(x) + v\varphi_2(x) = p(x)\psi(x) \tag{1}$$

can hold for at most one value $v \in \mathbb{R}$.

PROOF. Obviously, $\deg \psi(x) \leq 2$. Now, if $\psi(x) = x - \alpha$, then it is easy to see that equality (1) is satisfied only by one value of v, namely $v = -(\varphi_1(\alpha)/\varphi_2(\alpha))$.

If, however, $\psi(x) = (x - \beta)(x - \overline{\beta})$, where $\overline{\beta}$ is the complex conjugate of β, then it is easy to verify that equality (1) is satisfied in at most one case. The lemma is proved.

In [1] (p. 131) a theorem is proved on the semiscalar equivalence of matrices over $\mathbb{C}[x]$. It turns out that this result holds also for matrices over $\mathbb{R}[x]$.

THEOREM 2. *Let* $A(x)$ *be an* $m \times n$ *polynomial matrix* $(m \leq n)$ *over* $\mathbb{R}[x]$. *If* $\operatorname{rank} A(x) = m$, *then for* $A(x)$ *there exist a nonsingular matrix* P *and an invertible matrix* $Q(x)$ *over* $\mathbb{R}[x]$ *such that one has the relation*

$$F(x) = PA(x)Q(x) = \begin{Vmatrix} \varepsilon_1(x) & 0 & \cdots & 0 & \cdots & 0 \\ & \varepsilon_2(x) & \cdots & 0 & \cdots & 0 \\ * & & \ddots & & & \\ & & & \varepsilon_m(x) & \cdots & 0 \end{Vmatrix},$$

where $\varepsilon_i(x)$ *are the invariant polynomials of the matrix* $A(x)$.

PROOF. For the proof it is sufficient to make use of Lemma 2 and to carry out a reasoning similar to that used for the proof of Theorem 1 of [1, p. 131], replacing the expressions of the form $x - \alpha_i$ by $\psi_i(x)$, where $\psi_i(x)$ is a monic irreducible polynomial in $\mathbb{R}[x]$.

Assume that the polynomial matrix $A(x)$ has relatively prime elementary divisors. Then, on the basis of Theorem 2, there exist a nonsingular matrix P over $\mathbb{R}$ and an invertible matrix $Q(x)$ over $\mathbb{R}[x]$ such that

$$F(x) = PA(x)Q(x) = \left\| \begin{matrix} 1 & & & \\ & \ddots & & \\ & & 1 & \\ \varphi_1(x) & \cdots & \varphi_{n-1}(x) & \Delta(x) \end{matrix} \right\|, \tag{2}$$

where the blank spaces are occupied by zeros while $\Delta(x) = |A(x)|$. Taking the classical adjoints in the equality (2), we obtain

$$F_*(x) = Q_*(x)A_*(x)P_* = \left\| \begin{matrix} \Delta(x) & & & \\ & \ddots & & \\ & & \Delta(x) & \\ \psi_1(x) & \cdots & \psi_{n-1}(x) & 1 \end{matrix} \right\|,$$

where $\psi_i(x) = -\varphi_i(x)$. We denote the last row of the matrix $F_*(x)$ by $g(x)$:

$$g(x) = \| \psi_1(x) \quad \cdots \quad \psi_{n-1}(x) \ 1 \|.$$

Making use of the definition of the value of a polynomial matrix at the system of the zeros of a polynomial $\varphi(x)$ as well as of Statement 4 of [1, p. 29], we obtain

$$\operatorname{rank} M_{A_*(x)}(\varphi) = \operatorname{rank} M_{Q_*(x)A_*(x)P_*}(\varphi) = \operatorname{rank} M_{g(x)}(\varphi). \tag{3}$$

We note that an analogous equality has been obtained in [2] in the case of the field $\mathbb{C}$.

Let

$$A(x) = (Ex - B)D(x), \tag{4}$$

where $|Ex - B| = \varphi(x)$. On the basis of Theorem 1 of [1, p. 54], the matrix B is obtained as the solution of the matrix equation

$$M_{A_*(x)}(\varphi)X = M_{xA_*(x)}(\varphi).$$

Obviously, the equations

$$\begin{aligned} M_{F_*(x)}(\varphi)Y &= M_{xF_*(x)}(\varphi), \\ M_{g(x)}(\varphi)Y &= M_{xg(x)}(\varphi) \end{aligned} \tag{5}$$

are equivalent. Assume that a matrix B_1 satisfies these equations. Then we have the relation

$$F(x) = PA(x)Q(x) = (Ex - B_1)S(x),$$

and $|Ex - B_1| = \varphi(x)$. From here we have

$$A(x) = (Ex - P^{-1}B_1P)P^{-1}S(x)Q^{-1}(x).$$

By virtue of the uniqueness of the decomposition of the matrix $A(x)$ into factors, in which the left monic factor has characteristic polynomial $\varphi(x)$, we obtain that $B = P^{-1}B_1P$.

Taking into account the equalities (3) and Theorem 5 of [1, p. 52], we obtain

LEMMA 3. *The matrix* $A(x)$ *admits the decomposition* (4) *if and only if*

$$\operatorname{rank} M_{g(x)}(\varphi) = n,$$

and then $B = P^{-1}B_1P$, *where* B_1 *is the solution of equation* (5) *and* P *is the matrix from relation* (2).

We proceed to the proof of Theorem 1.

PROOF. The necessity is obvious.

Sufficiency. Since in $\mathbb{R}[x]$ only the polynomials of degree at most two are irreducible, it follows that $\varphi(x)$ can be represented in the following manner:

$$\varphi(x) = (x-\alpha_1)^{k_1}\cdots(x-\alpha_s)^{k_s}(x-\overline{\alpha}_1)^{k_1}\cdots(x-\overline{\alpha}_s)^{k_s}(x-\beta_1)^{p_1}(x-\beta_2)^{p_2},$$

where α_i, $\overline{\alpha}_i$ are complex conjugate numbers and $\beta_j \in \mathbb{R}$. Let

$$N = \left\| \begin{matrix} M_{g(x)}[\alpha_1^{(k_1)}] \\ \cdots\cdots\cdots \\ M_{g(x)}[\alpha_s^{(k_s)}] \end{matrix} \right\|, \qquad N_1 = \left\| \begin{matrix} M_{xg(x)}[\alpha_1^{(k_1)}] \\ \cdots\cdots\cdots \\ M_{xg(x)}[\alpha_s^{(k_s)}] \end{matrix} \right\|,$$

and let $\overline{N}$, $\overline{N}_1$ be the matrices conjugate to N and N_1, respectively. Then equation (5) can be written as

$$\left\| \begin{matrix} N \\ \overline{N} \\ K \end{matrix} \right\| Y = \left\| \begin{matrix} N_1 \\ \overline{N}_1 \\ K_1 \end{matrix} \right\|,$$

where K and K_1 are real matrices.

It is known that if a polynomial $f(x) \in \mathbb{R}[x]$ and $\alpha \in \mathbb{C}$, then from the equality $f(\alpha) = a$ there follows $f(\overline{\alpha}) = \overline{a}$. Taking this into account, it is clear that each complex row of the matrices $M_{g(x)}(\varphi)$ and $M_{xg(x)}(\varphi)$ has itself a conjugate and the conjugate rows in these matrices are situated at the same places. Therefore, applying Lemma 1, we obtain

$$|M_{g(x)}(\varphi)| = ai^s,$$

where $a \in \mathbb{R}$. In order to find the element y_{tj} of the matrix Y, we find the quotient

$$|M_{g(x)}(\varphi)|_{(tj)}/|M_{g(x)}(\varphi)|,$$

where $|M_{g(x)}(\varphi)|_{(tj)}$ is the determinant of the matrix, obtained by replacing the tth column of the matrix $M_{g(x)}(\varphi)$ by the jth column of the matrix $M_{xg(x)}(\varphi)$. Clearly, under this replacement, the structure of the obtained matrix remains unchanged, i.e., the rows which have been conjugate remain conjugate again. Therefore,

$$|M_{g(x)}(\varphi)|_{(tj)} = a_{tj} i^s ,$$

where $a_{tj} \in \mathbb{R}$. Consequently, $y_{tj} = a_{tj}/a$. Thus, the matrix B_1 consists of real numbers. Taking into account Lemma 3 and also the fact that the matrix P is real, we obtain that the matrix B is also real.

Bibliography

1. P. S. Kazimirskiĭ, *Decomposition of matrix polynomials into factors*, "Naukova Dumka", Kiev, 1981. (Ukrainian)

2. V. M. Petrichkovich, *Absolute decomposability of matrix polynomials*, Mat. Metody i Fiz.-Mekh. Polya No. 9 (1979), 37–41. (Russian)

Institute for Applied Problems of Mechanics and Mathematics
Academy of Sciences of the Ukrainian SSR
L′vov

Translated by E. DEUTSCH

Amer. Math. Soc. Transl.
(2) Vol. **149**, 1991

Subgroups of Symmetric and Alternating Groups

YA. G. BERKOVICH

§1. Introduction

We denote the set of all bijections of a set M by $S(M)$; the elements of $S(M)$ are called permutations. For elements $\sigma, \tau \in S(M)$, the product $\sigma\tau$ is defined by the equation $(\sigma\tau)(x) = \sigma(\tau(x))$, $x \in M$. $S(M)$ is a group under this operation, namely the symmetric group on M. If $M_1 \subset M$, there is a canonical injection of $S(M_1)$ into $S(M)$; thus, we shall allow ourselves the liberty of assuming that $S(M_1)$ is a subgroup of $S(M)$. If M is finite and of cardinality n, we write S_n for $S(M)$. There is a subgroup of index 2 in S_n, namely the alternating group A_n of degree n.

This paper contains a fairly complete survey of the subgroup structures of S_n and A_n. The theme has been developed comparatively little in the literature. We shall make almost no reference here to the important question of classifying the maximal subgroups of S_n and A_n. Notwithstanding the fact that this problem has been pursued intensively here and abroad (following on from the papers of L. A. Kaluzhnin and his students), the results obtained in this article are sporadic in nature. The representation theory of S_n is fully covered in the literature and is not touched upon here.

S_n and A_n are classical algebraic objects and have long attracted the attention of mathematicians. It is enough to quote here the proof, due to Galois, of the fact that A_n is nonabelian and simple for $n > 4$. Hölder showed that S_n is complete when $n > 2$ and $n \neq 6$. Schreier and Ulam carried this result over to the infinite case. The Sylow subgroups of S_n and their generalizations have been the subject of intensive study by Kaluzhnin, his students and successors. Phillip Hall [1] showed that S_n has nonprimary Hall subgroups only when $n = 3, 4, 5, 7$, and 8 in the solvable case; in the nonsolvable case, Thompson [2] proved that S_n has Hall subgroups only

1991 *Mathematics Subject Classification.* Primary 20B35, 20D06.

Translation of Severo-Kavkaz. Nauchn. Tsentra Vyssh. Shkoly Estestv. Nauk. **1984**, no. 1, 6–9.

when n is prime. Suprunenko [3] showed that maximal transitive nilpotent subgroups of S_n are conjugate (though the analogous result for A_n is false). Some series of maximal primitive subgroups have been constructed by Kaluzhnin, Klin, Ustimenko-Bakhumovskiĭ, and others. However, the status of this question is that we cannot exhibit, for each n, a maximal primitive subgroup of S_n different from A_n.

The present survey is entirely devoid of proofs, because of considerations of space. The specialist will have no difficulty in determining what has been done by the author and what by his predecessors. We recommend [4] to the reader in this connection. All notation and definitions are those standard to the discipline. Everywhere below, M is a finite set.

§2. Results

1. The maximal subgroups of S_n that are intransitive or imprimitive are listed in [5] and [6]. We give below a more general statement of the results of [5], [6].

1.1. Let $M = \bigcup_{\alpha\in\Lambda} M_\alpha$ be a partition with $|\Lambda| > 1$. We set $H_\alpha = S(M_\alpha)$ for $\alpha \in \Lambda$, and $H = \prod_{\alpha\in\Lambda} H_\alpha$. Let π be a permutation in $S(M)$ not normalizing H. If $|\Lambda| = 2$, then $\langle H, \pi\rangle = S(M)$; in particular, the normalizer of H is maximal in $S(M)$. Suppose now that $|\Lambda| > 2$, and let B be a subgroup acting 2-transitively on the system $\{M_\alpha\}_{\alpha\in\Lambda}$, $B < S(M)$ (clearly, B normalizes H). Then $\langle H, B, \pi\rangle = S(M)$; in particular, the normalizer of H is maximal in $S(M)$.

Assertions like 1.1 are also valid for A_n. A description of the maximal intransitive and imprimitive subgroups of S_n follows at once from 1.1. Little is known about the primitive maximal subgroups of S_n.

A subgroup G of $S(M)$ is said to be thin if $G \cap S(M\backslash M_1) = 1$ for every maximal normal subset M_1 of M such that $G \cap S(M_1) = 1$. It is not difficult to show that G is thin if it has no cycle of length 3. By a result of Jordan, every primitive subgroup not containing A_n is thin. Since regular groups are thin, there can be no description of the group structure of thin groups.

1.2. Let G be a transitive imprimitive thin subgroup of $S(M)$, set* $H = \langle \pi \in G \mid |T(\pi)| \leq 3\rangle$, and let $\{M_j\}_{j\in I}$ be the complete system of H-orbits. Then all M_j have the same cardinality and length not more than 4. Suppose that G contains a cycle of length 3. Then G contains no transpositions, so that $|M_1| \geq 3$, and $G = A(M_1) \operatorname{wr} D$, where D is a transitive subgroup of $S(J)$, M being identified with $M_1 \times J$.

Clearly, the index of a thin subgroup of S_n is not less than $[(n+1)/2]!$ This gives a bound for the index of a primitive subgroup of S_n (see [7]).

2. In this section we give the results of an analysis of certain invariants of S_n and A_n.

---

*$T(H)$ is the set of points permuted by the elements of H; that is, the support of H.

2.1. If $\alpha(n)$ is the maximum of the orders of abelian subgroups of S_n and $n = 3k + r$, $2 \leq r \leq 4$, then $\alpha(n) = 3^k r$ [8].

We note that the minimum of the orders of maximal abelian subgroups is $n - 1$ when $n > 2$. Let $p^{\alpha_p(n)}$ be the greatest of the orders of abelian p-subgroups of S_n, and $p^{\beta_p(n)}$ the minimum of the orders of maximal abelian p-subgroups of S_n. Then

2.2. $\alpha_p(n) = [n/p]$. If $n = a_0 + a_1 p + \cdots + a_t p^t$ is the p-adic decomposition of n, then $\beta_p(n) = a_1 + 2a_2 + \cdots + ta_t$.

Let $d_p(G)$ be the maximum of the ranks of p-subgroups of G and $e_{p^k}(G)$ the maximum of the orders of subgroups of exponent p^k in G. When $G = S_n$, we write $d_p(n)$ and $e_{p^k}(n)$ for $d_p(G)$ and $e_{p^k}(G)$ respectively.

2.3. Let $W = G \operatorname{wr} C_p$ be a finite p-group, where C_m denotes the cyclic group of order m.

a) If A is an abelian subgroup of W with $|A| > p^{1+\alpha_p(G)}$, then A lies in the base group of W. In particular, $\alpha_p(W) = p\alpha_p(G)$.

b) $e_p(W) = pe_p(G)$. In particular, $e_p(n) = [n/p]$.

2.4. $d_p(n) = [n/p]$. If $p > 2$ and H is a p-subgroup of rank $[n/p]$ in S_n, then H is elementary abelian.

A similar assertion valid for abstract groups follows from 2.2:

2.5. Suppose that the finite p-group G contains a subgroup H of order p^n such that $C_G(H) \leq H$. Then every abelian normal subgroup of G can be generated by p^{n-1} elements.

2.6. Suppose that $n = 2^m$, and let Σ_n be a Sylow 2-subgroup of S_n. Then Σ_n has subgroups $\Phi_1, \ldots, \Phi_{m-1}$ such that $\Phi(\Sigma_n) = \Phi_1 \leftthreetimes \Phi_2 \leftthreetimes \cdots \leftthreetimes \Phi_{m-1}$, where $\Phi_i \cong \Sigma_{2^i}$; here, $\Phi(G)$ is the Frattini subgroup of G. In particular, the rank of $\Phi(\Sigma_{2^m})$ is not more than $m(m-1)/2$.

2.7. The rank of a Sylow 2-subgroup of A_{2^m} is m.

2.8. If $n \geq p^k$, then $e_{p^k}(n) = [n/p] + \cdots + [n/p^k]$.

2.9. Let m and k be natural numbers such that $m > k + 1$ and $p > k$. Then all normal subgroups of exponent p^k of $\Sigma_{p^m} = \Sigma_{p^{m-1}} \operatorname{wr} C_p$ lie in the base group of the wreath product Σ_{p^m}.

For the general problem, it is important to establish the analogue of 2.4 for $p = 2$. The author conjectures that every 2-subgroup of S_n of rank $[n/2]$ is a homomorphic image of a direct product of dihedral groups of order 8.

3. The structure of the centralizer Z of a permutation in S_n is used in many arguments.

3.1. Let π be a permutation of type $(b_1, b_2, \ldots, b_n)$ in $S(M) = S_n$, M_i the set of points permuted by the cycles of length i in the canonical decomposition of π when $i > 1$, and set $M_1 = \{x \in M | \pi(x) = x\}$. We write $\sigma_i = \sigma|_{M_i}$ for every element σ of Z. Then $C_{S(M)}(\pi) = Z = Z_1 \times \cdots \times Z_n$, where $Z_i = \{\sigma_i | \sigma \in Z\}$. We have $T(Z_i) = M_i$, $Z_i \cong C_i \operatorname{wr} S_{b_i}$. The base group Z_i is generated by cycles in the decomposition of $\pi_i = \pi|_{M_i}$.

We get as a corollary to 3.1:

3.2. Let σ be a permutation of type $(b_1, \dots, b_n)$ and τ one of type $(c_1, \dots, c_n)$ in S_n; let Z^1 and Z^2 be their centralizers in S_n. If Z^1 and Z^2 are isomorphic as abstract groups, then $b_i = c_i$ for $i > 3$. If σ and τ are not conjugate in S_n, then one of the following holds:

a) $b_1 = 0, b_2 = 3, c_1 = 4, c_2 = 1, b_3 = c_3$;
b) $b_1 = 0, b_2 = c_2, b_3 = 2, c_1 = 3, c_3 = 1$.

It is easy to get a description of the maximal subgroups of S_n with nontrivial centers, using 3.1. There are two series of groups of this type: $S_2 \times S_{n-2}$ and $S_2 \operatorname{wr} S_{n-2}$.

4. In this section we give a "normal form" for the subgroups isomorphic to $S_r \operatorname{wr} S_t$ (in S_{rt}) and $S_r \times S_t$ (in S_{r+t}). We recall that G^m denotes the direct product of m copies of G.

4.1. Suppose that $n = rt$, $r \geq 3$, $t > 1$. Then A_n has a subgroup H isomorphic to A_r^t but no subgroup isomorphic to A_r^{t+1}. There exists a partition $M = \{1, 2, \dots, n\} = M_1 \cup \dots \cup M_t$ with $|M_j| = r$ and $H = A(M_1) \times \dots \times A(M_t)$. In particular, all subgroups of A_n isomorphic to H are conjugate in S_n.

The following similar results are corollaries of 4.1:

4.2. Suppose that $n = rt \neq 6$, $r > 1$, $t > 1$. Then S_n has a subgroup H isomorphic to $S_r \operatorname{wr} S_t$. All subgroups of S_n isomorphic to H are conjugate in S_n (in particular, they are maximal in S_n).

4.3. Suppose that $n = r + t$, $2 < r < t$. Then A_n contains a subgroup $H = H_1 \times H_2$ with $H_1 \cong A_r$, $H_2 \cong A_t$. All subgroups of A_n isomorphic to H are conjugate in S_n.

4.4. Suppose that $n = r + t \neq 6$, $1 < r < t$. Then S_n has a subgroup $H = H_1 \times H_2$ with $H_1 \cong S_r$, $H_2 \cong S_t$. All subgroups of S_n isomorphic to H are conjugate in S_n.

Let $C(r, n)$ be the number of conjugacy classes of subgroups of S_n isomorphic to A_r.

4.5. $C(r, n) \geq [n/r]$. We shall say that a pair $\{r, n\}$ is allowable if $C(r, n) = [n/r]$. We list below all allowable pairs $\{r, n\}$ with $r < n$:

a) $\{4, 5\}; \{6, 7\}; \{6, 8\}; \{6, 9\}; \{7, 8\}; \{7, 9\}; \{7, 10\}; \{7, 11\}; \{7, 12\}; \{7, 13\}; \{8, 9\}; \{8, 10\}; \{8, 11\}; \{8, 12\}; \{8, 13\}; \{8, 14\}$;
b) all pairs $\{r, n\}$ for $r > 8$ and $n < r(r-1)/2$.

5. The normalizers N_n of Sylow p-subgroups Σ_n of S_n are listed in 5.1.

5.1. Let $n = a_0 + a_1 p + \dots + a_t p^t$ be the p-adic decomposition of n. Then

a) $N_n = N_{a_0} \times N_{a_1 p} \times \dots \times N_{a_t p^t}$;

$$T(N_{a_i p^i}) \cap T(N_{a_J p^J}) = \varnothing, \quad i \neq J;$$

$$|T(N_{a_i p^i})| = a_i p^i;$$

b) if $1 < a < p$, then $N_{ap^m} \cong N_{p^m} \operatorname{wr} S_a$;

c) $N_{p^m}/\Sigma_{p^m} \cong C_{p-1}^m$ (again, C_t is the cyclic group of order t);

d) if $n = p^m$ and p is odd, then $[N_n, N_n] = \Sigma_n$ (it is always the case that $[N_n, N_n] \geq \Sigma_n$ if $p > 2$);

e) suppose that $m > 1$, $p > 2$, and $W = N_{p^{m-1}} \operatorname{wr} C_p$. If Q is a Sylow p-subgroup of the base group of the wreath product W and D is the diagonal subgroup, then $N_{p^m} = QDN_p$.

5.2. If $n > 5$, then every Sylow 2-subgroup of A_n is self-normalizing.

The proof of 5.2 is based on the fact that a Sylow 2-subgroup of A_n has a characteristic subgroup of order 2 when $n > 5$ (the centralizer of this subgroup is also determined; its structure is necessary for the proof).

5.3. Suppose that $\Pi = \{2, 3\}$ when $p = 2$, and that Π consists of the primes not exceeding p when $p > 2$. If G is a p-solvable subgroup of S_n whose index is not divisible by p, then $\Pi(G) \leq \Pi$. Here $\Pi(G)$ is the set of all prime divisors of $|G|$.

6. The following property is very useful in many arguments.

6.1. Let H be a subgroup of $S(M)$ such that $H = \{\pi \in H \,||T(\pi)| \leq 3\}$. Then $T(H)$ contains pairwise disjoint subsets M_α $(\alpha \in A)$ and N_β $(\beta \in B)$ such that

$$H = \left[\prod_{a \in A} S(M_a)\right] \times \left[\prod_{\beta \in B} A(N_\beta)\right].$$

6.1 has the following interesting corollary:

6.2. Let G be a transitive subgroup of $S(M)$ containing a transposition or a cycle of length 3. Then one of the following holds:

a) $G \geq A(M)$;

b) $M = M_1 \times M_2$, $|M_i| > 1$, $i = 1, 2$, $G = S(M_1) \operatorname{wr} H$, where H is a transitive subgroup of $S(M_2)$;

c) M_1 and M_2 are as in b), G is a transitive subgroup of $S(M_1) \operatorname{wr} S(M_2)$ containing the base group of the wreath product $A(M_1) \operatorname{wr} S(M_2)$.

These results carry over without difficulty to infinite groups. In particular, if a primitive subgroup of $S(M)$ (M arbitrary here) contains a transposition or a cycle of length 3, then it contains $A(M)$ (for finite M, this result is a theorem of Jordan).

7. In this section we describe the maximal subgroups of S_n whose indices are not divisible by 6.

7.1. Let G be a transitive subgroup of S_n whose index in S_n is not divisible by 6. Then one of the following holds:

a) $G \geq A_n$;

b) $G = S_r \operatorname{wr} H$, where $r = 2^k$, $k > 0$, $n = rt$, $t > 1$, and H is a transitive subgroup of degree t and odd index in S_t;

c) $G = S_r \operatorname{wr} S_2$, $n = 2r$, r is not a power of 2 or 3, but the coefficients of the 3-adic decomposition of r are 0 or 1;
d) G is a transitive subgroup of $S_r \operatorname{wr} S_t$ containing the base group of $A_r \operatorname{wr} S_t$, $r = 3^k$, $k > 0$, $n = rt$, $t > 1$, and the projection of G on the top group of the wreath product has index in it not divisible by 3.

7.2. Let G be a maximal subgroup of S_n whose index is not divisible by 6. Then one and only one of the following assertions holds:

a) $G = A_n$;
b) $G = S_r \operatorname{wr} S_t$, $r = 2^k$, $k > 0$, $t > 1$, $n = rt$;
c) $G = S_r \operatorname{wr} S_t$, $r = 3^k$, $k > 0$, $t > 1$, $n = rt$;
d) $G = S_r \operatorname{wr} S_t$, where r is as in 7.1c;
e) $G = S_r \times S_t$ and the index of G in S_{r+t} is not divisible by 6 (see 7.1a).

7.3. Let G be a regular elementary abelian subgroup of S_{2^n} and H its normalizer in S_{2^n}, where $n > 2$. Then

a) H is maximal in A_{2^n};
b) if F is a subgroup of S_{2^n} that is an extension of an elementary abelian subgroup of order 2^n by $\mathrm{GL}(n, 2)$ and F has trivial center, then F and H are conjugate in S_{2^n}.

7.3a) has been proved by many authors.

8. In this section we enumerate all the Hall subgroups of A_n. Our list is obtained as a very simple corollary of the more general Theorem 8.1. A decisive element of the proof of this theorem is the proof due to Freiman of a very nontrivial theorem about the divisibility of binomial coefficients. The theorem of Hall and Thompson concerning Hall subgroups of S_n stated in the introduction is a very special case of Theorem 8.1.

We say that a subgroup G of S_n is an $H(p)$-subgroup if the greatest common divisor of $|G|$ and $|S_n : G|$ is a power of p; G is a nontrivial $H(p)$-subgroup if it is nonprimary and different from S_n. Let $\Pi(G)$ be the set of all prime divisors of G. An $H(p)$-subgroup G is said to be an $H_0(p)$-subgroup in S_n if $G = F$ whenever $G \leq F < S_n$ and $\Pi(G) = \Pi(F)$. The enumeration of the $H_0(p)$-subgroups is not as cumbersome as that of the $H(p)$-subgroups.

8.1. Let G be a nontrivial $H_0(p)$-subgroup of S_n. Then one of the following assertions holds:

A. G is transitive.

A.1. $G = A_n$, n is bigger than 3, and $p = 2$.

A.2. $n = 20$, $p = 2$, $G = S_{10} \operatorname{wr} S_2$.

A.3. $G = S_{a_1} \operatorname{wr} S_{a_2} \operatorname{wr} \cdots \operatorname{wr} S_{a_m}$, $n = a_1 \cdots a_m$, where $a_1, \dots, a_{m-1}$ are 2 or 4 and a_m is not more than 4.

A.4. Let q be a prime such that $q \equiv 1(p)$, and G_0 an $H_0(p)$-subgroup of S_q with $\Pi(G_0) = \{p, q\}$ (G_0 is determined up to conjugacy in S_q). Then $n = q^s$, $s \geq 1$, and $G = G(0)$ is the wreath product of s groups isomorphic to G_0.

A.5. $p = 2$. Let q be a Mersenne prime. Choose a regular elementary abelian subgroup L of order $q + 1$ in S_{q+1}. Then L is normalized by a nonabelian subgroup F of S_{q+1} of order $2q$. Suppose that $G_1 = F \rightthreetimes L$ is a semidirect product. Then $G = G(1)$ is the wreath product of the group $G(0)$ of A.4 (constructed for $p = 2$) by G_1.

A.6. Let t be a power of p less than q and P a Sylow p-subgroup of S_t. Then $G = G(2) = G(0) \operatorname{wr} P$ or $G = G(3) = G(1) \operatorname{wr} P$ ($p = 2$ in the latter case).

B. The group G is intransitive.

B.1. If G is solvable, then it is a direct product of groups of type $G(0)$, $G(1)$, $G(2)$, $G(3)$ (the determination of which factors participate in this direct product, and in what numbers, is a fairly cumbersome number-theoretical calculation of no great value in our context).

B.2. $G = S_2 \times S_{n-2}$, $n = 2p^\alpha + 1$ is prime or $n - 1 = 2p^\beta - 1$ is prime.

B.3. $G = S_{n-1}$, n a power of p.

B.4. $G = S_{n-3} \times S_2$, $p = 3$, $n = 2 \cdot 3^\alpha + 1$ and $n - 2$ are both primes.

B.5. $n = 8$, $p = 2$, $G = S_3 \times S_5$.

B.6. $G = S_{n-3} \times S_3$, $n = 6p^\alpha + 1$ and $n - 2$ are both primes.

B.7. $n = 13$; $p = 5$, $G = S_4 \times S_9$.

B.8. $n = 14$, $p = 7$, $G = S_4 \times S_{10}$.

B.9. $n = 13$, $p = 3$, $G = S_5 \times S_8$.

B.10. $n = 17$, $p = 2$, $G = S_7 \times S_{10}$.

B.11. $n = 19$, $p = 2$, $G = S_9 \times S_{10}$.

We note that $S_{10} \times S_{10}$ is an $H(2)$-subgroup of S_{20}, but not an $H_0(2)$-subgroup.

Note too that a Hall subgroup of A_n is an $H(2)$-subgroup of S_n. Thus, we deduce from 6.1:

6.2. If H is a Hall subgroup of A_n, then S_n has a Hall subgroup H_0 such that $H = H_0 \cap A_n$.

§3. Open questions

1. Describe the maximal subgroups of A_n of odd index.
2. Show that every subgroup of S_n can be generated by $[n/2]$ elements when $n > 3$.
3. For which n does S_n have a subgroup of index $[(n + 1)/2]!$?
4. What is the greatest number of pairwise noncommuting elements of S_n (of A_n)?
5. A set T of permutations is said to be product-free if $\sigma_1, \sigma_2 \in T \Rightarrow \sigma_1\sigma_2 \notin T$. What is the greatest length of product-free set in S_n ?

6. Elucidate the structure of the Frattini subgroup of a Sylow p-subgroup of S_{p^m}, $p > 2$.
7. Let Π be a set of primes. Find the maximum of the orders of abelian Π-subgroups of S_n.
8. Find the maximum rank of maximal subgroups of Σ_{p^m}.
9. Determine the subgroups of S_n whose indices are of the form $p^a q^b$, where p and q are different primes.
10. Study the intransitive thin subgroups of S_n containing a ternary cycle.
11. Investigate the pairs $r < n$ such that S_n has a maximal subgroup isomorphic to S_r.

References

1. P. Hall, *Theorems like Sylow's*, Proc. London Math. Soc. **3** (1956), 286–304.

2. J. G. Thompson, *Hall subgroups of the symmetric groups*, J. Combinatorial Theory **1** (1966), 271–279.

3. D. A. Suprunenko, *Gruppy matrits*, Moscow, 1972. Translation: *Matrix groups*, Translations of Math. Monographs, Vol. 45, American Mathematical Society, Providence, RI, 1976.

4. D. Gorenstein, *Reviews on finite groups*, American Mathematical Society, Providence, RI, 1974.

5. R. W. Ball, *Maximal subgroups of the symmetric groups*, Trans. Amer. Math. Soc. **112** (1966), 398–407.

6. R. A. Baĭramov, *On the completeness problem in the symmetric semigroup of finite degree*, Diskret. Analiz. **8** (1966), 1–26.

7. H. Wielandt, *Finite Permutation Groups*, Academic Press, New York, London, 1964.

8. R. Bercov and L. Moser, *On abelian permutation groups*, Canad. Math. Bull. **8** (1965), 627–630.

9. * M. W. Liebeck and J. Saxl, *On the orders of maximal subgroups of the finite exceptional groups of Lie type*, Proc. London Math. Soc. (3) **55** (1987), no.2, 299–330.

Rostov-on-Don Civil Engineering Institute

Translated by JAMES WIEGOLD

* Added by the translator

Amer. Math. Soc. Transl.
(2) Vol. **149**, 1991

A Generalization of Burnside's Second Theorem on $\{p, q\}$-Groups

UDC 519.4

YA. G. BERKOVICH

§1.

Together with the theorem on the solvability of $\{p, q\}$-groups (in the sequel p and q are always different primes) Burnside also proved "another theorem on $\{p, q\}$-groups":

LEMMA 1 (Burnside [1]). *Let G be a group of order $p^a q^b$. If $q^b > p^a$ then $O_q(G) > 1$ (here $O_q(G) \in \mathrm{Syl}_q(F(G))$, where $F(G)$ is the Fitting subgroup of G), with the following possible exceptions:*

(i) *$q = 2$ and p is a Fermat prime;*
(ii) *$p = 2$ and q is a Mersenne prime;*
(iii) *$p = 7$ and $q = 2$.*

Let $G_p = \mathrm{Syl}_p(G)$. Put $p^{a_p(G)} = \max\{|H| : H \le G_p, \ \mathrm{cl}\, H \le 2\}$.
Glauberman gave an interesting and deep development of

LEMMA 2 (Glauberman [2]). *Let G be a $\{p, q\}$-group and let $p^{a_p(G)} < q^{a_q(G)}$. Then $O_q(G) > 1$.*

An interesting moment in Lemma 2 is the lack of restrictions on the primes p and q.

The following lemma appeared as Exercise 6.16 in [3].

LEMMA 3. *Suppose that P is a finite p-group, $p > 2$, $Q \in \mathrm{Syl}_q(\mathrm{Aut}\, P)$ with $q > 2$. Then $|Q| < \frac{1}{2}|P|$.*

In the proof of Lemma 1 one uses, as we will see, statements similar to Lemma 3. Therefore for strengthening of Lemma 1 one should strengthen

1991 *Mathematics Subject Classification.* Primary 20D20.
Translation of Izv. Severo-Kavkaz. Tsentra Vyssh. Shkoly Estestv. Nauk. **1985**, no. 4, 18–22; MR 87j:20042.

Lemma 3. This will be achieved due to Theorem 1, proved by G. A. Freĭman specifically for this paper.

We give additional notations (in addition to the ones already introduced).

Let $a > 1$ and n be natural numbers. Assume that

$$C(n, a) = (a-1)(a^2-1)\cdots(a^n-1).$$

Let q be a prime that does not divide a, and let $t_q(b)$ be the greatest power of q that divides a natural number b. Let $m = t_q(C(n, a))$. Let α denote the order of a (mod q), and suppose that $s = t_q(a^\alpha - 1)$, and $t = (a^s-1)/q^s$. To simplify our notation we did not reflect the dependence of m on a, n, and q. The same relates to the numbers α, s, and t. Recall that $t_q(n!) = [\frac{n}{q}] + [\frac{n}{q^2}] + \cdots$, where $[x]$ is the integer part of a real number x.

LEMMA 4. *Suppose that $B > 1$ is a natural number, $B \equiv 1 \pmod q$, and $M = 1 + B + \cdots + B^{r-1}$. If $q > 2$ or $q = 2$ and $B \equiv 1 \pmod 4$, then $t_q(M) = t_q(r)$.*

PROOF. For natural θ assume that

$$M(\theta) = 1 + B^\theta + B^{2\theta} + \cdots + B^{(q-1)\theta}.$$

Then

$$\begin{aligned} M(\theta) &= q + (B^\theta - 1) + \cdots + (B^{(q-1)\theta} - 1) \\ &= q + (B^\theta - 1)(1 + B^\theta + 1 + \cdots + B^{(q-2)\theta} + \cdots + B^\theta + 1) \\ &= q + (B^\theta - 1)\{q - 1 + (q-2)B^\theta + (q-3)B^{2\theta} + \cdots + B^{(q-2)\theta}\} \\ &= q + (B^\theta - 1)\{q - 1 + q - 2 + \cdots + 1 + (q-2)(B^\theta - 1) \\ &\quad + (q-3)(B^{2\theta} - 1) + \cdots + (B^{(q-2)\theta} - 1)\} \\ &= q + (B^\theta - 1)\left\{\frac{q(q-1)}{2} + (B^\theta - 1)N\right\} \equiv q \pmod{q^2}, \end{aligned}$$

if $q > 2$ or $q = 2$ and $B \equiv 1 \pmod 4$; here N is a natural number whose exact value we do not need.

Let $r = q^t v$, where $(v, q) = 1$ (so that $t = t_q(r)$). Then

$$\begin{aligned} M &= 1 + B + \cdots + B^{r-1} = 1 + B + \cdots + B^{vq^t - 1} \\ &= (1 + B + \cdots + B^{q^t - 1})(1 + B^{q^t} + \cdots + B^{(v-1)q^t}) \\ &= (1 + B + \cdots + B^{q-1})(1 + B^q + \cdots + B^{q(q-1)})\cdots \\ &\quad (1 + B^{q^{t-1}} + \cdots + B^{q^{t-1}(q-1)})(1 + B^{q^t} + \cdots + B^{(v-1)q^t}) \\ &= M(1)M(q)\cdots M(q^{t-1})(1 + B^{q^t} + \cdots + B^{(v-1)q^t}). \end{aligned}$$

We have proved that $t_q(M(q^i)) = 1$ for $0 \le i < t$. Clearly,

$$1 + B^{q^t} + \cdots + B^{(v-1)q^t} \equiv v \not\equiv 0 \pmod q.$$

Therefore $t_q(M) = t = t_q(r)$, Q.E.D.

In the sequel we assume that $A = a^\alpha$. Lemma 4 can be applied to A rather than B.

THEOREM 1 (G. A. Freĭman). (i) *If* $q > 2$ *or* $q = 2$ *and* $s > 1$ (*that is,* $a \equiv 1 \pmod 4$), *then*

$$m = t_q(C(n, a)) = s\left[\frac{n}{a}\right] + t_q\left(\left[\frac{n}{a}\right]!\right).$$

(ii) *If* $q = 2$ *and* $s = 1$ (*that is,* $a \equiv 3 \pmod 4$) *and* $s_1 = t_2(a+1)$, *then*

$$m = t_2(C(n, a)) = n + s_1\left[\frac{n}{2}\right] + t_2\left(\left[\frac{n}{2}\right]!\right).$$

PROOF. Obviously,

$$t_q(C(n, a)) = m = t_q((A-1)(A^2-1)\cdots(A^{[n/\alpha]}-1)).$$

However,

$$\begin{aligned}&(A-1)(A^2-1)\cdots(A^{[n/\alpha]}-1)\\&\quad = (A-1)^{[n/\alpha]}(A+1)(A^2+A+1)\cdots(A^{[n/\alpha]-1}+\cdots+A+1),\qquad (1)\\&\qquad t_q((A-1)^{[n/\alpha]}) = s[n/\theta].\end{aligned}$$

(i) Let either $q > 2$ or $q = 2$ and $s > 1$. Then, by Lemma 4, $t_q(A+1) = t_q(2)$, $t_q(A^2+A+1) = t_q(3), \ldots$; $t_q(1+A+\cdots+A^{[n/\alpha]-1}) = t_q([\frac{n}{\alpha}])$, so that $t_q((A+1)(A^2+A+1)\cdots(A^{[n/\alpha]-1}+A^{[n/\alpha]-2}+\cdots+A+1)) = t_q(2) + t_q(3) + \cdots + t_q([\frac{n}{\alpha}]) = t_q([\frac{n}{\alpha}]!)$, which proves statement (i).

(ii) Let $q = 2$ and let $s = 1$. Assume that $s_1 = t_q(a+1)$. In this case the product (1) takes the form

$$(a-1)^n(1+a)(1+a+a^2)\cdots(1+a+\cdots+a^{n-1}).\qquad (2)$$

Each of its factors with an odd number of summands is odd. On the other hand, the factors $1 + a + \cdots + a^{2^k v-1} = (1+a)(1+a^2) + \cdots + (1+a^{2^{k-1}})(1+a^{2^k} + \cdots + a^{(v-1)2^k})$ contain 2 precisely to the power $s_1 + k - 1 = s_1 + t_2(2^{k-1}v)$. Therefore, the right-hand side of (2) contains 2 precisely to the power

$$n + s_1\left[\frac{n}{2}\right] + \sum_{k=1}^{[n/2]} t_2(k) = n + s_1\left[\frac{n}{2}\right] + t_2\left(\left[\frac{n}{2}\right]!\right),$$

which completes the proof of (ii) and that of the theorem.

Observe that $|\mathrm{GL}(n, p^r)| = p^{rn(n-1)/2}C(n, p^r)$, so that the canonical decomposition of $|\mathrm{GL}(n, p^r)|$ may be deduced from Theorem 1.

§2.

Consider in detail the classical case corresponding to Lemma 3; namely, assume that

$$q^m > \frac{1}{2}a^n. \tag{3}$$

1. Let $a > 2$ and let $q > 2$. Then, by Theorem 1, $m = s[\frac{n}{\alpha}] + t_q([\frac{n}{\alpha}]!)$. However $2^s = \frac{1}{t}(a^\alpha - 1) < \frac{\alpha}{t}$, so that

$$q^m < \left(\frac{a^\alpha}{t}\right)^{[n/\alpha]} q^{[n/\alpha]1/q-1} \leq \left(\frac{q^{1/q-1}}{t}\right)^{[n/\alpha]} a^n,$$

and therefore, by (3),

$$\left(\frac{q^{1/q-1}}{t}\right)^{[n/\alpha]} > \frac{1}{2}. \tag{4}$$

Assume that $t \geq 2$. If $[\frac{n}{\alpha}] < q$, then $m = s[\frac{n}{\alpha}]$, and, by (3), $q^{s[n/\alpha]} > \frac{1}{2}a^n$, that is, $\left(\frac{a^{\alpha-1}}{t}\right)^{[n/\alpha]} > \frac{1}{2}a^n$, and $a^n \geq a^{[n/\alpha]\alpha} > \frac{1}{2}t^{[n/\alpha]}a^n$, whence $t^{[n/\alpha]} < 2$ which is a contradiction for $n \geq \alpha$. If $n < \alpha$ then $m = 0$ and $a^n < 2$, which is impossible for $a > 2$ and $n > 0$. Let $n > \alpha$. Then $[\frac{n}{\alpha}] \geq q$. By (4), $(\frac{q^{1/q-1}}{t})^{[n/\alpha]} > \frac{1}{2}$ and hence $t \geq 2$. Let $t = 2$. Then $2q^s = a^\alpha - 1 = (a-1)(1+a+\dots+a^{\alpha-1})$ and if $\alpha > 1$ then $a - 1 = 2$ and $a = 3$. However, in this case $q \geq 5$. If $q = 5$ then $t = 16 > 2$ which is a contradiction. It follows that $q > 5$ and hence (4) fails.

Thus $\alpha = 1$ and (4) becomes

$$\left(\frac{q^{1/q-1}}{t}\right)^n > \frac{1}{2}.$$

Since $q^{1/q-1} \leq \sqrt{3}$, we obtain $n \leq 4$. But $n \geq q \geq 3$. An immediate verification shows that this is impossible. Thus we have proved that $t = 1$, and so $q^s = a^\alpha - 1 = (a-1)(1 + a + \dots + a^{\alpha-1})$ which gives $\alpha = 1$ and $a = 1 + q^s$. In this case $m = sn + t_q(n!)$, and (3) becomes

$$2q^{t_q(n!)} > \left(1 + \frac{1}{q^s}\right)^n. \tag{5}$$

This inequality holds if and only if (3) holds. Inequality (5) definitely fails for $s = 1$, large q, and $n = q - 1$ (since its left-hand side is 2 while the right-hand side is close to e, the base of the natural logarithms).

2. $q = 2$ and $a \equiv 1 \pmod 4$, i.e., $s > 1$. In this case $\alpha = 1$, $m = sn + t_2(n!)$, and $2^s = \frac{a-1}{t}$.

Let $t \geq 3$. Then (4) becomes $(2/t)^n > \frac{1}{2}$, whence $t = 3$, $n = 1$, $m = s$, and $2^m = 2^s = \frac{1}{3}(a-1) < \frac{1}{2}a$, which is a contradiction. Thus $t = 1$, $a = 1 + 2^s$, and $2^m = (2^s)^n 2^{t_2(n!)} = (a-1)^n 2^{t_2(n)}$. If $n > 3$ then $t_2(n!) > n/2$

and hence $2^m > (a-1)^n 2^{n/2} = \{\sqrt{2}(a-1)\}^n > \frac{1}{2}a^n$, since $a \geq 5$. Clearly, (3) holds for $N = 1, 2, 3$.

3. $q = 2$ and $a \equiv 3 \pmod 4$, i.e., $s = 1$. Again $\alpha = 1$. Assume $s_1 = t_2(a+1)$, $2^{s_1} = (a+1)/t$, and $(t, 2a) = 1$. By Theorem 1, $m = n + s_1[\frac{n}{2}] + t_2([\frac{n}{2}]!) < n + \frac{s_1 n}{2} + \frac{n}{2} = \frac{3n}{2} + \frac{s_1 n}{2}$, $2^m < (2^{s_1} \cdot 8)^{n/2} = (\frac{8(a+1)}{t})^{n/2}$, and

$$\sqrt[n]{4} \cdot \frac{8(a+1)}{t} > \frac{1}{2}a^2. \tag{6}$$

Let $n \geq 2$. Then

$$\frac{16(n+1)}{t} > a^2, \tag{7}$$

and hence $a \leq 15$.

If $a = 15$ then $2^4 = \frac{15+1}{1}$ implies $s_1 = 4$ and $t = 1$ and, by (6), $128\sqrt[n]{4} > 225$. This is true only if $n = 2$. In this case $m = 6$, $2^m = 2^6 < \frac{1}{2} \cdot 15^2$, and (3) fails. If $a = 11$ then $s_1 = 2$, $t = 3$, and $32\sqrt[n]{4} > 121$, which fails for $n > 1$.

Let $a = 7$. Then $s_1 = 3$, $t = 1$, and

$$2 \cdot 2^m = 2^{n+3[n/2]+t_2([n/2]!)+1} > 7^n \tag{8}$$

must hold. However, (8) does not hold for all n; it fails, for example, for $n = 1, 3, 5, 7$, and it holds if and only if (3) holds.

Let $a = 3$. Then $t = 1$, $s_1 = 2$, $m = n + 2[n/2] + t_2([n/2]!)$, and the inequality

$$2^m = 2^{n+2[n/2]+t_2([n/2]!)} > \frac{1}{2} \cdot 3^n$$

must hold. It is not difficult to see that this inequality is not always true.

Finally, let $n = 1$. Then $m = 1$ and $2^m = 2 > \frac{1}{2}$ implies $a = 3$.

4. Let $a = 2$ and let $q > 2$. Inequality (4) holds. If $t \geq 3$ then $q \geq 5$ and (4) fails. Thus $t = 1$ and $q^s = 2^\alpha - 1$, whence $s = 1$ and $q = 2^\alpha - 1$ is a Mersenne prime. Inequality (3) holds if and only if

$$q^{[n/\alpha]+t_q([n/\alpha]!)} > 2^{n-1}. \tag{9}$$

All possible cases have been considered, and thus the following theorem has been proved.

THEOREM 2. *If $a > 1$ and n are natural numbers, q a prime which does not divide a, and $q^m > \frac{1}{2}a^n$, where $m = t_q(C(n, a))$, then exactly one of the following statements holds*:

1. $a = 1 + q^s$ *and* $2q^{tq(n!)} > (1 + \frac{1}{q^s})^n$, $a > 2$; $q > 2$.
2. $a = 1 + 2^s$; $q = 2$; n *is arbitrary.*
3. $a = 7$; $q = 2$; $2^{n+3[n/2]+t_2([n/2]!)+1} > 7^n$.
4. $a = 2$; $q = 2^\alpha - 1$, $q^{[n/\alpha]+t_q([n/\alpha]!)} > 2^{n-1}$.

Theorem 2 implies the following

COROLLARY. *Suppose that* p *and* q *are different primes,* n *is a natural number, and* $m = (t_q(C(n, p)))$. *If* $q^m > \frac{1}{2}p^n$, *then exactly one of the following statements holds*:

1. $q = 2$, p *is a Fermat prime,* n *is arbitrary.*
2. $q = 2$, $p = 7$, $2^{n+3[n/2]+t_2([n/2]!)+1} > 7^n$.
3. q *is a Mersenne prime,* $p = 2$, $q^{[n/2]+t_q([n/2]!)} > 2^{n-1}$.

In the sequel we use this corollary only.

§3.

Now we apply the results of the previous section to $\{p, q\}$-groups. We need certain elementary facts from the theory of finite groups (their proofs may be found in [4] and [5]).

LEMMA 5 (P. Hall). *Suppose that* G *is a finite p-group,* $\alpha \in \operatorname{Aut} G$, *and* α *induces the identity automorphism on* $G/\Phi(G)$, *where* $\Phi(G)$ *is the Frattini subgroup of* G. *Then the order of* α *is a degree of* p.

Let G be a finite p-group (in what follows only finite groups are considered). Assume that $p^{d(G)} = |G: \Phi(G)|$. The number $d(G)$ is called the *rank* of G. Lemma 5 shows that if $q \neq p$ and $Q \in \operatorname{Syl}_q(\operatorname{Aut} G)$, then Q is isomorphic to a subgroup of $\operatorname{Aut} G/\Phi|G| \cong \operatorname{GL}(D(G), p)$, so that $|Q| \leq q^{t_q(C(d(G), p))}$. However, as we will see, other invariants of G may be often substituted for $d(G)$ in this inequality.

If Π is a set of primes then $O_\Pi(G)$ denotes the greatest normal Π-subgroup of G.

LEMMA 6 (Hall-Higman). *Let* G *be a* Π*-solvable group and let* $O_{\Pi'}(G) = 1$. *Then* $C_G(O_\Pi(G)) \leq O_\Pi(G)$.

LEMMA 7 (Lemma 1.2 in [4]). *Let* G *be a p-group and let* $\alpha \in \operatorname{Aut}(G)$. *If* $H \leq G$ *and* $HC_G(H) \leq C_G(\alpha)$, *then the order of* α *is a degree of* p.

LEMMA 8 (Thompson). *An arbitrary p-group* G *contains a characteristic subgroup* H *with* $C_G(H) \leq H$, $\Phi(H) \leq Z(H)$, *and* $H/Z(H) \leq Z(G/Z(H))$.

LEMMA 9. *Suppose that* G *is a p-group,* $d(G) \leq 2$, *and*

$$\varepsilon(p) = \begin{cases} 1, & p > 2; \\ 2, & p = 2. \end{cases}$$

Then $G_{[\varepsilon(p)]} = \{x \in G | x^{p^{\varepsilon(p)}} = 1\}$ *is a subgroup of* G.

LEMMA 10. *Suppose that* G *is a p-group,* $\operatorname{cl} G \leq 2$, $\alpha \operatorname{Aut} G$, *and* $\alpha|_{G[\varepsilon(p)]} = \mathrm{id}$. *Then the order of* α *is a degree of* p.

A triple $\{p, q, n\}$ is called *special* if $q^{t_q(C(n, p))} > \frac{1}{2}p^n$. Special triples are described in a corollary to Theorem 2.

A subgroup H of a p-group G that satisfies the conditions of Lemma 8 is called a *Thompson subgroup* of G.

THEOREM 3. *Suppose that P is a p-group, $q \neq p$, and $Q \in \mathrm{Syl}_q(\mathrm{Aut}\, P)$.*

1. *If $|Q| > \frac{1}{2}p^{d(P)}$, then $\{p, q, d(P)\}$ is a special triple.*
2. *Let $r = \min\{d(H)\colon H$ is a Thompson subgroup of $P\}$. If $|Q| > \frac{1}{2}p^r$ then $\{p, q, r\}$ is a special triple.*
3. *Let $r = \min\{d(H_{[\varepsilon(p)]}\colon H$ is a Thompson subgroup of $P\}$. If $|Q| > \frac{1}{2}p^r$ then $\{p, q, r\}$ is a special triple.*
4. *Let $r = \min\{d(H)\colon H$ is a characteristic subgroup of P with $C_P(H) \leq H\}$. If $|Q| > \frac{1}{2}p^r$, then $\{p, q, r\}$ is a special triple.*

We prove claim 3 (other claims are proved either analogously or even more simply). Let H be a Thompson subgroup of P for which $d(H_{[\varepsilon(p)]}) = r$. By definition, Q faithfully acts on P. By Lemmas 7 and 8, Q faithfully acts on H. By Lemma 10, Q faithfully acts on $H_{[\varepsilon(p)]}$. By Lemma 5, Q faithfully acts on $H_{[\varepsilon(p)]}/\Phi(H_{[\varepsilon(p)]})$, and hence Q is isomorphic to a q-subgroup of $\mathrm{GL}(r, p)$. If $|Q| > \frac{1}{2}p^r$, then $\{p, q, r\}$ is a special triple by definition (and all such triples are described in the corollary).

THEOREM 4. *Suppose that $G = PQ$, $P \in \mathrm{Syl}_p(G)$, $Q \in \mathrm{Syl}_q(G)$ and p and q are different primes. Let $r = \max\{d(H)\colon H$ is a normal subgroup of P with $C_P(H) \leq H\}$. If $|Q| > \frac{1}{2}p^r$ then $O_q(G) > 1$ with the possible exception of the cases when $\{p, q, r_1\}$ is a special triple with $r_1 \leq r$.*

PROOF. Assume that $O_q(G) = 1$. Then $O_p(G) > 1$ because G is solvable. Since, by Lemma 6, $C_G(O_p(G)) \leq O_p(G)$, Q faithfully acts on $O_p(G)$. Let H be a Thompson subgroup of $O_p(G)$. Then $H \lhd P$ and Q faithfully acts on H. By Lemma 5, Q faithfully acts on $H/\Phi(H)$, so that Q is isomorphic to a subgroup of $\mathrm{GL}(d(H), p)$. By our condition, $d(H) \leq r$. Therefore $|Q| > \frac{1}{2}p^r \geq \frac{1}{2}p^{d(H)}$, so that $\{p, q, d(H)\}$ is a special triple. Theorem 4 is proved for $r_1 = d(H) \leq r$.

In Theorem 4 H can be replaced by $H_{[\varepsilon(p)]}$.

Analogously one can prove several stronger forms of Lemma 2. We give one of the related results. Recall that $a_p(G)$ was defined in Lemma 2.

THEOREM 5. *Let $G = PQ$ be a $\{p, q\}$-group, where p and q are different primes. Suppose that $P \in \mathrm{Syl}_p(G)$, $Q \in \mathrm{Syl}_q(G)$, $r = \max\{d(H) | H \trianglelefteq P$, $\mathrm{cl}\, H \leq 2$, $C_P(H) \leq H\}$. If $q^{a_q(G)} > p^r$ then $O_q(G) > 1$.*

PROOF. Let $O_q(G) = 1$. Then $O_p(G) > 1$ and $C_G(O_p(G)) \leq O_p(G)$, so that Q faithfully acts on $O_p(G)$. Let H be a Thompson subgroup of $O_p(G)$. By Lemmas 6, 7, and 8, Q faithfully acts on $H/\Phi(H)$. However, $q^{a_q(G)} > p^r \geq |H/\Phi(H)|$ and therefore, by Lemma 2, $O_q(QH/\Phi(H)) > 1$ which is a contradiction.

If the inequality $p^r < q^{a_q(G)}$ in Theorem 5 is replaced by $q^{a_q(G)} > \frac{1}{2}p^r$, we obtain the same conclusion with a few exceptions given by the corollary to Theorem 2; however, this is weaker than the results obtained.

§4.

In the previous section we saw the important role played by subgroups H with $C_G(H) \leq H$. These subgroups are important for studying nonprimary groups as well (if H is such a subgroup, then $\mathrm{Aut}\, H$ controls the factor group $N_G(H)/Z(H)$). Therefore, it is important to have sufficiently simple criteria for the existence of self-centralizers with a possibly uncomplicated structure. We axiomatize an approach to this problem.

Let $L(G)$ be the set of all subgroups of a group G. A function $f: L(G) \to L(G)$ is called *centralizing* if it satisfies the following conditions:

(SC1) $H \in L(G) \Rightarrow [H, f(H)] = 1$, i.e., $f(H) \leq C_G(H)$;
(SC2) $A, B \in L(G),\ A \leq B \Rightarrow f(A) \geq f(B)$;
(SC3) $\alpha \in \mathrm{Aut}\, G \Rightarrow f(H^\alpha) = f(H)^\alpha,\ H \in L(G)$.

A subgroup H is called *f-closed* if $f(H) \leq H$. If $f(H) = C_G(H)$, then an f-closed subgroup is called *closed*. Our goal is to indicate f-closed subgroups with sufficiently simple structure.

LEMMA 11. *$Hf(H)$ is an f-closed subgroup.*

Indeed, by (SC2), $f(Hf(H)) \leq f(H) \leq Hf(H)$.

Let $Z_k(G)$ be the kth term of the upper central series of G.

THEOREM 6. *Suppose that $A \leq \mathrm{Aut}\, G$, H is a nilpotent A-invariant subgroup of G, and $f(H)$ is nilpotent. Then between H and $Hf(H)$ there exists an A-invariant f-closed subgroup $F \trianglelefteq Hf(H)$ such that*

(a) *if H is abelian then* $\mathrm{cl}\, F \leq 2$;
(b) *if H is not abelian then* $\mathrm{cl}\, F = \mathrm{cl}\, H$.

PROOF. It follows from the axioms that $f(H)$ is A-invariant. If $\mathrm{cl}\, f(H) \leq \mathrm{cl}\, H$, then we assume that $F = Hf(H)$; then the result follows from Lemma 11. Further, suppose that $\mathrm{cl}\, f(H) > \mathrm{cl}\, H$. Assume that $H_1 = HZ_2(f(H))$. By (SC2), $f(H_1) \leq f(H)$. If H is not abelian then $\mathrm{cl}\, H_1 = \mathrm{cl}\, H$; if H is abelian then $\mathrm{cl}\, H_1 \leq 2$. If $f(H_1) \leq H_1$ then the theorem holds for $F = H_1$. Otherwise we assume that $H_2 = H_1 Z_2(f(H_1))$. Observe that $H \cap f(H) \leq Z(f(H))$ for every $H \in L(G)$. It follows from our assumptions that $H < H_1 < H_2$. Continuing in this way we construct a *strictly ascending chain*

$$H < H_1 < H_2 < \cdots < H_k$$

with $f(H_k) \leq H_k$. Assume that $F = H_k$. Then F is both f-closed and A-invariant. Since F is a central product of H and a subgroup whose class does not exceed 2, our theorem is proved.

REMARK. For $\operatorname{cl} H > 1$ the method of proof of Lemma 8 produces a subgroup which is either of class $\operatorname{cl} H$ or $\operatorname{cl} H + 1$.

COROLLARY. *Let G be a p-group and let H be maximal among characteristic subgroups of class $k > 1$ of G. Then $C_G(H) \leq H$.*

Indeed, $HC_G(H)$ is characteristic in G, and by Theorem 6, between H and $HC_G(H)$ there exists a characteristic subgroup F of class k such that $C_G(F) \leq F$. By our condition $F = H$.

Let G be a solvable group. Let $A_1(G)$ denote the set of all maximal elements of the set of characteristic abelian subgroups of G. If $T_1 \in A_1(G)$ and $T_2/T_1 \in A_1(G/T_1)$, then all such T_2 form $A_2(G)$. Analogously we define sets $A_3(G), A_4(G), \ldots$.

THEOREM 7. *Suppose that $A \leq \operatorname{Aut} G$, H is an A-invariant solvable subgroup of G, and $f(H)$ is solvable. Then between H and $Hf(H)$ there exists an f-closed and A-invariant subgroup F such that*

(a) *if H is abelian then the derived length of F does not exceed* 2;

(b) *if H is not abelian then H and F have the same derived length.*

Proof of this theorem is exactly the same, but, instead of $Z_2(f(H_1))$, one considers subgroups from $A_2(f(H_1))$.

COROLLARY. *Let G be a solvable group and let H be maximal among characteristic subgroups of G with derived length $k > 1$. Then $C_G(H) < H$.*

We notice that a group G in Theorem 6 is not assumed to be nilpotent, nor solvable in Theorem 7.

After this paper was finished V. A. Belonogov informed the author in a conversation that he had proved Theorem 1 too. However, as far as we know, his proof differs from ours.

REFERENCES

1. W. Burnside, *On groups of order $p^a q^b$*, Proc. London Math. Soc. **2 (1904)**, 388–392, 432–437.
2. G. Glauberman, *On Burnside's other $p^a q^b$-theorem*, Pacific J. Math. **56** (1975), 469–476.
3. I. M. Isaacs, *Character theory of finite groups*, N.Y., 1976.
4. B. Huppert and N. Blackburn, *Finite groups*, III, Berlin, 1982.
5. B. Huppert, *Endliche Gruppen*, Berlin, 1967.

Rostov Institute of Civil Engineering

Translated by BORIS M. SCHEIN

Amer. Math. Soc. Transl.
(2) Vol. **149**, 1991

Finite Groups of Large Degree

UDC 519.4

YA. G. BERKOVICH AND B. M. POGREBINSKIĬ

§1. Introduction

The Cayley Theorem, which is proved on the first pages of textbooks on the theory of groups, states that a group of order n is isomorphic to a subgroup of the symmetric group S_n of degree n. If we define

$$\delta(G) = \min\{n | G \cong H \leq S_n\}$$

and call $\delta(G)$ the degree of G, then the Cayley Theorem states that the degree of a group exists and always $\delta(G) \leq |G|$. If $G \in \{C(p^n), Q(2^n), C(2) \times C(2)\}$ where $C(m)$ is a cyclic group of order m and $Q(2^n)$ is the generalized quaternion group, then $\delta(G) = |G|$. It is interesting to note (and, as far as the authors know, B. M. Schein was the first to observe) that there are no other G with $\delta(G) = |G|$. Let

$$\Delta(G) = \frac{\delta(G)}{|G|}.$$

It turns out that it is better to handle $\Delta(G)$ rather than $\delta(G)$. The problem of complete classification of all groups G with $\Delta(G) \geq 1/2$ was raised in [1]. We solve a considerably more general problem here. Let p be the least prime divisor of $|G|$ and let $\Delta_0(G) = \frac{p+1}{p^3}$. G is called a group of large degree or a $(*)$-group if $\Delta(G) > \Delta_0(G)$. Our goal is to classify $(*)$-groups. We use elementary methods, but the proof is nontrivial.

In the sequel p, q, and r denote pairwise different primes; m and n are natural numbers; $D(2m)$ is the dihedral group of order $2m$; for $H \leq G$ the intersection $H_G = \bigcap_{x \in G} xHx^{-1}$ is called the core of H and G; $C_G(H)$ denotes the centralizer of a subset H in G; $G = A * B$ denotes the central product of A and B, that is, $G = AB$, and $[A, B] = 1$ (we always assume

1991 *Mathematics Subject Classification*. Primary 20D60; Secondary 20B10.

Translation of Izv. Severo-Kavkaz. Nauchn. Tsentra Vyssh. Shkoly Estestv. Nauk. **1986**, no. 1, 21–26; MR 88e:20011.

that $A \cap B = Z(A))$; $A \leftthreetimes B$ is a semidirect product of A and B with the kernel B.

We use standard notation and terminology from Huppert's book [2].

If G is a p-group then $G_{(m)} = \langle x \in G | x^{p^m} = 1 \rangle$, $G^{(m)} = \langle x^{p^m} | x \in G \rangle$.

Let G be a group of permutations of degree $\delta(G)$ of a set M; let $M_1, \dots, M_k$ be a complete set of G-orbits, let $x_i \in M_i$ and let G^i be the stabilizer of x_i in G. Then

$$\Delta(G) = \frac{1}{|G^1|} + \cdots + \frac{1}{|G^k|},$$
$$\delta(G) = |G: G^1| + \cdots + |G: G^k|.$$

$\delta(G)$ may be defined as

$$\min\left\{\sum_i |G: G^i| \mid G^i \leq G, (G^1 \cap \cdots)_G = 1\right\}.$$

In the sequel we always assume that

$$\delta(G) = \sum_{i=1}^k |G: G^i|, \quad (G^1 \cap \cdots \cap G^k)_G = 1.$$

However, note that k is not an invariant of G (it suffices to take $G = C(2) \times C(2)$ or $G = D(6) \times D(6))$.

If $D^1, \dots, D^s$ are subgroups of G such that

$$(D^1 \cap \cdots \cap D^s)_G = 1,$$

then

$$\sum_{i=1}^s |G: D^i| \geq \delta(G), \qquad \sum_{i=1}^s \frac{1}{|D^i|} \geq \Delta(G).$$

A useful lemma follows from this remark:

Lemma 1. *$H \leq G \Rightarrow \Delta(H) \geq \Delta(G)$. In particular, property $(*)$ is hereditary for subgroups.*

To compute $\delta(G)$ we use

Lemma 2. *Let $T_1 \times \cdots \times T_s$ be the socle of a group G, where T_i are minimal normal subgroups of G. Then $k \leq s$ (here k is the number of summands in the formula for $\delta(G)$).*

Lemma 3 (Johnson [3]). *If $(|A|, |B|) = 1$, then $\delta(A \times B) = \delta(A) + \delta(B)$ (see also §1 of [1]).*

Lemma 4 (Lyubich [4] and [5]). *If A and B are nilpotent, then $\delta(A \times B) = \delta(A) + \delta(B)$.*

In particular, if an abelian group is of type $(m_1, \dots, m_t)$, where all m_i are powers of primes, then its degree is $m_1 + \cdots + m_t$ (A. Ya. Povzner).

Additional information on the function d can be found in the mentioned papers.

§2. Classification of $(*)$-groups which are p-groups

In this section G is a p-group which is a $(*)$-group. In particular, if H is a subgroup of order p^2 of G, then $H_G > 1$.

LEMMA 5. *If G is abelian, then it is of one of the following types:* (p^n), (p^n, p), (p^2, p^2), (p, p, p), (p^2, p, p), $(2, 2, 2, 2)$.

Since the groups of types (p^3, p^2), (p^3, p, p), (p^2, p^2, p), $(4, 2, 2, 2)$ are not $(*)$-groups, the proof, by Lemma 1, is reduced to verification of the fact that the groups listed in Lemma 5 have the $(*)$ property.

LEMMA 6. *G does not contain a subgroup A of type (p, p, p, p).*

The intersection of any two nonincident subgroups C and B of orders p^2 and p^3 with $B, C < A$, respectively, is a central subgroup, and $C_G(A) > A$, whence G contains an abelian subgroup $D > A$, which contradicts Lemmas 5 and 1.

LEMMA 7. *If G contains a subgroup A of type (p, p, p), then* $\exp G \leq p^2$.

Indeed, if B is a cyclic subgroup of order p^3 in G, and $\Lambda = (A \cap B) \times A^1$, then $\frac{1}{|B|} + \frac{1}{|A^1|} \leq \Delta_0(G)$ and G is not a $(*)$-group.

LEMMA 8. *G does not contain a subgroup of type (p^2, p, p).*

PROOF. It follows from a remark made before Lemma 1 that $A_{(1)} \leq Z(G)$ where A is a subgroup of type (p^2, p, p) in G. If x is an element of order p^2 in $G - A$ and $x^p \notin A^{(1)}$, then

$$A = (\langle x \rangle \cap A)) \times A^1, \qquad |A^1| \geq p^3,$$

$$\frac{1}{|\langle x \rangle|} + \frac{1}{|A^1|} \leq \Delta_0(G),$$

although $\langle x \rangle \cap A^1 = 1$, and hence G is not a $(*)$-group. Thus $G^{(1)} \leq A^{(1)}$ so that $\exp G/A^{(1)} = p$.

Let $A_{(1)} = A^{(1)} \times A_1$ and let H/A_1 be a subgroup of order p in G/A_1 different from $A_{(1)}/A_1$. Then $A = C(p^2) \times A_1$ and $C(p^2) \cap H = 1$, $\frac{1}{|C(p^2)|} + \frac{1}{|H|} = \Delta_0(G)$, so that G is not a $(*)$-group. Therefore, G/A_1 contains the only subgroup of order p and is not cyclic (because $A_1 < Z(G)$). Thus $G/A_1 \cong Q(8)$. But $G/A^{(1)}$ is elementary, and hence $\Phi(A) = \Phi(G)$. Since $Z(G)$ is elementary, $G = G^1 \times A_0$, $A_0 \cong C(2) \times C(2)$, and $\Delta(G) \leq \Delta_0(G)$, which is a contradiction. Lemma 8 is proved.

LEMMA 9. *If* $|G| = p^4$, *then* $G \not\geq C(p) \operatorname{wr} C(p)$ *and* $G \neq G^1 \times C(p)$ *for* $p > 2$, *and if* $p = 2$, *then* $G \neq D(8) \times C(2)$.

It is easily seen that other groups G of order p^4 are $(*)$-groups.

In the sequel $|G| > p^4$.

LEMMA 10. *Let* G *contain a subgroup* A *of type* (p, p, p). *Then* $|G| = p^6$ *and* $Z(G)$ *is the only* G*-admissible subgroup of order* p^2 *in* A, *where* $\exp G/Z(G) = p$. *If* $p = 2$, G *is special.*

PROOF. By Lemmas 6 and 8, $C_G(A) = A$. It follows from the remark made before Lemma 5 that $|Z(G)| = p^2$. If B is a subgroup of order p^3 of G and $B \not\geq Z(G)$, then $A = (B \cap Z(G)) \times A^1$ implies $\frac{1}{|R|} + \frac{1}{|A^1|} \leq \Delta_0(G)$, and so G is not a $(*)$-group. In particular, all subgroups of G of order p^3 contain $Z(G)$ and hence are abelian, and $Z(G) \leq \Phi(G)$. Furthermore, $\exp G/Z(G) = p$ (since $\exp G \leq p^2$ and we may apply the remark made before Lemma 5 to cyclic subgroups of order p^2). If $x \notin A - Z(G)$, then $A = C_G(\langle x, Z(G)\rangle) = C_G(x)$, and hence $Z(G)$ is the only G-admissible subgroup of order p^2 in A. If $p = 2$, we have already proved that $Z(G) = \Phi(G)$. If $G^1 < Z(G)$ then $Z(G)$ is not the only G-admissible subgroup of order 4 in A, and hence $G^1 = Z(G) = \Phi(G)$ and G is special. If $p > 2$ then, again by a theorem of Blackburn [6] (see also [1]), we may assume that $A \triangleleft G$. Then $C_G(x) = A$ for all $x \in A - Z(G)$ implies $|G| = p^5$. This equality holds for $p = 2$ as well. Lemma 10 is proved.

We have already seen that if G contains a subgroup A of type (p, p, p), then all subgroups of G of order p^3 are abelian. Conversely, if G contains a subgroup such as A, then , for $p > 2$, it is a $(*)$-group, provided G satisfies all conditions found in Lemma 10. By Lemma 2, $k \leq 2$. If $k = 1$, then $G^1 \cap Z(G) = 1$ and $G^1 \times Z(G)$ is abelian of rank exceeding 2 and order exceeding p^3 which contradicts Lemmas 6 and 8. Let $k = 2$ and let $|G{:}G^1| \leq |G{:}G^2|$. If G is not a $(*)$-group, then $\delta(G) \leq p^3 + p^2$ and $|G^1| \geq p^3$. Since $Z(G) \leq G^1$, we see that $|G{:}G^1| = p^2$, G^1 is abelian by our assumption, and $G^1 Z(G)$ is abelian of order p^4 and rank exceeding 2, which is a contradiction. (Here we used the inequality $\exp G \leq p^2$.) Thus, G is a $(*)$-group.

In the sequel we omit analogous verifications which are almost always similar to this one.

LEMMA 11. *Let a 2-group contain an abelian subgroup* A *of type* $(2, 2, 2)$. *Then it does not contain a normal cyclic subgroup of order* 4.

PROOF. By Lemma 10 G is special of order 32. Let $L \cong C(4)$ and normal in G. It follows from $LZ(G) \leq Z(C_G(L))$ and $|G{:}C_G(L)| = 2$ that $C_G(L)$ is abelian of type (4, 4) (apply Lemmas 6 and 8). Since G/L is not abelian and AL/L is of type $(2, 2)$, we see that $G/L \cong D(8)$. Let

$C_G(L) = L \times L_1$. Consider a noninvariant subgroup H/L of order 2 in G/L. Then $C_G(L) \cap H = L$, and hence $H \cap L_1 = 1$, $\frac{1}{|H|} + \frac{1}{|L_1|} = \Delta_0(G)$ and G is not a $(*)$-group. This contradiction proves Lemma 11.

Conversely, if G does not contain a normal cyclic subgroup of order 4 and G is a group from Lemma 10 (without the assumption that it is a $(*)$-group), then G is a $(*)$-group.

In the sequel we assume that G does not contain a subgroup of type (p, p, p).

LEMMA 12. *Let G contain a subgroup A of type (p^2, p^2). Then $p = 2$ and G is special with $G_{(1)} = A_{(1)} = Z(G)$ and $|G| < 2^6$.*

(a) *If $|G| = 2^5$ then G does not contain $Q(8)$.*

(b) *If $|G| = 2^6$ then G contains a subgroup C of order 2 such that $G/C \cong Q(8) * D(8)$.*

Conversely, the groups described in (a) *and* (b) *are $(*)$-groups.*

PROOF. By Lemmas 1 and 5, $C_G(A) = A$. By our condition (see the remark preceding Lemma 5) $A_{(1)} \leq Z(G)$. Since G does not contain a group of type (p, p, p), $A_{(1)} = G_{(1)}$. As above, $\exp G = p^2$. If $p > 2$ then G is metacyclic by a theorem of Blackburn [6], and $\exp G = p^2 \Rightarrow |G| < p^4$, which is a contradiction. Thus $p = 2$.

Every subgroup of G of order 8 contains $A_{(1)} \leq Z(G)$ and hence is abelian. If x is an element of order 4 in G, then $\langle X\rangle_G \neq 1$, whence $x^2 \in A_{(1)}$. Since $\exp G = 4$, we see that $G/A_{(1)}$ has exponent 2 and so is abelian. Let C be a subgroup of G of order 2; assume that $\overline{G} = G/C$. Since every subgroup of $\overline{G}$ of order 4 contains $A_{(1)}/C = \Phi(\overline{G})$, we see that $\overline{G}$ is not abelian, for its rank exceeds 2 and this produces $G' = \Phi(G) = A_{(1)}$.

Let $|G| = 32$. Then $|G| = 2|Z(G)| \times |G'|$ implies $|Z(G)| = 4$, and it is clear now that G is special.

Let $|G| > 32$. By Lemma 1, all proper subgroups of G are $(*)$-groups. Let $A < H < G$ and let $|H| = 32$. By what we have proved above, $|Z(H)| = 4$, since $Z(G) < A < H$, $Z(G) \leq Z(H)$, and $|Z(G)| = 4$. It is clear now that in this case G is special. Now we prove that $|G| = 64$. Let B be a G-admissible subgroup of A of order 8. Then $G/C_G(B)$ is isomorphic to a subgroup of $\operatorname{Aut} B \cong D(8)$, so that $|G: C_G(B)| \leq 4$. If $C_G(B) = A$, we are done. Let $C_G(B) > A$. Then, if D is a G-admissible subgroup of order 32 contained between A and $C_G(B)$, then its center has order exceeding 4 and so D is not a $(*)$-group, which is impossible. Thus $|G| = 2^6$.

Let T be a minimal nonabelian subgroup of G, $|G| = 64$. It follows from the structure of G that $|T| = 16$. It is known that $|T'| = 2$, so that T has a subgroup C of order 2 different from T'. Assume that $\overline{G} = G/C$, $\overline{T} = T/C$, and $\overline{L} = L_G(\overline{T})$. Suppose that $\overline{L}$ is abelian. Since every subgroup of $\overline{L}$ of order 4 contains $\Phi(\overline{G})$, we see that $\overline{L}$ must be of type $(4, 2)$, $\overline{L} = \overline{Z}^1 \times \overline{Z}^2$, $\overline{Z}' \cong C(4)$, $\overline{Z}^2 \cong C(2)$. Thus $\overline{G} = (\overline{T} * \overline{Z}^1) \times \overline{Z}^2$. If $\overline{C}_0$ is a subgroup of order 2 in $\overline{T} * \overline{Z}^1$ that is different from $\Phi(\overline{G})$, then $\overline{C}_0 \times \overline{Z}^2$ does not contain $\Phi(\overline{G})$ which contradicts the property of G established above. Thus $\overline{L}$ is nonabelian of order 8. Since $H = Q(8) * Q(8) \cong D(8) * D(8)$ of order 32 has a subgroup of order 4 that does not contain $\Phi(H)$, we obtain $\overline{G} \cong Q(8) * D(8)$. It is easy to verify that a group G with these properties is a $(*)$-group. Lemma 12 is proved.

In the sequel we assume that G does not contain a subgroup of type (p^2, p^2). We can assume that G has no cyclic subgroup of index p (these are $(*)$-groups). In such a case G contains a noncyclic minimal abelain subgroup $A \lhd G$. It is clear now that A is of the type (p^n, p) for $n > 1$. Assume that $A = Z^n \times Z^1$, $Z^n \cong C(p^n)$, $Z^1 \cong C(p)$. Let $A_0 = A_{(2)}$; then A_0 is of type (p^2, p). Since $(C_G(A_0))_{(2)} = A_0$, we obtain $C_G(A_0) = A$.

Let $n > 2$. Then every subgroup of G of order p^2 intersects Z^n (for otherwise G is not a $(*)$-group). Therefore $C_G(Z^1)/Z^1$ contains only one subgroup of order p.

Let $p > 2$. Since in this case $C_G(Z^1)/Z^1$ is cyclic, it is abelian and coincides with A. Moreover, $|G: A| = p$. By a theorem of Blackburn [6] either G is metacyclic or $|G_{(1)}| = p^3$, $G_{(1)}$ is not abelian, and $G/G_{(1)}$ is cyclic. In the second case $\Delta(G) = \frac{1}{p} > \Delta_0(G)$ and G is a $(*)$-group. Let G be metacyclic; then G/B is cyclic for a suitable cyclic subgroup B. We can assume that B is maximal cyclic possessing this property. If $|B| < |G/B|$, then considering $C_G(B)$ leads to a contradiction. Thus $|B| \geq |G/B|$ and it is easy to show that $G = B \leftthreetimes D$. Since $|D| > p$, we see that $D_G > 1$ and $G_{(1)} \leq Z(G)$ which is a contradiction. The case of $p > 2$ has been completely considered. Let $n > 2$ and $p = 2$. Then $C_G(Z^1)/Z^1$ is either cyclic or a generalized quaternion group and $|G: C_G(Z^1)| \leq 2$.

Suppose that $G = C_G(Z^1)$. Then $G/Z^1 \cong Q(2^{n+1})$ and obviously $k \leq 2$. Since $G_{(1)} = Z(G)$ we obtain that $k = 2$ (see Lemma 2 for notations). Then $|G^1 \cap Z(G)| = 2$. We can assume that $G^1 \geq Z^1$. Then $G^2 \ngeq Z^1$. It follows from the structure of G that $G^1 = Z^1$ and G is a $(*)$-group.

Let $|G: C_G(Z^1)| = 2$. First we assume that $C_G(Z^1)/Z^1$ is cyclic. Then $Z(G)$ is cyclic, and hence $k = 1$. Now it is obvious that $G^1 \cap Z^n = 1$, whence $|G^1| \leq 4$. Assume that $|G^1| = 4$. Then $G^1 \cap A > 1$ and $C_G(G^1 \cap A) = G$, which contradicts $G^1_G = 1$. Thus $|G^1| = 2$ and G is a $(*)$-group. Finally, let $C_G(Z^1)/Z^1 \cong Q(2^{n+1})$, $C_G(Z^1) < G$; then $|G| = 2^{n+3}$. Again $k = 1$ by Lemma 2. Therefore $G^1 \cap Z^n = 1$ and $|G^1| \leq 8$. If $G^1 \cap C_G(Z^1) = 1$, then $|G^1| = 2$ and G is a $(*)$-group. Thus $G^1 \cap C_G(Z^1) > 1$, whence, if $|G^1| > 2$ then $G^1_G > 1$. It follows that $|G^1| = 2$ and G is a $(*)$-group. The case of

$n > 2$ is considered completely. In the sequel $n = 2$.

Let $n = 2$ and $p > 2$. By the Blackburn theorem [6] we easily see that G is not a $(*)$-group.

Now $n = 2$ and $p = 2$. Then $C_G(A) = A$ and $A \triangleleft G$ imply that G/A is isomorphic to a subgroup of $D(8)$. Furthermore, $Z(G) < A$.

Let $\Phi(G)$ be cyclic of order greater than 4. If M is a normal subgroup of G of type $(2, 2)$, then $\Phi(G)M$ has order greater than 8 and it is not cyclic; we have already considered this case. Therefore we may assume that A and $\Phi(M)$ are incident.

Let $\Phi(G) > A$. Then $G/A \cong D(8)$ and $|Z(G)| = 2$, $k = 1$, and $Z(G) = \Phi(A)$. By a theorem of Taussky, $|G\colon G'| > 4$. By our assumption, $G' \neq A$. If G' is abelian of type $(4, 2)$ then we can replace A by G' and not consider this situation. If G' is cyclic of order greater than 2, then we can replace A by $G'A_{(1)}$ and not consider this situation. Clearly, $|G'| > 2$. Thus G' is of type $(2, 2)$. In this case G' centralizes $A < \Phi(G)$ and $G'A$ is of type $(2, 2, 2)$, which is a contradiction.

Let $\Phi(G) = A$. Then G/A is of type $(2, 2)$, so that $|G| = 32$. Since $\Phi(G) = \langle x^2 | x \in G\rangle$, G contains an element x of order 8. Let $A \leq \langle a, b | a^2 = b^4 = [a, b] = 1\rangle$; then $\langle a, x\rangle$ is not of the maximal class because we may assume that $x^2 = b$, in which case $A < \langle a, x\rangle$. Now $\langle a, x\rangle$ is not abelian by assumption. G contains an element y whose square is a direct complement to $\langle x^2\rangle$ in A. In this case $\frac{1}{|\langle x\rangle|} + \frac{1}{|\langle y\rangle|} = \Delta_0(G)$ and G is not a $(*)$-group. Thus $\Phi(G) < A$.

Assume that $\Phi(G) = \langle a\rangle$ is of order 4. Then G contains an element x such that $x^2 = a$, where $A = \langle a\rangle \times \langle b\rangle$, and $|\langle b\rangle| = 2$. Since G cannot be decomposed in a nontrivial direct product because it does not contain a noncyclic abelian subgroup of order 16, the subgroup $Z(G)$ is cyclic. It follows that $A_{(1)}$ is the only subgroup of order 4 which contains b, such that $C_G(b)/\langle b\rangle$ contains only one subgroup of order 2 and hence is isomorphic to $Q(8)$. Then $|G| = 32$ and $k = 1$. But $G^1 \cap \Phi(G) = 1$, so that $\exp G^1 = 2$. Since $G^1\Phi(G)_{(1)}$ is elementary, G is a $(*)$-group.

Let $\Phi(G) = A_{(1)}$. Again $|G\colon A| = 4$ and $\exp G = 4$. Let $\Phi(G) = Z(G)$. Then $G_{(1)} = Z(G)$ and $k = 2$. We obtain $|G^1 \cap Z(G)| = 2$ for $i = 1, 2$. In the sequel G^i does not contain a subgroup of type (2, 2) (it suffices to consider the product of such a subgroup and $Z(G)$). Therefore G^i is either cyclic or $Q(8)$. Let $|G^1| \geq |G^2|$. Clearly, $G^1 \cap G^2 = 1$. Assume that G is not a $(*)$-group. Then $|G^1| = 8$, $|G^2| = 4$, $G^1 \cong Q(8)$, $G^2 \cong C(4)$, $G = G^1G^2$. Suppose that $G^1 \ntrianglelefteq G$. Then $G^1_G \cong C(4)$ (because $G/G^1_G \cong D(8)$), and $C_G(G^1_G)$ is an abelian subgroup of order 16, which is a contradiction. Thus $G^1 \triangleleft G$ and then $G = G^1 * C_G(G^1)$. Since $C_G(G^1)$ is abelian, $LC_G(G^1)$, where L is a subgroup of order 4 in G^1, is abelian of order 16, which is a contradiction. Thus G is a $(*)$-group. Now let $\Phi(G) \neq Z(G)$. Then $Z(G)$ is cyclic. Suppose that $|G^1| \geq 4$. It follows from $G^1_G = 1$ that $|G^1| = 4$, and consideration of $G^1 Z(G)$ shows that G^1 is cyclic. If $G^1\Phi(G)$ is abelian, then it is of type $(4, 2)$ and $G^1_G = (G^1\Phi(G))^1 > 1$, which is a contradiction. If $G^1\Phi(G)$ is not abelian, it is isomorphic to $D(8)$. In this case G^1 is a characteristic subgroup of $G^1\Phi(G) \triangleleft G \Rightarrow G^1 \triangleleft G$, which is a contradiction. Thus $|G^1| = 2$ and G is a $(*)$-group.

Finally, let $\Phi(G) = \Phi(A)$. Again, as above, $Z(G)$ is cyclic and $|G| = 32$. It is known that $|G{:}\,Z(G)|$ is a perfect square, and hence $|Z(G)| = 2$ and G is extra special. It is clear now that $G = D(8) * Q(8)$ (since $D(8) * D(8) \cong Q(8) * Q(8)$ contains a subgroup of type $(2, 2, 2)$ and, obviously, is not a $(*)$-group).

This completes the classification of primary $(*)$-groups. The results obtained will be stated as a theorem.

THEOREM 1. *Let a $(*)$-group G be a p-group. Then it is isomorphic to one and only one of the following groups:*

1. $C(p^n)$; $C(p^n) \times C(p)$; $n > 1$; $C(p^2) \times C(p^2)$; $C(p^2) \times C(p) \times C(p)$; $C(p) \times C(p) \times C(p)$; $C(2) \times C(2) \times C(2) \times C(2)$.
2. *A nonabelian group of order* p^3 *and exponent* $p > 2$.
3. *A nonabelian group of order* p^4 *that is isomorphic neither to* $D(8) \times C(2)$ *nor (for* $p > 2$*) to a subgroup of* $C(p)\mathrm{wr}\,C(p)$.
4. *A nonabelian group that contains a cyclic subgroup of index* p.
5. $p > 2$, $|G| = p^5$; $Z(G)$ *is of type* (p, p) *and is contained in all subgroups of order* p^3; $G_{(1)} > Z(G)$.
6. $p = 2$; $|G| = 32$, G *is special with* $|Z(G)| = 4$; G *does not contain a normal cyclic subgroup of order* 4, $G_{(1)} > Z(G)$.
7. $p = 2$; $|G| \leq 64$, G *contains a normal abelian subgroup* A *of type* $(4, 4)$; $G_{(1)} = A_{(1)} = Z(G)$; G *is special.*
   (a) *If* $|G| = 32$ *then all subgroups of order* 8 *contain* $Z(G)$.
   (b) *If* $|G| = 64$ *then* G *contains a subgroup* C *of order* 2 *such that* $G/C \cong D(8) * Q(8)$.
8. $p > 2$, $G_{(1)}$ *is nonabelian of order* p^3 *and exponent* p; $G/G_{(1)}$ *is cyclic.*
9. $p = 2$, G *is an extension of* $C(2)$ *by* $Q(2^{n+1})$, $n \geq 2$.
10. $p = 2$, G *contains a subgroup* L *of order* 2 *such that* $C_G(L)/L \cong C(2^n)$, $n > 2$; $[G{:}\,C_G(L)| = 2$, *and* G *does not contain a subgroup of type* $(2, 2, 2)$.
11. $p = 2$, G *contains a subgroup* L *of order* 2 *such that* $C_G(L)/L \cong Q(2^{n-1})$, $n > 2$; $|G{:}\,C_G(L)| = 2$.
12. $p = 2$; $\Phi(G) \cong C(4)$; $|G| = 32$; G *does not contain an abelian subgroup of order* 16 *and type* $(2, 2, 2)$; $Z(G)$ *is cyclic.*

13. $p = 2$; $\Phi(G) = Z(G) \cong C(2) \times C(2) = G_{(1)}$; $\exp G = 4$; *G does not contain an abelian subgroup of order* 16; $|G| = 32$.
14. $p = 2$; $\Phi(G) \cong C(2) \times C(2)$; $|G| = 32$; $Z(G)$ *is cyclic; G does not contain an abelian subgroup of order* 16 *and a subgroup of type* $(2, 2, 2)$.
15. $D(8) * Q(8)$ *is of order* 32.

§3. Classification of nonprimary $(*)$-groups

In this section G is a nonprimary $(*)$-group and p is the least prime divisor of $|G|$. Not to extend the size of this paper too much, almost no proofs are given in this section.

LEMMA 14. *G possesses an ordered Sylow tower of the same type as a supersolvable group.*

PROOF. Let G be a counterexample of the minimal order. Then $G = P \leftthreetimes Q$, $P \in \mathrm{Syl}_p(G)$, $Q \in \mathrm{Syl}_q(G)$, G is a minimal nonnilpotent group, and $p < q$. Suppose that $|P| = p^n$, $|Q| = q^m$, and let b be the order of $p (\mathrm{mod}\, q)$. Obviously, $b > 1$. If b is odd, then P is an elementary minimal normal subgroup of G, and hence, if G^1 is a subgroup of order p^2 in P, then $G^1_G = 1$ and G is not a $(*)$-group. If $b = 2$ then $p = 2$ and $q = 3$; if $m = 1$ we have $\frac{1}{|Q|} < \Delta_0(G)$; if $m > 1$, then $\frac{1}{|P|} + \frac{1}{|Q|} < \Delta_0(G)$ and G is not a $(*)$-group. In the sequel $b \geq 4$ is even. Then P is elementary by Theorem 1, and again G is not a $(*)$-group. Lemma 14 is proved.

Since a description of nonprimary $(*)$-groups is quite long we subdivide the statement of the theorem into parts.

LEMMA 15. *If $p = 2$ and G is abelian, then it is of one of the types:* $(2, q^b)$; $(4, 3)$; $(4, 5)$; $(4, 7)$; $(8, 3)$; $(16, 3)$; $(2, 2, 3)$; $(2, 2, 5)$; $(2, 2, 7)$.

LEMMA 16. *If $p = 2$ and G is not abelian, then* $G \in \{Q(8) \times C(3); Q(16) \times C(3)\}$.

LEMMA 17. *If $p = 2$, $P \in \mathrm{Syl}_2(G)$, $|P| = 2$ and G is not nilpotent, then* $G = D(2q^b)$.

LEMMA 18. *Suppose that $p = 2$, $P \in \mathrm{Syl}_2(G)$, $|P| > 2$, and G is not nilpotent. Then G is isomorphic to one of the following groups:*

1. $D(12)$.
2. $C(4) \leftthreetimes C(3)$.
3. $C(4) \leftthreetimes C(5)$, $Z(G) \cong C(2)$.
4. $C(4) \leftthreetimes C(7)$.
5. $C(8) \leftthreetimes C(3)$.
6. $C(16) \leftthreetimes C(3)$.

7. $Q(8) \leftthreetimes C(3)$.
8. $Q(16) \leftthreetimes C(3)$.

THEOREM 2. *If* $p = 2$, *then Lemmas* 15–18 *contain a complete list of nonprimary* $(*)$*-groups of even order.*

In the sequel we assume that $p > 2$. Suppose that $G = P \leftthreetimes H$, $P \in \mathrm{Syl}_p(G)$, and $|P| = p^n$.

LEMMA 19. *If* $n = 1$ *and* $G = P \times H$ *then one of the following statements holds*:

1. $H = C(q^m)$.
2. $H = C(q) \times C(q)$, $(p + 2q)p^2 > (p + 1)q^2$.
3. $H = C(q^2) \times C(r)$, $r \neq q$, $p^2(p + r + q^2) > q^2r(p + 1)$.
4. $H = C(q^2) \times C(q)$, $p^2(p + q + \sigma^2) > q^3(p + 1)$.
5. $H = C(q) \leftthreetimes C(r)$ *is not abelian,* $p^2(p + r) > qr(p + 1)$.

THEOREM 20. *If* $n = 1$, $C \neq P \times H$, *and* H *is not primary, then* $H = C(qr)$, $p^2(q + r) > qr(p + 1)$.

LEMMA 21. *If* $n = 1$, $p \ntriangleleft G$, *and* H *is primary, then* G *is isomorphic to one of the following groups*:

1. $G = C(p^2) \leftthreetimes C(q)$, $Z(G) \cong C(p)$, $q \equiv 1 (\mathrm{mod}\ p)$, $q < p^3$.
2. $H = C(qr)$, $q \neq r$, $Z(G) \cong C(p)$, $qr < p^3$.
3. $G = P \times C(q)$, $p^3 > q$.

LEMMA 23. *Let* $P = C(p) \times C(p)$. *Then* G *is isomorphic to one of the following groups*:

1. $P \times C(q)$, $q < 2p^2$.
2. $P \times C(q^2)$, $q^2 < 2p^2$.
3. $C(p) \times (C(p) \leftthreetimes C(q))$: *(parentheses enclose a nonabelian group),* $q < p^2$.

LEMMA 24. *If* $n > 2$ *then* $G = C(p^n) \leftthreetimes C(q)$ *is nonabelian,* $q \equiv 1 (\mathrm{mod}\ p)$, $p \mid |z(G)|$, $\frac{1}{p^n} + \frac{1}{q} > \frac{p+1}{p^3}$.

THEOREM 3. *If* $p > 2$ *is the least prime divisor of a* $(*)$*-group* G, *then* G *is one of the groups described in Lemmas* 19–24.

Thus, groups G with $\delta(G) > \frac{p+1}{p^3}|G|$ have been completely classified. An examination of proofs (omitted in this paper) of Theorems 2 and 3 shows that, in fact, they produce a classification of nonprimary groups G with $\Delta(G) \geq \frac{p+1}{p^3}$. However, a classification of p-groups G with $\Delta(G) = \frac{p+1}{p^3}$ is a very difficult matter.

Next we introduce a generalization of $(*)$-groups.

Let p be the least prime divisor of $|G|$. We say that G is a $(**)$-group if $\Delta(G) > \frac{1}{p^2}$. Clearly, every $(*)$-group is a $(**)$-group. The following Lemma may serve as a foundation for a classification of $(**)$-groups.

LEMMA 25. *Let p be the least prime divisor of the order of a $(**)$-group G.*

1. *If $p > 2$ then G has an invariant p-complement. Since the property $(**)$ is hereditary for subgroups, G has an ordered Sylow tower.*
2. *If $p = 2$ and if 8 does not divide $|G|$, then either G has an invariant 2-complement or G is a minimal nonnilpotent group of order $2^2 \cdot 3^m$.*
3. *If $p = 2$ and $8 \mid |G|$ but G has no invariant 2-complement, then $|G| = 3 \cdot 2^n$.*

PROOF. 1. Let $p > 2$ and let G have no invariant p-complement. Then* $G = P \leftthreetimes Q$ is a minimal nonnilpotent group, $P \in \mathrm{Syl}_p(G)$, $Q \in \mathrm{Syl}_q(G)$, $|P| = p^n|Q| = q^m$. Let b be the order of $p(\mathrm{mod}\ q)$ and let $m = a + b$. Since $p > 2$, we see that $b > 2$. If b is odd, then P is elementary and, as in Lemma 14, G is not a $(**)$-group. Now let P be not elementary. Then $b \geq 4$ is even, $a > 0$, and $\exp P = p$. If $a = 1$, then P contains a subgroup A of type (p, p, p), $A > Z(P)$, $A = A^1 \times Z(P)$, $|A^1| = p^2 A^1_G = 1$, and G is not a $(**)$-group. Let $a > 2$. Choose in P a minimal nonabelian subgroup T; $|T| = p^3$ and $|T \cap Z(P)| = p$. Therefore $Z(P)$ contains a subgroup L such that $L \cap T = 1$ and $|L| = p^2$. It follows from $\frac{1}{|T|} + \frac{1}{|QL|} < \frac{1}{p^2}$ that G is not a $(**)$-group. Thus $a = 2$. Let $A > Z(P)$ be a subgroup of P of order p^3. Then A is of type (p, p, p) and $A \triangleleft P$. Since $b \geq 4$, we see that $C_p(A) > A$, and hence $C_p(A)$ has a subgroup B of type (p, p, p, p) which contains A, $B = B^1 \times Z(P)$, $|B^1| = p^2$, and $B^1_G = 1$, so that G is not a $(**)$-group. This contradiction shows that G has an invariant p-complement.

2. Suppose that $p = 2$, that 8 does not divide $|G|$, and G has no invariant 2-complement. Then $P \in \mathrm{Syl}_2(G)$ has order 4 and type $(2, 2)$. In this case $P_G > 1$ implies $P \triangleleft G$, whence $G = P \leftthreetimes R$. If $|R|$ is a prime then $|R| = 3$ and $G = A_4$ is the minimal nonnilpotent group. Now let $|R|$ be a composite number. Assume that R is not primary. By the above, $R = Q \leftthreetimes S$, where $Q \in \mathrm{Syl}_3(R)$ (observe that $Q > 1$). It follows from $\frac{1}{|PQ|} + \frac{1}{|R|} < \frac{1}{4}$ that $(PQ \cap R)_G Q_G > 1$. Since $Q \ntriangleleft G$, we see that $|Q| \geq 3^2$. Since $(PS \cap Q)_G = 1$, we obtain $\frac{1}{|PS|} + \frac{1}{|Q|} > \frac{1}{4}$, which is impossible in view of $|Q| \geq 9$ and $|S| \geq 5$. Thus R is primary, and hence a 3-group, $|R| = 3^m$, $m > 1$. If A and B are nonincident subgroups of R of orders 3 and 9, then $\frac{1}{|PA|} + \frac{1}{|B|} < \frac{1}{4}$ and G is not a $(**)$-group. Therefore R has no subgroups of type $(3, 3, 3)$ and $(9, 3)$. Analogously we prove that all subgroups of R of order 3 are normal in G. All this shows that R is cyclic. Since $|G : C_G(R)| = 3$, we see that G is a minimal nonnilpotent group.

* We use induction on $|G|$.

3. Suppose that $p = 2$, $8 \mid |G|$, $|P| = 2^n$, and $n > 2$. Then $\frac{1}{|P|} + \frac{1}{|R|} > \frac{1}{4}$ implies $|R| < 8$. But then $|R| = q$ is prime, $q > 2$. If $q > 3$ then $R \triangleleft G$, so that R is a normal 2-complement in G. Thus if $R \ntriangleleft G$ then $|R| = 3$. Lemma 25 is proved.

Possibly the problem of classification of nonprimary $(**)$-groups is not too difficult.

References

1. Ya. G. Berkovich, *On minimal representations of finite groups by permutations*, Izv. Severo-Kavkaz. Nauchn. Tsentra Vyssh. Shkoly Estestv. Nauk. **1** (1982), 8–15 [Russian].

2. B. Huppert, *Endliche Gruppen*, Bd. 1, Berlin, 1967, 739 pp.

3. D. L. Johnson, *Minimal permutation representations of finite groups*, Amer. J. Math. **93** (1971), 857–866.

4. M. Yu. Lyubich, *On a logarithmic property of a degree of a finite group*, in *Problems of the Theory of Groups and Homological Algebra*, Yaroslavl′, Vyp. **2** (1979), 120–130 [Russian].

5. ——, *On a logarithmic property of a degree of a finite group*, Dokl. Akad. Nauk SSSR **247** (1979), no. 4, 791–794 [Russian].

6. N. Blackburn, *Generalizations of certain elementary theorems on p-groups*, Proc. London Math. Soc. **11** (1961), 1–21.

Rostov Civil Engineering Institute

Translated by BORIS M. SCHEIN

Amer. Math. Soc. Transl.
(2) Vol. **149**, 1991

Maximal Subgroups of Chevalley Groups Containing a Maximal Split Torus

UDC 513.6

N. A. VAVILOV

Abstract. We will classify, to within conjugacy, the maximal subgroups of Chevalley groups of normal types over a field that contain a maximal split torus, in those cases where the description of the subgroups containing it is standard.

Bibliography: 21 titles.

§1. Standard description

Suppose Φ is a reduced irreducible system of roots of rank ℓ, K is a field, $G = G(\Phi, K)$ is a Chevalley group of type Φ over K, and $T = T(\Phi, K)$ is a maximal split torus in this group. For a root $\alpha \in \Phi$ we denote by $\mathfrak{X}_\alpha$ the corresponding unipotent root subgroup in G. We denote by $N = N(\Phi, K)$ the subgroup of G generated by all elements $w_\alpha(\varepsilon)$, where $\alpha \in \Phi$, $\varepsilon \in K^*$ [12]. It is known that the factor group N/T is canonically isomorphic to the Weyl group $W = W(\Phi)$ of the system of roots Φ. For $w \in W$ we denote by n_w some preimage of it in N.

Suppose $S \subseteq \Phi$ is a closed set of roots (i.e., if $\alpha, \beta \in S$ and $\alpha + \beta \in \Phi$, then $\alpha + \beta \in S$). Let $E(S)$ denote the subgroup of G generated by the $\mathfrak{X}_\alpha$, $\alpha \in S$, $G(S)$ the product $T \cdot E(S)$, and $N(S)$ the subgroup generated by $G(S)$ and those n_w for which $wS = S$. It is known [20] that if the field K contains at least five elements, then $N(S)$ is equal to the normalizer of $G(S)$ in G. A subgroup H of G such that $G(S) \leq H \leq N(S)$ for some closed set of roots S will be called standard.

We say that for a group G there is a standard description of the subgroups containing T if each subgroup of G containing T is standard. A necessary condition for such a description is $\operatorname{char} K \neq 2$ for $\Phi = B_\ell, C_\ell, F_4$

1991 *Mathematics Subject Classification*. Primary 20D06, 20E28.

Translation of Rings of modules, Limit theorems of probability theory, No. 1, Leningrad Univ., Leningrad, 1986, pp. 67–75; MR 88b:20067.

and $\operatorname{char} K \neq 2, 3$ for $\Phi = G_2$, which we will always assume. A standard description has been established in the following cases:

1) The field K is algebraically closed [3].
2) The field K is finite, $\operatorname{char} K \neq 2$, $|K| \geq 13$ [18].
3) The field K is infinite and $\Phi \neq C_\ell$ (recall that $A_1 = C_1$, $B_2 = G_2$, $D_2 = C_1 \times C_1$) [6].
4) $\Phi = C_\ell$ and the field K satisfies the following condition: $|K^*| > |K^*/K^{*2}|$, [11], [7], [1].

Suppose $\mathbb{F}_q$ is a finite field of q elements. It is known that the fields $\mathbb{F}_q$, $q \leq 11$, are exceptions for a standard description in symplectic groups [7], [11], and the fields $\mathbb{F}_q$, $q \leq 5$, in all groups [10]. Thus, after the appearance of the papers [3], [18], [6], [7], and [10] there remained unsettled only the following cases:

5) $\Phi = C_\ell$ and K is an infinite field such that $|K^*| = |K^*/K^{*2}|$.
6) $\Phi \neq C_\ell$ and K is a finite field of characteristic 2 or $K = \mathbb{F}_7$, $\mathbb{F}_9$, $\mathbb{F}_{11}$. In the case of exceptional characteristics, a description of the subgroups containing T can also be obtained, but in terms of quasiclosed, rather than closed, sets of roots [3].

We mention a small modification of this problem for which a complete solution has been obtained. Suppose $\overline{G} = \overline{G}(\Phi, K)$ is an extension of the Chevalley group G in which all diagonal automorphisms [12] become inner (e.g., GL_n for SL_n, GSp_n for Sp_n, and GO_n for SO_n; a general construction of such extensions was given in [13]). Suppose also that $\overline{T} = \overline{T}(\Phi, K)$ is a maximal split torus in $\overline{G}$. The subgroups obtained from $G(S)$ and $N(S)$ by replacing T by $\overline{T}$ are denoted by $\overline{G}(S)$ and $\overline{N}(S)$. The concepts of standard subgroup and standard description have to be accordingly modified. For extended Chevalley groups the situation is much simpler than for the ordinary ones. In [5] the standardness of the description was announced for any field K containing at least seven elements (and, of course, satisfying the above-mentioned conditions on the characteristic). For classical groups there are published proofs: for GL_n in [1], for GSp_n in [7], and for GO_n in [21].

§2. Main reduction

Suppose S is a closed set of roots. Then $S = S^r \cup S^u$, where $S^r = \{\alpha \in S \mid -\alpha \in S\}$ and $S^u = \{\alpha \in S \mid -\alpha \notin S\}$. The group $W(S) = W(S^r)$ is contained in the subgroup $X(S)$ consisting of those $w \in W$ for which $wS = S$. The following assertion is obvious.

Lemma 1. *The inclusion $N(S_1) \subseteq N(S_2)$ holds if and only if $S_1 \subseteq S_2$ and $X(S_1) \subseteq X(S_2)$.*

Thus, the problem of describing the maximal standard subgroups reduces to the purely combinatorial problem of determining, to within conjugacy via elements of the Weyl group, all maximal pairs among the $(S, X(S))$, where S is a closed subset of Φ.

We now fix an order on Φ and let Φ^+ and $\Pi = \{\alpha_1, \dots, \alpha_\ell\}$ be the sets of positive and simple roots, respectively. Recall that a standard parabolic set is a closed set containing Φ^+, and the parabolic sets are those conjugate to a standard parabolic set. For each parabolic set P we have $X(P) = W(P)$. We denote by P_r, $1 \le r \le \ell$, the smallest closed set of roots containing Φ^+ and $-\alpha$, where $\alpha \in \Pi \backslash \{\alpha_r\}$. Two sets P_r and P_s, $r \ne s$, are not conjugate. This gives us exactly ℓ conjugacy classes of maximal parabolic subsets. Thus the $G(P_r)$, $1 \le r \le \ell$, are representatives of the ℓ conjugacy classes of maximal parabolic groups. The following result is a direct consequence of the Borel–Tits Theorem [4].

THEOREM 2. *Suppose S is a set of roots such that $S^u \ne \varnothing$. Then the group $N(S)$ is contained in a proper parabolic subgroup.*

Therefore in what follows we can limit ourselves to the case where $S = S^r$, i.e., where S is a subsystem of roots in Φ. All subsystems of roots were classified in [9, Tables 9 and 11]. For exceptional systems we can also use Tables 7–10 of [14], it being necessary only to eliminate from them all graphs containing cycles and to add the subsystems $B_2 + 2A_1$ in F_4 and $D_5 + A_3$ in E_8. For exceptional systems there are the following numbers of proper (different from Φ and from $\varnothing$) subsystems of roots in Φ: 4 for G_2, 22 for F_4, 19 for E_6, 45 for E_7, and 75 for E_8.

§3. Statement of the theorem

We will employ the following notation: $\Delta + \Sigma$ is an orthogonal sum of systems of roots, $k\Delta$ is the sum of k copies of the system Δ, $A_0 = D_1 = \varnothing$ is the empty system of roots, $\widetilde{\Delta}$ is an embedding of Δ into Φ onto the short roots, and $N(\varnothing) = N$.

THEOREM. *A Chevalley group $G = G(\Phi, K)$ has the following conjugacy classes of nonparabolic maximal standard subgroups*:

A_ℓ $N\left(\frac{\ell+1}{k+1} A_k\right)$, $(k+1) \mid (\ell+1)$, $k \ne \ell$;

B_ℓ $N(B_k + D_{\ell-k})$, $0 \le k \le \ell - 1$;

C_ℓ $N(k/\ell C_k)$, $k \mid \ell$, $k \ne \ell$; $N(A_{\ell-1})$; $G(C_k + C_{\ell-k})$, $1 \le k < \ell/2$;

D_ℓ $N(k/\ell D_k)$, $k \mid \ell$, $k \ne \ell$; $N(A_{\ell-1})$; $N(D_k + D_{\ell-k})$, $1 \le k < \ell/2$;

G_2 $N(A_2)$, $G(A_1 + \widetilde{A}_1)$;

F_4 $G(B_4)$, $N(D_4)$, $G(A_1 + C_3)$, $N(A_2 + \widetilde{A}_2)$;

E_6 $G(A_5 + A_1)$, $N(3A_2)$, $N(D_4)$, N;

E_7 $G(D_6 + A_1)$, $N(A_7)$, $N(A_5 + A_2)$, $N(D_4 + 3A_1)$, $N(7A_1)$, $N(E_6)$, N;

E_8 $G(E_7+A_1)$, $G(D_8)$, $N(E_6+A_2)$, $N(A_8)$, $N(2D_4)$, $N(2A_4)$, $N(4A_2)$, $N(8A_1)$, N;

In the case where the description of subgroups of G containing T is standard, this theorem provides a description, to within conjugacy, of all maximal subgroups containing T. This question is relevant to various problems, in particular to the problem of describing the maximal subgroups of finite simple groups.

Of course, the theorem carries over without any changes to extended Chevalley groups, it being necessary only to replace the groups $G(S)$ and $N(S)$ by $\overline{G}(S)$ and $\overline{N}(S)$, respectively. For the case of the full linear group (i.e., $\Phi = A_\ell$) this theorem was established earlier in [2].

If the field K is infinite, then the subgroups $G(S)$ and only these among the standard subgroups are connected in the Zariski topology. As already stated, parabolic subgroups are connected. The theorem provides, in particular, a classification of the connected maximal standard subgroups. For this purpose we need only choose those Δ for which $G(\Delta)$ figures in the theorem.

An explicit list of the maximal standard subgroups is interesting also because the subgroups not contained in any proper standard subgroup play the same role in the theory of Chevalley groups as do the primitive irreducible subgroups in the theory of linear groups. In particular, the theorem of the present paper can be used to shorten the proof of the results announced in [5], [6].

§4. Method of proof

Note that the proof of the theorem for classical types is not very difficult, since in these cases the action of the Weyl group can easily be described in terms of permutations. Most of the calculations needed in the proof pertain to the types F_4, E_6, E_7, E_8.

Let us make some obvious remarks. First, if Δ is a subsystem of roots that is maximal as a closed subset, then the pair $(\Delta, X(\Delta))$ is maximal. These subsystems are listed in Table 12 of [9] (as was mentioned in [19], this table contains five mistakes: the subsystems $A_3+\widetilde{A}_1$ in F_4, $2A_3+A_1$ in E_7, and A_3+D_5, A_7+A_1, $A_5+A_2+A_1$ in E_8 are not maximal; a correct table can also be found in [8]). Second, if Δ is a maximal subsystem of roots, contained, however, in a proper parabolic set [9, Table 13], then the pair $(\Delta, X(\Delta))$ is maximal if and only if $X(\Delta) \neq W(\Delta)$. Third, if Δ contains roots of different lengths, then the pair (Φ_ℓ, W), where Φ_ℓ is the set of long roots of the system Φ, is maximal, and no other pair $(\Delta, X(\Delta))$, where Δ consists entirely of long roots, is maximal. If all roots in Φ have the same length, then the pair $(\varnothing, W)$ is maximal. These simple ideas already enable us to prove the maximality of all subgroups listed in the theorem except for $N(D_4)$ in E_6, $N(D_4+3A_1)$, $N(7A_1)$ in E_7, and $N(2D_4)$, $N(4A_2)$, $N(8A_1)$ in E_8.

Suppose Φ is a system of roots in Euclidean space V of dimension ℓ. Recall that we denote by $A(\Phi)$ the group of orthogonal transformations of V that fix Φ. The distinction between $A(\Phi)$ and $W(\Phi)$ is connected only with the presence of nontrivial automorphisms of the Dynkin diagram of Φ and arises for irreducible systems only in the following cases: $A(\Phi)/W(\Phi) \cong \mathbb{Z}/2\mathbb{Z}$ for $\Phi = A_\ell$, $\ell \geq 2$, D_ℓ, $l \geq 5$, E_6 and $A(\Phi)/W(\Phi) \cong S_3$ for $\Phi = D_4$. For all other irreducible systems of roots, $A(\Phi) = W(\Phi)$. If the system Φ is an orthogonal sum of isotypical irreducible subsystems, i.e., $\Phi = k\Delta$, then $A(\Phi)$ is the wreath product of the symmetric group S_k and $A(\Delta)$. Finally, if $\Phi = k_1\Delta_1 + \cdots + k_t\Delta_t$, where $\Delta_1, \ldots, \Delta_t$ are irreducible systems of different types, then $A(\Phi) = A(k_1\Delta_1) \times \cdots \times A(k_t\Delta_t)$.

It is well known that a subgroup of W consisting of the elements fixing certain given roots is generated by the reflections relative to the roots orthogonal to the given ones. This implies almost immediately the following result, which is essentially Proposition 2.8 of [14]. Let $\Delta^\perp$ denote the set of roots in Φ orthogonal to Δ. It is closed and symmetric and is therefore a system of roots. Even though the system $\Delta + \Delta^\perp$ is, in general, not closed in Φ, it is nevertheless true that $W = W(\Phi)$ contains the Weyl subgroup $W(\Delta) \times W(\Delta^\perp)$.

LEMMA 3. *Suppose* $\Delta \leq \Phi$. *Then* $X(\Delta) = (A(\Delta) \cap W(\Phi)) \leftthreetimes W(\Delta^\perp)$.

Thus we have to calculate $\Delta^\perp$ and $A(\Delta) \cap W(\Phi)$. The calculation of $\Delta^\perp$ is done as follows. If Δ and $\Delta^\perp$ are strongly orthogonal (i.e, for any $\alpha \in \Delta$ and $\beta \in \Delta^\perp$ the sum $\alpha + \beta$ is not a root) and if, in addition, Φ has only one conjugacy class of subsystems of type Δ, then $\Delta^\perp$ can be found directly from a table of subsystems of Φ. In fact, $\Delta^\perp$ is simply the (unique) largest subsystem such that $\Delta + \Delta^\perp$ is contained in Φ. The case where Δ and $\Delta^\perp$ are not strongly orthogonal is possible only for $\Phi = G_2, F_4$, and here it is easy to do all the calculations directly. Finally, if Φ has nonconjugate subsystems $(\Delta)'$, $(\Delta)''$ of type Δ (this is possible in the systems E_7, E_8 for certain subsystems of the form $A_{k_1} + \cdots + A_{k_t}$, where all of the k_i are odd), we have to explicitly construct $(\Delta)'$, $(\Delta)''$ and look for the roots orthogonal to them. It usually (but not always) turns out that the orthogonal complements of $(\Delta)'$ and of $(\Delta)''$ have a different type. For systems of type E_ℓ we actually carried out all calculations not in the usual realization, but in the much more convenient realization of these systems in the pseudo-Euclidean space suggested in [11].

The calculation of the factor group $(A(\Delta) \cap W(\Phi))/W(\Delta)$ is done as follows. Suppose $w \in A(\Delta) \cap W(\Phi)$. Then there exists an element $w_0 \in W(\Delta)$ such that $w_0 w$ fixes certain simple roots of Δ (e.g., the roots $\alpha_1, \alpha_3, \alpha_5, \ldots$ for A_ℓ, the roots $\alpha_1, \alpha_2, \ldots, \alpha_{\ell-2}$ for D_ℓ, and the roots α_1, α_4, α_6 for E_6). If we denote by Σ the subsystem generated by these roots, we can conclude that $w_0 w \in W(\Sigma^\perp)$. It remains to consider whether the group $W(\Sigma^\perp)$

realizes outer automorphisms on the system Δ. For the case where Δ is equal to D_4 and contains some isotypical components or some components for which the group of automorphisms is not equal to the Weyl group it is necessary to repeat the above argument several times and to see that each time we may assume that a larger number of simple roots remains fixed. Particularly difficult are the cases where the group $A(\Delta)/W(\Delta)$ is very large, e.g., $\Delta = 7A_1$ in E_7 or $\Delta = 2D_4$ in E_8.

Once we know the group $X(\Delta)$, it is usually rather easy either to find a set S (a subsystem of roots or a parabolic subset) such that the pair $(S, X(S))$ is strictly larger than the pair $(\Delta, X(\Delta))$ or to prove by elementary considerations (the order type of $X(\Delta)$, transitivity of the action on isotypical components, etc.) the maximality of $(\Delta, X(\Delta))$. The author has prepared for publication a paper containing the calculations, with all details, that were described here in a general way. In particular, that paper contains tables in which for each subgroup $N(\Delta)$ that is not maximal there are indicated the standard subgroups in which it is contained.

After the present paper had been completed, the author learned of the publications [15]–[17], in which the analogous question was considered for semisimple complex Lie groups. Since the main calculations are connected only with systems of roots and their Weyl groups and do not depend in any way on the ground field, the present paper has a very large intersection with [15]–[17]. Nevertheless, it seems to us that it is not completely covered by them. First, independent calculations by another method and in a different realization have enabled us to discover three mistakes in the tables of [16] and [17]. Namely, the subgroup $N(D_5) = G(D_5)$ is not maximal in $G(E_6, K)$ since it is contained in a proper parabolic subgroup; the subgroup $N(3A_2)$ is not maximal in $G(E_7, K)$, since it is contained in $N(E_6)$; the system E_7 has no subsystem $D_5 + A_2$, but it does have a subsystem $A_5 + A_2$. Second, along the way we obtained considerable additional information on the groups $X(\Delta)$, $N(\Delta)$ and the inclusions among them, which will be used in future publications.

Finally, the author thanks Z. I. Borevich for his interest in this research.

BIBLIOGRAPHY

1. Z. I. Borevich, *A description of the subgroups of the full linear group that contain the group of diagonal matrices*, Zap. Nauch. Seminar. Leningrad. Otd. Mat. Inst. im. V. A. Steklov., Akad. Nauk SSSR **64** (1976), 12–29.

2. ——, *On some subgroups of the full linear group*, Zap. Nauch. Seminar. Leningrad. Otd. Mat. Inst. im. V. A. Steklov., Akad. Nauk SSSR **71** (1977), 42–46.

3. A. Borel and J. Tits, *Reductive groups*, Matematika **11** (1967), No. 1, 43–111; No. 2, 3–31.

4. ——, *Unipotent elements and parabolic subgroups of reductive groups*. I, Matematika **16** (1972), 3–12.

5. N. A. Vavilov, *On subgroups of extended Chevalley groups containing a maximal torus*, XVI All-Union Algebra Conference, Abstracts of Reports, Part 1, Leningrad, 1981, pp. 26–27.

6. ——, *Subgroups of Chevalley groups over a field that contain a maximal torus*, XVII All-Union Algebra Conference, Abstracts of Reports, Part 1, Minsk, 1983, pp. 38–39.

7. N. A. Vavilov and E. V. Dybkova, *On subgroups of the full symplectic group that contain the group of diagonal matrices.* I, II, Zap. Nauch. Seminar. Leningrad. Otd. Mat. Inst. im. V. A. Steklov., Akad. Nauk SSSR **103** (1980), 31–47; **132** (1983), 44–56.

8. M. Goto and F. Grosshans, *Semisimple Lie Algebras*, Moscow, 1981.

9. E. B. Dynkin, *Semisimple subalgebras of semisimple Lie algebras*, Mat. Sb. **30** (1952), 349–462.

10. V. A. Koïbaev, *Examples of nonmonomial linear groups without transvections*, Zap. Nauch. Seminar. Leningrad. Otd. Mat. Inst. im. V. A. Steklov., Akad. Nauk SSSR **71** (1977), 153–154.

11. Yu. I. Manin, *Cubic Forms*, Moscow, 1975.

12. R. Steinberg, *Lectures on Chevalley Groups*, Moscow, 1975.

13. S. Berman and R. V. Moody, *Extensions of Chevalley groups*, Israel J. Math. **22** (1975), 42–51.

14. R. W. Carter, *Conjugacy classes in the Weyl group*, Compositio Math. **25** (1972), 1–59.

15. M. Golubitsky, *Primitive actions and maximal subgroups of Lie groups*, J. Differential Geom. **7** (1972), 175–191.

16. M. Golubitsky and B. Rothschild, *Primitive subalgebras of exceptional Lie algebras*, Bull. Amer. Math. Soc. **77** (1971), 983–986.

17. ——, *Primitive subalgebras of exceptional Lie algebras*, Pacific J. Math. **39** (1971), 371–393.

18. G. M. Seitz, *Subgroups of finite groups of Lie type*, J. Algebra **61** (1979), 16–27.

19. J. Tits, *Sous-algèbres des algèbres de Lie semi-simples* (*d'après V. Morozov, A. Malcev, E. Dynkin, et F. Karpelevitch*), Séminaire Bourbaki, 7-e année 1954/1955, Exposé 119, pp. 1–18, Paris, 1959.

20. ——, *Groupes semi-simples isotropes*, Coll. Théorie des Groupes Algébriques (Bruxelles, 1962), pp. 137–147, Paris, 1962.

21. N. A. Vavilov, *Subgroups of split orthogonal groups in even dimensions*, Bull. Acad. Polon. Sci., Sér Sci. Math. **29** (1981), 425–429.

Translated by G. A. KANDALL

Amer. Math. Soc. Transl.
(2) Vol. **149**, 1991

The Corona Theorem in Algebras of Bounded Analytic Functions

V. A. TOLOKONNIKOV

§1. Introduction

In 1962 L. Carleson [1] proved the following theorem: if the functions $f_1, f_2, \dots, f_n$ are bounded and analytic in the open unit disk $\mathbf{D}$ (briefly, they are functions in the class H^∞) and if

$$\inf\left\{\sum_1^n |f_i(z)|^2 : z \in \mathbf{D}\right\} > 0 \tag{1}$$

then there are functions $g_1, g_2, \dots, g_n \in H^\infty$, such that

$$\sum f_i g_i = 1. \tag{2}$$

This assertion is known as the "Corona Theorem." Let us agree to call a a system of functions $\{f_1, f_2, \dots, f_n\}$ in H^∞ satisfying (1) *data* (for the corona problem), and to call a system $\{g_1, g_2, \dots, g_n\}$ satisfying (2) a *solution* for the corona problem (for the data $\{f_1, f_2, \dots, f_n\}$).

We shall say that the corona problem is solvable in a subset $B \subset H^\infty$ (more briefly, the corona problem is B-solvable) if for all data lying in B there is a solution $\{g_1, g_2, \dots, g_n\}$ to the corona problem with $g_i \in B$ $(i = 1, \dots, n)$.

It is known that the corona problem need not be B-solvable even when B is a closed subalgebra of H^∞ [2]. The basic result of the present paper is the description of a large class of subalgebras of H^∞ in which the corona problem is solvable.

There is an interesting quantitative aspect to the problem of B-solvability of the corona problem. To describe this we need some notation. To each normed space X we associate the space $l^2(X)$ of all sequences $f = \{f_i\}_1^\infty$

1991 *Mathematics Subject Classification.* Primary 30H05, 46J15.

0065-9290/91 $1.00 + $.25 per page

of elements of X such that

$$\|f\|^2_{l^2(X)} \stackrel{\text{def}}{=} \sum_1^\infty \|f_i\|^2_X < \infty .$$

Let $a = \{a_i\}_1^\infty$, $b = \{b_i\}_1^\infty$ be two sequences of complex numbers. If $\sum |a_i b_i| < \infty$, then we write

$$\langle a, b\rangle = \sum a_i b_i \quad \text{and} \quad |a| = \langle a, \overline{a}\rangle^{1/2},$$

where the bar denotes complex conjugation. Thus if $X \subset H^\infty$, $z \in \mathbf{D}$, and $f \in l^2(X)$, then $|f(z)|^2 = \sum |f_i(z)|^2$.

We return now to the corona problem and consider a more general formulation. We shall allow not only finite, but also countable, sets of "data" with a corresponding change in the concept of solution.

Let $B \subset H^\infty$, and let δ be a positive number. By *vector data* (more precisely, (B, δ)-vector data) we mean an element f of the space $l^2(B)$ that satisfies the conditions:

$$\|f\|^2_{l^2(X)} \le 1; \qquad \inf\{|f(z)| : z \in \mathbf{D}\} \ge \delta .$$

The set of all (B, δ)-vector data will be denoted by $D(B, \delta)$. In this connection we shall always assume that the embedding of B in H^∞ is continuous (see the remark after Axiom M), and thus $|f(z)|$ is finite, for f in $l^2(B)$ and $z \in \mathbf{D}$. By a B-solution of the corona problem corresponding to the vector $f \in l^2(B)$ we mean a vector $g \in l^2(B)$ such that

$$\langle f(z), g(z)\rangle = 1 \quad (z \in \mathbf{D}).$$

Finally, let

$$C(f) \stackrel{\text{def}}{=} \inf\{\|g\|_{l^2(B)} : \langle g(z), f(z)\rangle = 1, \ z \in \mathbf{D}\},$$

$$C(B, \delta) \stackrel{\text{def}}{=} \sup\{C(f) : f \in D(B, \delta)\}.$$

The finiteness of $C(f)$ means that the corona problem is B-solvable for the given vector data f. The quantitative aspect of our work referred to above involves the study of $C(B, \delta)$ (in particular, estimating it as $\delta \to 0$). In [3]–[7] various authors studied quantities close to $C(H^\infty, \delta)$.

The present paper is based on T. Wolff's method of solving the corona problem [3].

We prove the B-solvability of the corona problem, and find estimates for $C(B, \delta)$ for algebras B satisfying certain axioms. Section 2 is devoted to these axioms, and to solving the corona problem "in general form." In §3 we show that many concrete algebras B satisfy our axioms, and thus we obtain the B-solvability of the corona problem for these algebras, together with estimates for $C(B, \delta)$ from the general considerations in §2. Among such algebras B is the algebra of all those functions $f \in H^\infty$ with a finite Dirichlet integral:

$$\int_{\mathbf{D}} |f'|^2 \, dm_2 < \infty .$$

The question of the B-solvability of the corona problem for this algebra was raised in [8]. Section 4 is devoted to this and to similar algebras.

To each normed space X of analytic functions in the open unit disk $\mathbf{D}$ there is associated its algebra of multipliers:

$$M(X) \stackrel{\text{def}}{=} \{f \in \mathscr{A} : fX \subset X\},$$

where $\mathscr{A}$ denotes the set of all analytic functions in $\mathbf{D}$. In §5 we solve the corona problem for many multiplier algebras. In particular, we give a partial solution to one of the questions posed in [20]. The sixth paragraph studies closed subalgebras of H^∞, with the supremum norm. The seventh paragraph contains some additional comments.

Notation. Let letters $\mathbf{C}$, $\mathbf{D}$, $\mathbf{T}$ denote, respectively, the complex plane, the open unit disk, and the unit circle; $\mathbf{D}^- = \mathbf{D} \cup \mathbf{T}$ denotes the closed unit disk.

The letter c will denote a real constant depending only on the space being considered.

Spaces of analytic and harmonic functions. We use $\mathscr{A}$ ($\mathscr{H}$) to denote the set of all analytic (harmonic) functions in $\mathbf{D}$, with the topology of uniform convergence on compact subsets.

We shall often identify, without express notice, functions in $\mathscr{H}$ with their boundary values on $\mathbf{T}$ (in the sense of generalized functions), and we shall identify functions on $\mathbf{T}$ with their harmonic extension into $\mathbf{D}$.

We consider also the following spaces: $\mathscr{H}_{1+\varepsilon}$, $\varepsilon > 0$, is the space of all bounded harmonic functions in $\{z : |z| < 1 + \varepsilon\}$, and

$$\mathscr{H}(\mathbf{D}^-) \stackrel{\text{def}}{=} \bigcup_{\varepsilon>0} \mathscr{H}_{1+\varepsilon}$$

is the space of all functions harmonic on $\mathbf{D}^-$ (that is, in a neighborhood of $\mathbf{D}^-$, the neighborhood depending on the function). We denote the subspaces of analytic functions by

$$\mathscr{A}_{1+\varepsilon} \stackrel{\text{def}}{=} \mathscr{A} \cap \mathscr{H}_{1+\varepsilon} \quad \text{and} \quad \mathscr{A}(\mathbf{D}^-) \stackrel{\text{def}}{=} \mathscr{A} \cap \mathscr{H}(\mathbf{D}^-).$$

The set of all Laurent polynomials (that is, finite linear combinations of powers of z and powers of $\bar{z}$) we denote by $\mathscr{P}$ (restricted to $\mathbf{T}$ these are just the trigonometric polynomials).

If X is a space of functions in $\mathbf{D}$ then we denote its analytic part by $X_A \stackrel{\text{def}}{=} X \cap \mathscr{A}$; thus $\mathscr{P}_A$ denotes the polynomials in z. Also

$$X_0 \stackrel{\text{def}}{=} \{f \in X : f(0) = 0\}$$

denotes the subspace of functions vanishing at the origin.

If X has a topology, then we denote the closure of the polynomials in X by

$$X_c \stackrel{\text{def}}{=} \mathrm{cl}_X(\mathscr{P} \cap X).$$

If X and Y are Banach spaces, then the Banach space of all continuous linear operators from X to Y will be denoted by $L(X, Y)$, and by $L(X)$ when $Y = X$.

If X is a Banach space then we introduce the space of square summable sequences of vectors of X:

$$l^2(X) \stackrel{\text{def}}{=} \left\{ f = \{f_i\}_1^\infty : f_i \in X, \|f\|^2_{l^2(X)} \stackrel{\text{def}}{=} \sum_1^\infty \|f_i\|^2_X < \infty \right\},$$

and the corresponding space of doubly-indexed "sequences":

$$S_2(X) \stackrel{\text{def}}{=} \left\{ f = \{f_{ij}\}_{i,j=1}^\infty : f_{ij} \in X, \|f\|^2_{S_2(X)} \stackrel{\text{def}}{=} \sum_{i,j=1}^\infty \|f_{ij}\|^2_X < \infty \right\}.$$

For the nonnormable space $\mathscr{H}(\mathbf{D}^-)$ we make the following definition:

$$l^2(\mathscr{H}(\mathbf{D}^-)) \stackrel{\text{def}}{=} \bigcup_{\varepsilon > 0} l^2(\mathscr{H}_{1+\varepsilon}).$$

The spaces $l^2(\mathscr{A}(\mathbf{D}^-))$, $S_2(\mathscr{H}(\mathbf{D}^-))$, and $S_2(\mathscr{A}(\mathbf{D}^-))$ are defined similarly. Also, by definition

$$l^2(\mathscr{P}) \stackrel{\text{def}}{=} \{f = \{f_i\}_1^\infty : f_i \in \mathscr{P}, \text{ and } f_i \neq 0 \text{ for only finitely many } i\}.$$

If X is a Banach space of functions in $\mathbf{D}$ then we define a bilinear and a quadratic form in $l^2(X)$ as follows:

$$\langle f, g\rangle(z) \stackrel{\text{def}}{=} \sum f_i(z) g_i(z),$$
$$|f|^2(z) = \sum |f_i(z)|^2.$$

Spaces of measurable functions. Normalized Lebesgue measure on $\mathbf{T}$ will be denoted by dm_1, and normalized Lebesgue area measure on $\mathbf{D}$ by dm_2; $S(dm_1)$ and $S(dm_2)$ will denote the spaces of all measurable functions for the measures dm_1 and dm_2.

A Banach space $E \subset S(dm_1)$ will be called a *solid Banach space*([1]) on the circle (denoted SBS($\mathbf{T}$)) if $f \in E$, $g \in S(dm_1)$, and $|g| \leq |f|$ almost everywhere $[dm_1]$ always implies that $g \in E$ and $\|g\|_E \leq \|f\|_E$. Analogously one defines solid Banach spaces in the open disk: SBS($\mathbf{D}$).

We shall say that an SBS E is dual to an SBS F, with respect to a measure μ, if E is the set of all those measurable functions f such that $\int |fg|\, d\mu < \infty$ for every function $g \in F$. Notation: $E = F'$.([2])

Derivatives. Let $f \in \mathscr{H}$ have the (formal) Fourier series

$$f(\xi) \sim \sum_{k=-\infty}^\infty \hat{f}(k)\xi^k, \quad |\xi| = 1.$$

([1])[Translator's note. Called a *Banach ideal space* in the original (see [12], Chapter IV, §3.1).

([2])Editor's note. Following [12], Chapter VI, §1.1, F' is here called the dual of E.

This means that $f(r\xi) = \sum \hat{f}(k) r^{|k|} \xi^k$, $0 \le r < 1$. By the symbol $f^{(n)}$, $n = \pm 1, \pm 2, \dots$, we denote the harmonic function with the Fourier series:

$$f^{(n)}(\xi) \sim \sum_{\substack{k=-\infty \\ k \neq 0}}^{\infty} \hat{f}(k) k^n \xi^k .$$

In particular, if $f \in \mathscr{A}$, then $f^{(1)}(z) = z f'(z)$ for $z \in \mathbf{D}$.

The space of functions with continuous Nth derivative on $\mathbf{T}$, respectively in $\mathbf{D}^-$, will be denoted by C^N, respectively $C^N(\mathbf{D}^-)$. Note that if $f \in C^1$, then

$$f^{(1)}(e^{i\theta}) = \frac{1}{i}\frac{d}{d\theta} f(e^{i\theta}), \quad 0 \le \theta < 2\pi .$$

If $X \subset \mathscr{H}$ is a class of harmonic functions and n is a whole number, then by definition: $X^{(n)} = \{f \in \mathscr{H} : f^{(n)} \in X\}$.

In what follows we shall frequently use different variants of Green's formula and integration by parts. These are listed in the following lemma. We use the following notations:

$$\bar{\partial} = \frac{1}{2}\left(\frac{\partial}{\partial x} + i\frac{\partial}{\partial y}\right), \qquad \partial = \frac{1}{2}\left(\frac{\partial}{\partial x} - i\frac{\partial}{\partial y}\right), \qquad \Delta = 4\partial\bar{\partial} .$$

LEMMA 1. *Let $d\tilde{m}_1$ denote normalized Lebesgue measure on the circle $\gamma = \{|z| = \frac{1}{2}\}$, and let $\Gamma = \{z : \frac{1}{2} \le |z| \le 1\}$. Then*

1. $\int_{\mathbf{D}} \partial a \, dm_2 = \int_{\mathbf{T}} a z \, dm_1$, $a \in C^1(\mathbf{D}^-)$.
2. $\int_\Gamma \bar{\partial} a \, dm_2 = \int_{\mathbf{T}} a z \, dm_1 - \int_\gamma a z \, d\tilde{m}_1$, $a \in C^1(\Gamma)$.
3. $\int_\Gamma b(\frac{1}{z} - \bar{z})^{n-1} \, dm_2 = \int_\Gamma \frac{1}{n} \bar{\partial} b (\frac{1}{z} - \bar{z})^n \, dm_2 - \int_\gamma \frac{1}{n} b (3\bar{z})^n \, d\tilde{m}_1$, $b \in C^1(\Gamma)$.
4. $\int_{\mathbf{T}} a \, dm_1 = \frac{1}{2} \int_{\mathbf{D}} \Delta a \ln \frac{1}{|z|} \, dm_2$, $a \in C^2(\mathbf{D}^-)$, $a(0) = 0$.
5. $\int_{\mathbf{T}} ab \, dm_1 = \int_{\mathbf{T}} a^{(k)} b^{(-k)} \, dm_1$, $a \in C^k$, $b \in C$, *and either* $a(0) = 0$ *or* $b(0) = 0$; $k = 1, 2, \dots$.

We shall denote the end of a proof by □.

§2. Axioms. "Abstract" estimates for $C(B, \delta)$

Let X be a set of functions in $\mathbf{D}$. By X_A we denote its "analytic part," that is, the set $X \cap \mathscr{A}$ (recall that $\mathscr{A}$ is the set of all holomorphic functions in $\mathbf{D}$).

We shall assume that B is an algebra (with respect to pointwise multiplication), representable in the form

$$B = X_A$$

where X is some Banach space of complex-valued harmonic functions in $\mathbf{D}$. The conditions that we place on B are expressed in terms of axioms on X; these axioms are arranged in groups.

As above the symbol $\mathscr{H}$ denotes the space of all harmonic functions in $\mathbf{D}$; both it and the space $\mathscr{A}$ have the topology of uniform convergence on compact subsets of $\mathbf{D}$. Recall that we do not distinguish between a (generalized) function f on $\mathbf{T}$, and its harmonic extension to $\mathbf{D}$. In keeping with this we regard spaces of functions given on $\mathbf{T}$ as being defined in $\mathbf{D}$, whenever this is convenient.

Axioms $(\mathscr{H})$. General properties of a Banach space X of harmonic functions.

Axiom $(\mathscr{H}1)$. The embedding $X \subset \mathscr{H}$ is continuous.

It follows from the closed graph theorem that $(\mathscr{H}1)$ is equivalent to the continuity on X of all point evaluation functionals:

$$f \to f(\lambda) \quad (\lambda \in \mathbf{D}).$$

Axiom $(\mathscr{H}2)$. $\mathscr{H}(\mathbf{D}^-) \subset X$.

Recall that $\mathscr{H}(\mathbf{D}^-)$ is the space of functions harmonic on $\mathbf{D}^-$. It follows from $(\mathscr{H}1)$ that the embedding $\mathscr{H}(\mathbf{D}^-) \subset X$ is continuous.([3])

The next axiom describes a connection between X and the space M_1 of all (complex-valued) Borel measures on $\mathbf{T}$. Each $h \in \mathscr{H}$ has a "Fourier transform" $\hat{h} : \mathbf{Z} \to \mathbf{C}$, where

$$\hat{h}(n) = r^{-|n|} \int_{\mathbf{T}} h(rz)\bar{z}^n \, dm_1(z) \quad (|z| = 1;\ 0 < r < 1).$$

If $h \in \mathscr{H}$ and $\mu \in M_1$, then by $h * \mu$ we mean the harmonic function

$$(h * \mu)(re^{i\theta}) = \sum_{n=-\infty}^{\infty} \hat{h}(n)\hat{\mu}(n) r^{|n|} e^{in\theta} \quad (\theta \in \mathbf{R};\ 0 < r < 1),$$

where $\hat{\mu}(n) = \int_{\mathbf{T}} \bar{z}^n \, d\mu(z)$.

Axiom $(\mathscr{H}3)$. If $f \in X$, $\mu \in M_1$, then $f * \mu \in X$, and

$$\|f * \mu\| \le \|f\|_X \|\mu\|_{M_1}.$$

It follows from this axiom that the space X is invariant under rotations of $\mathbf{D}$. The converse is false. For an example one may take the space of measures of the form $fm_1 + \mu$, where $f \in L^2(\mathbf{T})$ and μ is singular with respect to m_1, with the norm

$$\|fm_1 + \mu\| \stackrel{\text{def}}{=} \|f\|_2 + \|\mu\|_{M_1}.$$

If X satisfies axioms $(\mathscr{H}1)$–$(\mathscr{H}3)$, then we write $X \in (\mathscr{H})$. These are the only spaces we shall consider in what follows. Note that these spaces are close to the homogeneous Banach spaces on the line, introduced in [25].

We now consider the space X^* conjugate to X, $X \in (\mathscr{H})$. We denote the identity (or position) function on $\mathbf{T}$ by z. Thus z^k may be taken as a

([3])Translator's note. By this is meant that if $\{f_n\}$ is a sequence of functions harmonic in an open set $U \supset \mathbf{D}^-$, and converging to a function f uniformly on each compact subset of U, then $f_n \to f$ in the norm of X.

function in $\mathbf{D}$, for $k = 0, 1, 2, \dots$. This is also true for $k = -1, -2, \dots$, but then we have in mind the function

$$re^{i\theta} \to r^{|k|}e^{ik\theta} \quad (r \in [0, 1), \theta \in \mathbf{R}).$$

Every functional $g \in X^*$ has a "Fourier transform" $\hat{g}$:

$$\hat{g}(k) \stackrel{\text{def}}{=} g(z^{-k}), \quad k \in \mathbf{Z}.$$

Also, $|\hat{g}(k)| \le c\|z^{-k}\|_X$. From $(\mathscr{H}1)$ and $(\mathscr{H}2)$ it follows that

$$\varlimsup_{k\to\infty} (\|z^k\|_X + \|z^{-k}\|_X)^{1/k} = 1,$$

so that the functional g corresponds to a function $j(g) \in \mathscr{H}$: $j(g) \stackrel{\text{def}}{=} \sum \hat{g}(k)z^k$ (the summation is $-\infty < k < \lambda$). If the set $\mathscr{P}$ of all Laurent polynomials is dense in X (that is, $X = X_c \stackrel{\text{def}}{=} \mathrm{cl}_X \mathscr{P}$), then the operator j is one-to-one, and thus X^* may be identified with the set of functions $j(X^*) \subset \mathscr{H}$. And in every case it is easy to give an explicit formula for $g|\mathscr{H}(\mathbf{D}^-)$. Indeed, if $g \in X^*$ and $f \in \mathscr{H}(\mathbf{D}^-)$ then

$$(g, f) \stackrel{\text{def}}{=} g(f) = g\left(\sum_{k=-\infty}^{\infty} \hat{f}(k)z^k\right) = \sum_{k=-\infty}^{\infty} \hat{f}(k)\hat{g}(-k).$$

(The Fourier series of f converges in the space $\mathscr{H}(\mathbf{D}^-)$, and hence also in X.) If, in addition, $g \in L^1 \cap X^*$, then

$$(g, f) = \int_{\mathbf{T}} gf\,dm_1.$$

Thus we have a "Cauchy duality" between the spaces X^* and X.

To each space $X \in (\mathscr{H})$ we associate (along with X_c) another space X_m. This will help us to characterize those $X \in (\mathscr{H})$ that are dual (with respect to the Cauchy duality) to a given space of class $(\mathscr{H})$.

DEFINITION. 1. Given a space $X \in (\mathscr{H})$ we say that a sequence $\{x_n\} \subset X$ w-converges to an element $f \in \mathscr{H}$, if $\sup \|x_n\|_X < \infty$, and $x_n(z) \to f(z)$ for all $z \in \mathbf{D}$.

2. If $X \in (\mathscr{H})$, then we define the space X_m as follows. Here $f_r(z) = f(rz)$, $z \in \mathbf{D}$.

$$X_m \stackrel{\text{def}}{=} \left\{ f \in \mathscr{H} : \|f\|_{X_m} \stackrel{\text{def}}{=} \sup_{0<r<1} \|f_r\|_X < \infty \right\}.$$

The following lemma connects w-convergence and the space X_m.

From Part 1 of the lemma it follows that X_m is obtained by completing X with respect to w-convergence. From Parts 2 and 5 it follows that X is the "Cauchy dual" of some other space if and only if $X = X_m$. Also, if $X = Y^*$, then $Y = (((Y^*)_c)^*)_c = ((X_c)^*)_c$ (see 5), so that Y is uniquely determined by X. We shall call Y the *predual* of X, and write $Y = X_*$.

LEMMA 2. *Let* $X \in (\mathscr{H})$. *Then*

1. $x \in X_m \Leftrightarrow \exists \{x_n\} \subset X, x_n \xrightarrow{w} x$; *if* $x \in X_m$, *then we may take* $x_n = x_{r_n}, r_n \uparrow 1$.
2. *If* $X = Y^*$ *with respect to the Cauchy duality, then* $x_n \xrightarrow{w} x$ *if and only if* $x_n \to x$ *in the weak** *topology.*
3. $X_m \in (\mathscr{H})$.
4. $X_{cm} = X_m$, *and* $X_{mc} = X_c$.
5. $X_m = (((X_c)^*)_c)^*$ *(isometric isomorphism).*

PROOF. We only prove 5. If $Z = ((X_c)^*)_c$, then for $p \in \mathscr{P}$ we have

$$\begin{aligned}\|p\|_{X_c} &= \sup\{|(p, y)| : \|y\|_{(X_c)^*} \le 1\}\\ &= \sup\{|(p, y)| : \|y\|_Z \le 1\} = \|p\|_{Z^*}.\end{aligned}$$

Thus Z is w-dense in $(X_c)^*$ and so $X_c = (Z^*)_c$, and

$$X_m = X_{cm} = (Z^*)_{cm} = Z^*. \quad \square$$

The following axioms concern multiplication in the space X.

Axiom (M). If $f \in \mathscr{P}$, $g \in \mathscr{P}_A$, then $\|fg\| \le c\|f\|_X\|g\|_X$, where c depends on X.

Extending this bilinear map by continuity, we obtain a partial multiplication in X_c,that is, a bilinear map

$$\rho : X_c \times X_{cA} \to X_c.$$

In particular, if $X = X_c$, then X_A is an algebra. This is also true when $X = X_m$.

LEMMA 3. *If* $X \in (\mathscr{H})$, (M) *and if* $X = X_m$, *then* X_A *is an algebra.*

PROOF. If $f, g \in \mathscr{P}_A$ and $z \in \mathbf{D}$, then $\rho(f, g)(z) = f(z)g(z)$, and thus the bilinear form $\rho|(X_{cA} \times X_{cA})$ can be extended, using w-continuity in each argument. $\square$

It is easy to see that if $X \in (\mathscr{H}1)$ and if X_A is a Banach algebra, then $X_A \subset H^\infty$. This fact is easily established by "Landau's trick": by $(\mathscr{H}1)$ we have $|f(z)| \le c(z)\|f\|_X$ $(z \in \mathbf{D}, f \in X_A)$. Applying this to f^n we have: $|f(z)| \le c(z)^{1/n}\|f^n\|_X^{1/n}$, $n \ge 1$.(4) If the norm of the bilinear map ρ of Axiom (M) is equal to 1, then the embedding $X_A \subset H^\infty$ is contractive.

We note that most of the time the following stronger form of Axiom (M) will be satisfied:

$$X \text{ is a subalgebra of } L^\infty. \qquad (\widetilde{\mathrm{M}})$$

(4)Translator's note. We may assume that f has norm one; letting $n \to \infty$ we obtain $|f(z)| \le 1$ in $\mathbf{D}$. This result may also be obtained as follows, without assuming $(\mathscr{H}1)$. If not all functions in X_A vanish at z, then evaluation at z is a nontrivial multiplicative linear functional on X_A, and hence is bounded and has norm one.

The continuity of multiplication follows from the closed graph theorem (the embedding $X \subset L^\infty$ is continuous).

The following Axiom (C) has a more special character and is more closely related to the corona problem.

Before formulating Axiom (C) we note that if $f, g \in l^2(X_A)$, if $X \in (\mathscr{H}) \cap M$, and if

$$\langle f, g\rangle \equiv 1, \tag{3}$$

then for all $z \in \mathbf{D}$,

$$1 \le |f(z)|\,|g(z)| \le |f(z)|\,\|g\|_{l^2(H^\infty)} \le c|f(z)|\,\|g\|_{l^2(X_A)},$$

where c depends only on X. Thus the estimate

$$\inf\{|f(z)| : z \in \mathbf{D}\} \ge \delta, \tag{4}$$

for some positive δ, is necessary for the solvability of equation (3) (with respect to the vector-function g).

In other words, if (3) is solvable for g in X_A, then f forms (X_A, δ)-vector data (see the Introduction): $f \in D(X_A, \delta)$.

If $f \in D(X_A, \delta) \cap l^2(\mathscr{A}(\mathbf{D}^-))$, then it makes sense to speak of the vector-valued function $\overline{f}\cdot|f|^{-2}$ and the matrix-valued function $u_f = \{(u_f)_{ij}\}$ on $\mathbf{D}$, where

$$(u_f)_{ij} \stackrel{\text{def}}{=} (\overline{f'_i(z)}\cdot\overline{f_j(z)} - \overline{f_i(z)}\cdot\overline{f'_j(z)})\cdot|f|^{-4}(z). \tag{5}$$

Axiom (C). Properties of X-data.

Axiom (C1). For every $\delta \in (0, 1)$

$$C_1(\delta) \stackrel{\text{def}}{=} \sup\{\|\overline{f}|f|^{-2}\|_{l^2(X)} : f \in D(X_A, \delta) \cap l^2(\mathscr{A}(\mathbf{D}^-))\} < +\infty.$$

Axiom (C2). For every $\delta \in (0, 1)$

$$C_2(\delta) \stackrel{\text{def}}{=} \sup\left\{\left|\int_{\mathbf{D}} \langle u_f, y\rangle\, dm_2\right| : f \in D(X_A, \delta) \cap l^2(\mathscr{A}(\mathbf{D}^-)),\right.$$
$$\left. y \in S_2(\mathscr{P}_A),\ \|zy\|_{S_2(Y_A)} \le 1\right\} < +\infty.$$

Here $Y = (X_m)_*$ is the space predual to X_m, and

$$\langle u_f, y\rangle(z) \stackrel{\text{def}}{=} \sum_{i,j} (u_f)_{ij}(z) y_{ij}(z) \quad (z \in \mathbf{D}).$$

Since $\overline{f}|f|^{-2}$ is not harmonic in $\mathbf{D}$, the computation of its norm is understood according to our convention: the norm of the harmonic extension of the restriction of $\overline{f}|f|^{-2}$ to $\mathbf{T}$ is computed. It is very essential that in Axiom (C2) we *do not use* this convention: the integral on the area is computed for a nonharmonic function.

Everything is now ready for formulating our "abstract" corona theorem.

THEOREM 1. *If* $X \in (\mathscr{H}) \cap (\mathrm{M}) \cap (\mathrm{C})$, *then*

$$C(X_{cA}, \delta) = C(X_{ma}, \delta) \leq C_1(\delta) + \|\rho\| C_2(\delta) \quad (0 < \delta < 1),$$

where $\|\rho\|$ *is the norm of the bilinear mapping* ρ *in Axiom* (**M**).

Before proving the theorem we discuss the axioms introduced.

The axioms (M) and (C) are satisfied if there is a description of the space X via the boundary values in terms, for example, of the L^p-modulus of continuity or of membership of the derivative in some solid Banach space (SBS) on $\mathbf{T}$. In this case a stronger axiom $(\widetilde{\mathrm{C}1})$ holds.

Axiom $(\widetilde{\mathrm{C}1})$. X is a symmetric full subalgebra of L^∞ with an estimate, i.e., $X \in (\widetilde{\mathrm{M}})$, $X = \overline{X}$, and for every $\delta \in (0, 1)$

$$\widetilde{C}_1(\delta) \stackrel{\text{def}}{=} \sup\{\|h^{-1}\|_X : \|h\|_X \leq 1, |h| \geq \delta \text{ a.e. on } \mathbf{T}\} < +\infty.$$

Indeed, if $f \in D(X_A, \delta)$, then $|f|^2 \in X$ and $\| |f|^2 \|_X \leq \|\rho\|$; therefore $\| |f|^{-2} \| \leq \widetilde{C}_1(\delta^2 \cdot \|\rho\|^{-1}) \|\rho\|$, and

$$C_1(\delta) \leq \widetilde{C}_1(\delta^2 \|\rho\|^{-1}) \|\rho\|^2.$$

The following theorem gives a typical case in which Axiom (C2) can be verified.

THEOREM 2. 1) *If* E *and* F *are in* SBS($\mathbf{D}$), $E = F'$ *is the dual with respect to the measure* dm_2, $X = E_A^{(1)} \in (\mathscr{H})$, $X \subset L^\infty$, *and* $Y \stackrel{\text{def}}{=} X_* = F_{\mathscr{H}}$, *then Axiom* (C2) *is satisfied, and* $C_2(\delta) \leq c\delta^{-4}$.

2) *If, moreover,* $F_{\mathscr{H}} = F_A \oplus F_{\overline{A0}}$, *i.e.,* $Y = Y_A \oplus Y_{\overline{A0}}$, *then* $X \in (\widetilde{\mathrm{M}})$, (C1), *and* $C_1(\delta) \leq c\delta^{-4}$.

PROOF OF THEOREM 2. 1) f and g are the same as in Axiom (C2). Then

$$\begin{aligned}\left|\int_{\mathbf{D}} \langle u_f, y\rangle \, dm_2\right| &\leq \int_{\mathbf{D}} \sum_{i,j} (|f_i| \cdot |f'_j| + |f'_i| \cdot |f_j|) \cdot |f|^{-4} |y_{ij}| \, dm_2 \\ &\leq \sum_{i,j} \delta^{-4} \left(\|f_i\|_\infty \int_{\mathbf{D}} |f'_j| \cdot |y_{ij}| \, dm_2 + \|f_j\|_\infty \int_{\mathbf{D}} |f'_i| \cdot |y_{ij}| \, dm_2 \right) \\ &\leq c \sum_{i,j} \delta^{-4} (\|f_i\|_X \cdot \|f'_j\|_E + \|f_j\|_X \cdot \|f'_i\|_E) \|y_{ij}\|_F \leq c\delta^{-4}.\end{aligned}$$

2) Let us estimate $C_1(\delta)$. Each element y with $\|y\|_Y \leq 1$ can be decomposed as $y = y_1 + \overline{y}_2$: $\|y_1\|_{Y_{A0}}, \|y_2\|_{Y_A} \leq c$; therefore, for $f \in$

$D(X_A, \delta) \cap l^2(\mathscr{A}(\mathbf{D}^-))$

$$\|\overline{f}|f|^{-2}\|_{l^2(X)} = \sup\left\{\left|\int_{\mathbf{T}} \langle \overline{f}|f|^{-2}, y\rangle\, dm_1\right| : y \in l^2(\mathscr{H}(\mathbf{D}^-)), \|y\|_{l^2(Y)} \le 1\right\}$$
$$\le c\cdot \sup\left\{\left|\int_{\mathbf{T}} \langle \overline{f}|f|^{-2}, y_1\rangle\, dm_1\right| + \left|\int_{\mathbf{T}} \langle \overline{f}|f|^{-2}, y_2\rangle\, dm_1\right| : \right.$$
$$\left. y_1, y_2 \in l^2(A(\mathbf{D}^-)); \|y_1\|_{l^2(Y)}, \|y_2\|_{l^2(Y)} \le 1\right\}$$
$$\stackrel{\text{def}}{=} c \sup(I_1 + I_2).$$

We use Lemma 1, part 1:

$$I_1 \le \left|\int_{\mathbf{D}} \overline{\partial}\left\langle \overline{f}|f|^{-2}, \frac{y_1 - y_1(0)}{z}\right\rangle dm_2\right| + |\langle \overline{f}|f|^{-2}, y_1\rangle(0)|$$
$$\le \left|\int_{\mathbf{D}} \left\langle \overline{f}|f|^{-2} - \overline{f}\langle f, \overline{f}'\rangle|f|^{-4}, \frac{y_1 - y_1(0)}{z}\right\rangle dm_2\right| + c\delta^{-2}$$
$$= \int_{1>|z|\ge\frac12} + \int_{|z|\le\frac12} + c\delta^{-2}.$$

The integral over $\{1 > |z| \ge \frac12\}$ can be estimated as in the first part of the proof, and the second integral can be estimated with the help of the imbedding $X, Y \subset \mathscr{H}$. The integral I_2 can be estimated similarly with the use of the operator ∂ $(\partial f \stackrel{\text{def}}{=} \overline{\partial}(\overline{f}))$ instead of $\overline{\partial}$.

Verification of $(\widetilde{\mathrm{C}})$. Suppose that $f, g \in \mathscr{P}$ are polynomials on $\mathbf{T}$, extended harmonically to $\mathbf{D}$, and consider the decomposition $y = y_1 + \overline{y}_2 : y_1, y_2 \in \mathscr{P}_A$. Then $\int_{\mathbf{T}} f g y_1\, dm_1 = \int_{\mathbf{D}}((\overline{\partial} f)g + f(\overline{\partial} g))\cdot\frac{y_1 - y_1(0)}{z}\, dm_2 + f(0)\cdot g(0)\cdot y_1(0)$, and we easily get the necessary estimate. □

Proof of Theorem 1. Suppose that $X = X_m$ and $f \in D(X_A, \delta) \cap l^2(\mathscr{A}(\mathbf{D}^-))$. We consider a matrix-valued function v with the properties:

1. $v \in C^1(\mathbf{D}^-)$, $v|_{\mathbf{T}} \in \mathscr{H}(\mathbf{D}^-)$;
2. $\overline{\partial} v = u_f$, where u_f is given by (5);
3. $v^{\mathrm{T}} = -v$, where $(v^{\mathrm{T}})_{ij} \stackrel{\text{def}}{=} v_{ji}$ is the transposed matrix-valued function.

Such functions v exist; for example,

$$v_0(\xi) \stackrel{\text{def}}{=} \int_{\mathbf{D}} \frac{u_f(z)}{\xi - z}\, dm_2(z), \quad \xi \in \mathbf{D}^-.$$

We verify only the property 1:

$$v_0(\xi) = \frac{1}{\pi}\int_0^1 \left(\int_0^{2\pi} \frac{u_f(re^{i\theta})}{(\xi/2 - e^{i\theta})} d\theta\right) dr = -\int_0^1 p_-(u_r)\left(\frac{\xi}{r}\right)\frac{r\,dr}{2\xi},$$

where $|\xi| = 1$, $u_r(\eta) \stackrel{\text{def}}{=} u_f(r\eta)$, $|\eta| = 1$, and $p_-(w)(z) = \sum_{k=0}^\infty \hat{w}(-k) z^{-k}$, $|z| > 1$. The integrand is a continuous $\mathscr{H}(\mathbf{D}^-)$-valued function on $[0, 1]$; therefore, $v_0|\mathbf{T} \in \mathscr{H}(\mathbf{D}^-)$.

Define the vector-valued function $g \overset{\text{def}}{=} \overline{f}|f|^{-2} + vf$. Clearly, $g|_{\mathbf{T}} \in l^2(\mathscr{A}(\mathbf{D}^-))$ and $\langle f, g\rangle = 1$. Indeed, $\langle g, f\rangle = 1 + \langle vf, f\rangle = 1$, since $v = -v^{\mathrm{T}}$. Moreover, in $\mathbf{D}$

$$\overline{\partial} g = \overline{f'}|f|^{-2} - \overline{f}\langle f, \overline{f'}\rangle + \overline{\partial} v f = 0.$$

Let us estimate the norm of g:

$$\|g\|_{l^2(X)} \le \|\overline{f}|f|^{-2}\|_{l^2(X)} + \|vf\|_{l^2(X)} \le C_1(\delta) + \|vf\|_{l^2(X)}.$$

We estimate the second term; vf is the result of applying the smooth matrix-valued function v pointwise to the smooth vector-valued f, and we have that

$$\begin{aligned}\|vf\|_{l^2(X)} &= \sum_i \left\|\sum_j \rho(v_{ij}, f_j)\right\|_X^2 \le \|\rho\|^2 \sum_i \left(\sum_j \|v_{ij}\|_X \|f_j\|_X\right)^2 \\ &\le \|\rho\|^2 \cdot \left(\sum_{i,j} \|v_{ij}\|_X^2\right) \cdot \left(\sum_j \|f_j\|_X^2\right) = \|\rho\|^2 \cdot \|v\|_{S_2(X)},\end{aligned}$$

since $f \in D(X_A, \delta)$, and thus $\|f\|_{l^2(X)} \le 1$. Accordingly,

$$\|g\|_{l^2(X)} \le C_1(\delta) + \|\rho\| \cdot \|v\|_{S_2(X)}. \tag{6}$$

It remains for us to estimate

$$\begin{aligned}\Delta &\overset{\text{def}}{=} \inf\{\|v\|_{S_2(X)} : \text{ the properties } 1, 2, \text{ and } 3 \text{ hold for } v\} \\ &= \inf\{\|v_0 - a\|_{S_2(X)} : a \in S_2(\mathscr{A}(\mathbf{D}^-)), a = -a^{\mathrm{T}}\} \\ &= \inf\left\{\left\|v_0 - \frac{b - b^{\mathrm{T}}}{2}\right\|_{S_2(X)} : b \in S_2(\mathscr{A}(\mathbf{D}^-))\right\}.\end{aligned}$$

But $v_0 = -v_0^{\mathrm{T}} = \frac{v_0 - v_0^{\mathrm{T}}}{2}$; therefore,

$$\left\|v_0 - \frac{b - b^{\mathrm{T}}}{2}\right\|_{S_2(X)} \le \frac{\|v_0 - b\| + \|v^{\mathrm{T}} - b^{\mathrm{T}}\|}{2}$$

and

$$\begin{aligned}\Delta &= \inf\{\|v_0 - b\|_{S_2(X)} : b \in S_2(\mathscr{A}(\mathbf{D}^-))\} \\ &= \inf\{\|v_0 - b\|_{S_2(X)} : b \in S_2(X_A)\}.\end{aligned} \tag{7}$$

The last equality follows from $v_0 - b_r \xrightarrow[r \to 1]{} v_0 - b$, $b \in S_2(X_A)$ (weak* convergence).

If $Y = X_*$, then

$$\mathrm{cl}_Y \mathscr{P}_{AC} = Y_{A0} \tag{8}$$

(recall that if $z \in \mathscr{H}$, then $z_0 = \{f \in z : f(0) = 0\}$, and cl_Y is the closure in Y). Indeed, $Y = Y_c$; therefore, $\lim_{\xi \to 1} f_\xi = f$, $f \in Y$ (here $f_\xi(z) \overset{\text{def}}{=} f(\xi z)$,

$z \in \mathbf{D})$. From this equality it is easy to conclude that the Fejér polynomials of an arbitrary function $f \in Y$ converge to it in Y, and this implies (8). This relation implies, in turn, that $Y_{A0}^{\perp} = \mathscr{P}_{A0}^{\perp} = X_A$. Then $S_2(Y_{A0})^{\perp} = S_2(X_A)$, but $S_2(Y)^* = S_2(X)$; therefore, $S_2(Y_{A0})^* = S_2(X)/S_2(X_A)$. We now get from (7) that

$$\Delta = \|v_0\|_{S_2(X)/S_2(X_A)} = \sup\left\{\left|\int_{\mathbf{T}} \langle v_0, g\rangle\, dm_1\right| : g \in S_2(\mathscr{P}_A),\ \|g\|_{S_2(Y_{A0})} \le 1\right\}.$$

By Lemma 1, part 1, $\int_{\mathbf{T}} \langle v_0, g\rangle\, dm_1 = \int_{\mathbf{D}} \bar\partial(\frac{\langle v_0, g\rangle}{z})\, dm_2 = \int_{\mathbf{D}} \langle \bar\partial v_0, \frac{g}{z}\rangle\, dm_2$, and by the definition of $C_2(\delta)$ we get that $\Delta \le C_2(\delta)$, which together with (6) and the definition of the constant Δ gives the necessary lower estimate of the corona problem for $f \in l^2(\mathscr{A}(\mathbf{D}^-))$.

Let $f \in D(X_A, \delta)$. Then $f \in D(X_A, \delta) \cap S_2(\mathscr{A}(\mathbf{D}^-))$ for $r \in (0, 1)$, and by the part of the theorem already proved, there exists a function $g_r \in l^2(X_A)$, $\|g_r\|_{l^2(X_A)} \le C_1(\delta) + \|\rho\| C_2(\delta)$, for which $\langle f_r, g_r\rangle = 1$.

Since $l^2(Y)$ is separable, the weak* topology in $l^2(X)$ is metrizable on every ball, and there exists a sequence $\{g_{r_n}\}$, $r_n \uparrow 1$, that weak* converges to some function g. Then $f_{r_n}(z) \to f(z)$ in the l^2-norm and $g_{r_n}(z) \to g(z)$ weakly in l^2; therefore, $\|g\|_{l^2(X)} \le \varlimsup_{n\to\infty} \|g_{r_n}\|_{l^2(X)} \le C_1(\delta) + \|\rho\| C_2(\delta)$, and

$$\langle f(z), g(z)\rangle = \lim_{n\to\infty} \langle f_{r_n}(z), g_{r_n}(z)\rangle = 1.$$

It remains to verify that $C(X_{cA}, \delta) \le C(X_A, \delta)$ (recall that $X = X_m$). If $f \in D(X_{cA}, \delta)$, then there exists a solution g of the corona problem in X_A with the estimate $C(X_A, \delta)$. If $0 < r < 1$ and $\varepsilon(r) \overset{\text{def}}{=} \|f - f_r\|_{l^2(X_A)}$, $h_r \overset{\text{def}}{=} \langle f_r, g_r\rangle$, then $\|h_r - 1\|_{X_{cA}} = \|\langle f, g_r\rangle - \langle f_r, g_r\rangle\|_{X_{cA}} \le c\cdot\varepsilon(r)$.

As $r \to 1$ we have that $\varepsilon(r) \to 0$; therefore, for $r > r_0$ the element h_r^{-1} exists in X_{cA}, and $\|h_r^{-1}\|_{X_{cA}} \le (1 - c\varepsilon)^{-1}$. Suppose that $\tilde g \overset{\text{def}}{=} g_r h_r$ for some $r > r_0$. Then

$$\langle f, \tilde g\rangle = 1, \quad \tilde g \in l^2(X_{cA}), \quad \|\tilde g\|_{l^2(X_{cA})} \le C(X_A, \delta)(1 - c\varepsilon)^{-1}. \quad \square$$

We now dwell on the next natural generalization of the corona problem: describe the functions belonging to the ideal generated by a given collection of elements of the algebra (the corona theorem itself means that the ideal generated in X_A by the data contains *all* functions in X_A).

The following sufficient condition for membership in an ideal can be proved in a completely similar way. We introduce the axiom $(\mathrm{I}(k))$, where k is a positive integer. It will be said that $f \in D(X_A, h)$ if $h \in X_A$, $f \in l^2(X_A)$, $\|f\|_{l^2(X_A)} \le 1$, $\|h\|_{X_A} \le 1$, and

$$|f(z)| \ge |h(z)|, \quad z \in \mathbf{D}.$$

Axiom (I(k)) (k a positive integer).
Membership of a function in an ideal.
Axiom (I1(k)).

$$L_1 \overset{\text{def}}{=} \sup\{\|\overline{f}h^k|f|^{-2}\|_{l^2(X)} : f \in D(X_A, h) \cap l^2(\mathscr{A}(\mathbf{D}^-)), h \in \mathscr{A}(\mathbf{D}^-)\} < +\infty.$$

Axiom (I2(k)).

$$L_2 \overset{\text{def}}{=} \sup\left\{\left|\int_{\mathbf{D}} \langle u_f, y\rangle h^k \, dm_2\right| : f \in D(X_A, h) \cap l^2(\mathscr{A}(\mathbf{D}^-)), \right.$$
$$\left. h \in \mathscr{A}(\mathbf{D}), y \in S_2(\mathscr{P}_A), \|zy\|_{S_2(Y)} \le 1\right\} < +\infty$$

(the matrix-valued function u_f is given by (5)).

THEOREM 3. *If* $X \in \mathscr{H} \cap (\mathrm{M}) \cap (\mathrm{I}(k))$ *and* $f \in D(X_{mA}, h)$, *then* $h^k = \langle f, g\rangle$, *where* $g \in l^2(X_{mA})$ *and* $\|g\|_{l^2(X_{mA})} \le L_1 + \|\rho\| \cdot L_2$.

The proof is just as earlier: if $f \in D(X_{mA}, h)$, then $f_r \in D(X_A, h_r)$ for $r \in (0, 1)$, we find a function $g_r : h_r^k = \langle f_r, g_r\rangle$, and we pass to the limit as $r \to 1$.

Note that if $X \in (\mathrm{I}(k))$, then $X \in (\mathrm{C})$ and $C_i(\delta) \le L_i \delta^{-k}$, $i = 1, 2$.

Two important remarks. In all the theorems in this paper in which it is proved that $X \in (\mathrm{C})$ and $C_i(\delta) \le C_i \cdot \delta^{-k}$, $i = 1, 2$, it can be verified that $X \in \mathrm{I}(k)$ (see, for example, the proof of Theorem 2).

Moreover, there exist a positive integer $k_1 < k$ and, for each $n = 1, 2, \dots$, a constant $\widetilde{L}_n$ such that if $f \in D(X_{mA}, h)$ and $f_l = 0$ for $l = n+1, n+2, \dots$, then there is a function $g \in l^2(X_{mA})$ such that $h^{k_1} = \langle f, g\rangle$ and $\|g\|_{l^2(X_{mA})} \le \widetilde{L}_n$ (i.e., in the most important case of finitely generated ideals of the algebra X_A we can ensure the occurrence in the ideal of the function h to a *smaller* power, but with an estimate depending on the number of generators of the ideal).

We can sharpen the estimates of the solution of the corona problem. Let

$$C_n(X_A, \delta) \overset{\text{def}}{=} \sup\{C(f) : f \in D(X_A, \delta), \ f_l = 0, \ l = n+1, n+2, \dots\}, \quad n = 1, 2, \dots.$$

Then

$$C_n(X_{cA}, \delta) = C_n(X_{mA}, \delta) \le \widetilde{L}_n \delta^{-k_1} \quad (\delta > 0).$$

For example, under the conditions of Theorem 2 we can take $k_1 = 3$ and $\widetilde{L}_n = cn^{1/2}$. Indeed, we verify, for example, the first part of this theorem:

$$\left|\int_{\mathbf{D}} \langle u_f, y\rangle h^3\, dm_2\right| \le \int_{\mathbf{D}} \sum_{i,j} (|f_i| \cdot |f'_j| + |f_j| \cdot |f'_i|) \cdot |f|^{-4} \cdot |h|^3 |y_{ij}|\, dm_2$$

$$\le \int_{\mathbf{D}} \sum_{i,j} (|f'_j| \cdot |f|^{-3} \cdot |h|^3 |y_{ij}| + |f'_i| \cdot |f|^{-3} \cdot |h|^3 \cdot |y_{ij}|)\, dm_2 \le c \cdot n^{1/2}. \quad \square$$

§3. Some examples

By using Theorems 1 and 2 it is not hard to give directly many examples of spaces consisting of functions "of first order of smoothness" and admitting an estimate of the solution of the corona problem: it suffices simply to "substitute" known examples of spaces having preduals in these theorems (see, for example, [18]–[20]). Spaces consisting of functions of a higher degree of smoothness pose greater difficulties: in verifying Axiom (C) here one must first transform the corresponding integrals with the use of Lemma 1. We carry out corresponding computations in the following two examples.

It will be necessary to estimate the product of several derivatives. Such estimates are prepared in Lemmas 4 and 5.

EXAMPLE 1. The Besov spaces $B_s^{p,q}$.

We begin with a summary of facts relating to these spaces (see, for example, [13]). Denote by f_ξ, where $\xi \in \mathbf{D}^-$ and $f \in \mathscr{H}$, the function $z \to f(\xi z)$ $(z \in \mathbf{D})$; the symbol $f^{(n)}$, where $n \in \mathbf{Z}$ and $f \in \mathscr{H}$, denotes the function

$$z \to \sum_{\substack{k\in\mathbf{Z}\\ k\neq 0}} \hat{f}(k) \cdot k^n \cdot z^k \quad (z \in \mathbf{D}).$$

The symbol $\|\cdot\|_p$ denotes the norm in the space

$$L^p(\mathbf{T}, m_1), \quad p \in [1, +\infty].$$

DEFINITION. For $s \in \mathbf{R}$ let

$$B_s^{p,q} \stackrel{\text{def}}{=} \left\{ f \in \mathscr{H} : \|f\|_{B_s^{p,q}} \stackrel{\text{def}}{=} \left(\int_0^1 \|(f^{(n)})_r (1-r)^{n-s}\|_p^q \frac{dr}{1-r} \right)^{1/q} + |f(0)| < +\infty \right\}, \quad n = [s]+1$$

(we have in mind the usual modification in the case $q = \infty$).

The following relations hold between these classes:

$$q_1 \ge q \Rightarrow B_s^{p,q} \subset B_s^{p,q_1}, \tag{9}$$

$$s - s_1 \ge \frac{1}{p} - \frac{1}{p_1} \Rightarrow B_s^{p,q} \subset B_{s_1}^{p_1,q}. \tag{10}$$

The class $B^{p,q}_{sA}$ forms an algebra (with respect to pointwise multiplication on the circle $\mathbf{T}$) if and only if

$$s > \frac{1}{p} \quad \text{or } s = \frac{1}{p}, \quad q = 1. \tag{11}$$

The condition (11) is equivalent to the inclusion $B^{p,q}_s \subset L^\infty$. The class $B^{p,q}_s$ forms an algebra if and only if (11) holds and $s > 0$, i.e., if $B^{p,q}_{sA}$ is an algebra, then so is $B^{p,q}_s$, with the exception of the case $B^{\infty,1}_0$. Moreover,

$$(B^{p,q}_s)_c = B^{p,q}_s \Leftrightarrow q < \infty,$$
$$B^{p,q}_s \simeq ((B^{p',q'}_{-s})_c)^*,$$

where p' and q' are the exponents conjugate to p and q, respectively (we have in mind the isomorphism determined by the bilinear form $(f, g) \to \sum_{k\in\mathbf{Z}} \hat{f}(k)\cdot\hat{g}(-k)$, where f is a trigonometric polynomial).

THEOREM 4. *If the exponents p, q, and s satisfy condition* (11) *and $s > 0$ (i.e., if $B^{p,q}_s$ is an algebra), then*

$$C(B^{p,q}_{sA}, \delta) \le c(p, q, s)\cdot\delta^{-(4+2[s])} \quad (\delta \in (0, 1)).$$

The proof amounts to a verification of Axioms $(\mathscr{H})$, (M), and (C) for the space $X = B^{p,q}_s$. For the first two this is trivial. Let us check (C). For this we need

LEMMA 4. *Suppose that $s \ge \frac{1}{p}$, $\{f_i\}_{i\in I} \subset B^{p,q}_{sA}$, $k_i \ge 1$ $(i \in I)$, are positive integers, and $n \overset{\text{def}}{=} \sum_I k_i > s$. Then $\|\prod_I f_i^{(k_i)}\|_{B^{p,q}_{s-n}} \le c\cdot\prod_I \|f_i\|_{B^{p,q}_s}$, where c does not depend on $\{f_i\}_{i\in I}$.*

PROOF. Let $q_i \overset{\text{def}}{=} \frac{n}{k_i}q$, $p_i \overset{\text{def}}{=} \frac{n}{k_i}p$, $s_i = (s-n)\frac{k_i}{n}$, and $f_i(0) = 0$. Then $\frac{1}{q} = \sum_I \frac{1}{q_i}$, $\frac{1}{p} = \sum_I \frac{1}{p_i}$, $s - n = \sum_I s_i$, and

$$\begin{aligned}
\left\|\prod_I f_i^{(k_i)}\right\|_{B^{p,q}_{s-n}} &= \left(\int_{\mathbf{D}} \left\|\prod_I (f_i^{(k_i)})_r (1-r)^{n-s}\right\|_p^q \frac{dr}{1-r}\right)^{1/q} \\
&\le \left(\int_{\mathbf{D}} \left(\prod_I \|(f_i^{(k_i)})_r (1-r)^{-s_i}\|_{p_i}\right)^q \frac{dr}{1-r}\right)^{1/q} \\
&\le \prod_I \left(\int_{\mathbf{D}} \|(f_i^{(k_i)})_r (1-r)^{-s_i}\|_{p_i}^{q_i} \frac{dr}{1-r}\right)^{1/q_i} = \prod_I \|f_i^{(k_i)}\|_{B^{p_i,q_i}_{s_i}} \\
&\le c\prod_I \|f_i^{(k_i)}\|_{B^{p_i,q}_{s_i}} \le c\cdot\prod_I \|f_i^{(k_i)}\|_{B^{p,q}_{s-k_i}} = c\prod_I \|f_i\|_{B^{p,q}_s}.
\end{aligned}$$

We use Hölder's inequality twice, and then the inclusions (9) and (10), since $q_i \ge q$ and

$$s - k_i - s_i = s\left(1 - \frac{k_i}{n}\right) \ge \frac{1}{p}\left(1 - \frac{k_i}{n}\right) = \frac{1}{p} - \frac{1}{p_i}. \quad \square$$

PROOF OF THE THEOREM. Let $n = [s]+1$. We first verify (C2). Part 3 of Lemma 1 is used n times to do this:

$$\int_{\mathbf{D}} \langle u_f, g\rangle\, dm_2 = \int_{|z|\le\frac12} + \int_{\frac12\le|z|\le1}$$
$$= \int_{|z|\le\frac12} - \sum_{t=0}^{n-1}\int_{|z|=\frac12} \frac{1}{t!}(3\overline{z})^{t+1}\langle \overline{\partial}^t u_f, g\rangle\, d\tilde{m}_1$$
$$+ \int_{\frac12\le|z|\le1} \frac{1}{n!}\langle \overline{\partial}^{n-1} u_f, g\rangle \left(\frac{1}{z} - z\right)^{n-1} dm_2$$
$$\stackrel{\text{def}}{=} I_1 + I_2 + I_3. \tag{12}$$

Since $u_f = (\overline{f'}\otimes\overline{f} - \overline{f}\otimes\overline{f'})\cdot|f|^{-4}$, it follows that

$$\overline{\partial}^t u_f = \sum_{\substack{\alpha_m\ge0\\ \sum\alpha_m=t+1}} a_\alpha \prod_{m\ge3}\langle \overline{f^{(\alpha_m)}}, f\rangle \overline{f^{(\alpha_1)}}\otimes\overline{f^{(\alpha_2)}} \tag{12'}$$

(use the Leibnitz formula for differentiating a product along with the formula $\overline{\partial}^k(|f|^2) = \overline{\partial}^k(\langle\overline{f}, f\rangle) = \langle\overline{f^{(k)}}, f\rangle)$.

Here the a_α are functions on $\mathbf{D}$, and

$$|a_\alpha(z)| \le c_t\delta^{-(4+2t)}, \quad z\in\mathbf{D},\ t = 0, 1, \ldots.$$

We substitute (12′) in (12) and estimate the modulus of this expression. In I_1 and I_2 let us replace the derivatives of the function by their maxima in the disk $|z|\le 1/2$. To estimate I_3 we use Lemma 4 and the inequalities

$$\int_{\mathbf{D}} |h|\cdot|g|\cdot(1-|z|)^{n-1}\, dm_2(z) \le \|h\|_{B^{p,q}_{s-n}}\cdot\|g\|_{B^{p',q'}_{-s}}$$

(this estimate follows from Hölder's inequality) and

$$\|f\|_\infty \le c\|f\|_{B^{p,q}_s} \quad (\text{see } (11)).$$

Thus,

$$I_3 \le c\sum_{i,j,\alpha}\sup_{z\in\mathbf{D}}|a_\alpha(z)|\cdot\prod_{m\ge3}\sum_l(\|f_l\|_{B^{p,q}_s}\cdot\|f_l\|_\infty)$$
$$\cdot\|f_i\|_{B^{p,q}_s}\cdot\|f_j\|_{B^{p,q}_s}\cdot\|g_{ij}\|_{B^{p',q'}_{-s}} \le c\delta^{-(4+2(n-1))}$$

if $f\in D(B^{p,q}_{sA}, \delta)$ and $\|g\|_{S_2(B^{p',q'}_{-sA})} \le 1$. Therefore, $C_2(\delta)\le c\delta^{-(4+2[s])}$ (since $n=[s]+1$).

Let us verify Axiom (C1). Since the Riesz projection acts in a Besov space, we can proceed as in the proof of the second part of Theorem 2. We decompose a function $g\in(B^{p,q}_s)_*$ into the sum of an analytic and a co-analytic function. Part 3 of Lemma 1 is used n times in the two integrals obtained in estimating $\|\overline{f}|f|^{-2}\|_{l^2(B^{p,q}_s)}$, with the operators ∂ and $\overline{\partial}$, respectively. Then we carry the moduli under the integration sign and estimate with the help of Lemma 4. □

REMARK. If p, q, and s satisfy (11) and $s>0$, then

$$C_n(B^{p,q}_{sA},\delta)\le c(p,q,s)\cdot n^{([s]+1)/2}\cdot\delta^{-(3+[s])}.$$

Consider the following example.

A Banach space E of functions Lebesgue-measurable on $\mathbf{T}$ is said to be a totally SBS (TSBS) if $f\in E$, $g\in S(dm_1)$, $|g|\le|f|\Rightarrow g\in E$ and $f\in E$, $\xi\in\mathbf{T}\Rightarrow f_\xi\in E$, $\|f_\xi\|=\|f\|$ (recall that $f_\xi(z)\overset{\text{def}}{=}f(\xi z)$).

EXAMPLE 2. Spaces with derivative in a TSBS.

If E is a TSBS and $E\subset L^1$, then each function in E will be identified with its harmonic extension to the disk $\mathbf{D}$. Let k be an integer, and let $E^{(k)}\overset{\text{def}}{=}\{f\in\mathscr{H}:f^{(k)}\in E\}$ (recall that $f^{(k)}$ is defined by the formula $\hat f^{(k)}_{(n)}=n^k\hat f(n)$, $n\in\mathbf{Z}\backslash\{0\}$, $\hat f^{(k)}_{(0)}=0$).

THEOREM 5. *Suppose that E and F are TSBS's, E is the dual of F, F is separable, $L^\infty\subset E$ and $F\subset L^1$ (for this it suffices that $\mathbf{1}\in E$, F). Then $E_A^{(k)}$ is a Banach algebra (with respect to pointwise multiplication of functions on $\mathbf{T}$), and*

$$C(E_A^{(k)},\delta)\le c\delta^{-(2k+4)}\quad(\delta\in(0,1),\ k=1,2,\dots).$$

PROOF. The fact that $E_A^{(k)}$ forms a Banach algebra follows from the Leibnitz formula for the derivative of a product and from the inclusions $E^{(k)}\subset L\subset L^\infty$. Verification of Axioms ($\mathscr{H}$), ($\widetilde{\mathrm{M}}$), and ($\widetilde{\mathrm{C1}}$) is not difficult. Since F is separable, the dual of F coincides with the space conjugate to F ([12], Chapter IV, §3.2 and Chapter VI, §1.2); therefore, $F^*=E$ and $(F^{(-k)})^*=E^{(k)}\overset{\text{def}}{=}X$; consequently, $X=X_m$. It remains to estimate $C_2(\delta)$. Using Lemma 1, we get that

$$\int_{\mathbf{D}}\langle u_f,g\rangle\,dm_2=\int_{\mathbf{T}}\langle v_0,zg\rangle\,dm_1=\int_{\mathbf{T}}\langle v_0^{(k)},(zg)^{(-k)}\rangle\,dm_1,$$

$$v_0^{(k)}=(-i)^k\frac{d^k}{d\theta^k}v_0=(z\partial-z\bar\partial)^kv_0=((z\partial)^k+Q(z,\partial,\bar\partial)\cdot\bar\partial)v_0,$$

where Q is a differential operator of order k. Therefore,

$$\int_{\mathbf{D}}\langle u_f,g\rangle\,dm_2=\int_{\mathbf{T}}\langle Q(u_f),(zg)^{-k}\rangle\,dm_1+\int_{\mathbf{D}}\left\langle\bar\partial(z\partial)^kv_0,\frac{(zg)^{(k)}}{z}\right\rangle dm_2.$$

In estimating the modulus of the first expression we use the fact that

$$f_i^{(k)}\in E,\ f_i^{(k-1)}\in H^\infty,\ (zg)_{ij}^{(-k)}\in F,\quad i,j=1,2,\dots;$$

therefore, $\int_{\mathbf{T}}|f_i^{(k)}|\cdot|(zg)_{ij}^{(-k)}|\,dm_1\le c\|f_i\|_{E^{(k)}}\cdot\|g_{ij}\|_{F^{(-k)}}$. In estimating the second expression we pass to integration over the disk. Note that

$$\partial\bar\partial v_0=\partial u=-e\langle\bar f,f'\rangle|f|^{-6}(\overline{f'}\otimes\bar f-\bar f\otimes\overline{f'});$$

therefore, the order of the highest derivative in the expression $\bar\partial(z\partial)^k v_0 = (z\partial)^{k-1}(z\partial\bar\partial)v_0$ is equal to k. These relations suffice for an estimate when $k \geq 2$.

For $k = 1$ we use one more inequality:

$$\int_{\mathbf{D}} |f_i^{(t)}|^2 |g_{ij}^{(-1)}|\, dm_2 \leq c\|f_i\|^2_{E^{(1)}} \cdot \|g_{ij}\|_{F^{(-1)}},$$

which follows from the next lemma.

LEMMA 5.

$$\int_{\mathbf{D}} |u|^2 \cdot |v|\, dm_2 \leq c\|u\|_E^2 \cdot \|v\|_F \quad (u \in E_A,\ v \in F_A).$$

PROOF. Let $u \in E_A$, $v \in F_A$, and $r \in (0, 1)$. Then

$$\int_{\mathbf{T}} |u_r|\, |v_r|\, dm_1 \leq c\|u_r\|_E \cdot \|v_r\|_F \leq c\|u\|_E \cdot \|v\|_F,$$

because E is dual to F, and the spaces E and F are invariant under rotations. Therefore, $h \stackrel{\text{def}}{=} uv$ belongs to the Hardy class H^1, and, by the Hardy–Littlewood theorem [14],

$$\int_{\mathbf{D}} |h|^2\, dm_2 = \sum_{n=0}^{\infty} \frac{|\hat h(n)|^2}{n+1} \leq \|h\|_{H^1} \cdot \sum_{n=0}^{\infty} \frac{|\hat h(n)|}{n+1} \leq \pi \cdot \|h\|^2_{H^1}.$$

Further, for an arbitrary constant $A > 0$

$$|u|^2 |v| \leq \frac{1}{2A} |u|^2 \cdot |v|^2 + \frac{A}{2} |u|^2,$$

and

$$\begin{aligned}\int_{\mathbf{D}} |u|^2 |v|\, dm_2 &\leq \frac{\pi}{2A} \|uv\|^2_{H^1} + \frac{\pi A}{2} \|u\|^2_{H^1} \\ &\leq c\|u\|_E^2 \cdot \left(\frac{1}{A}\|v\|_F + A\|1\|_F^2\right).\end{aligned}$$

Set $A = \|v\|_F$ to finish the proof of the lemma, and with it the theorem. □

REMARKS. 1. The conditions of the theorem are satisfied, in particular, by the space of all functions f analytic in $\mathbf{D}$ with $f^{(k)} \in H^p$, $1 \leq p \leq \infty$. Strictly speaking, the case $p = 1$ is not covered by Theorem 5, because here E is the space M_1 of measures on the circle, but all the computations are valid also in this case.

2. If under the condition of the theorem the spaces E and F are not only TSBS's but also symmetric [12], then $L^\infty \subset E$ and $F \subset L^1$.

3. Under the conditions of the theorem we have the estimate

$$C_n(E_A^{(k)}, \delta) \leq c \cdot n^{(k+1)/2} \delta^{-(3+k)} \quad (k, n = 1, 2, \dots)$$

(C_n was defined in the remark after Theorem 3).

§4. The spaces $B^{p,p}_{1/p} \cap H^\infty$, $1 < p < \infty$

In the examples above X_A was a subalgebra not only of H^∞, but also of C_A. In this case the corona theorem for a finite collection of functions (but without an estimate for the norm of the solution!) has an easy proof.

THEOREM 6. *If the Banach algebra B is continuously imbedded in C_A as a full subalgebra of C_A ($\overset{\text{def}}{\Leftrightarrow} \forall f \in B, f^{-1} \in C_A \Rightarrow f^{-1} \in B$), then the maximal ideal space of B coincides with $\mathbf{D}^-$.*

This is a special case of a theorem in [15], Chapter III, §11.7. □

Since it is easy to check that the subalgebras of H^∞ considered in this paper are full, the algebras not contained in the disk algebra C_A are of greatest interest.

EXAMPLE 3. $X_A = B^{p,p}_1 \cap H^\infty$, $1 < p < \infty$.

In this case $X = B^{p,p}_{1/p} \cap L^\infty$, and $Y = L^1 + B^{p',p'}_{-1/p}$, $\frac{1}{p} + \frac{1}{p'} = 1$, with norm $\|f\|_Y = \inf\{\|h\|_{L^1} + \|g\|_{B^{p',p'}_{-1/p}} : f = g + h\}$. Denote the L^p-modulus of continuity of a function f by $\omega_p(f, h) \overset{\text{def}}{=} \sup\{\|f - f_{e^{it}}\|_p : t \in (-h, h)\}$. Then $X = \{f \in L^\infty : \int_0^1 \omega_p^p(f, t)\frac{dt}{t^2} < \infty\}$. Since $\omega_p(fg, t) \le \|f\|_\infty \omega_p(g, t) + \|g\|_\infty \omega_p(f, t)$, the multiplication axiom $(\widetilde{\mathrm{M}})$ holds for X. If $f \in L^\infty$ and $|f| \ge \delta > 0$ a.e. on $\mathbf{T}$, then $\omega_p(\frac{1}{f}, t) \le \delta^{-2}\omega_p(f, t)$, and $X \in (\widetilde{\mathrm{C1}})$.

Verification of Axiom (C2) by the methods in §3 involves great difficulties, since here the space Y consists of two very different summands. Therefore, we use the following theorem of Peller and Khrushchev [16].

THEOREM 7.

$$(L^1 + B^{p',p'}_{-1/p})_A = B^{p',p'}_{-1/p,A}, \quad \frac{1}{p} + \frac{1}{p'} = 1, \quad 1 < p < \infty.$$

In verifying Axiom (C2) we can now estimate the supremum in (12) over the ball in the space $S_2(B^{p',p'}_{-1/p,A})$ by the same computations as in the proof of Theorem 3. Accordingly, we have proved

THEOREM 8.

$$C(B^{p,p}_{1/p} \cap H^\infty, \delta) \le c_p\delta^{-4}, \quad 1 < p < \infty.$$

For $p = 2$ (the algebra of bounded analytic functions with finite Dirichlet integral) this theorem gives a positive answer to a question in [8]. For a finite collection of functions we can give the estimate

$$C_n(B^{p,p}_{1/p} \cap H^\infty, \delta) \le n^{1/2} \cdot c_p \cdot \delta^{-3}, \quad 1 < p < \infty.$$

§5. Multipliers

In contrast to the previous sections, here the letter X will denote a Banach space consisting of functions analytic in $\mathbf{D}: X \subset \mathscr{A}$.

With each such space X we associate the algebra $M(X)$ of its multipliers:

$$M(X) \overset{\text{def}}{=} \{f \in A : fX \subset X\},$$

equipped with the norm $\|f\|_{M(X)} \overset{\text{def}}{=} \sup\{\|fg\|_X : g \in X, \|g\|_X \leq 1\}$.

The goal of this section is the "corona theorem with estimates" for the algebra $M(X)$. To apply Theorem 1 to $M(X)$ we must begin by determining the space of harmonic functions on $\mathbf{D}$ that contains $M(X)$ (in accordance with Axiom $(\mathscr{H})$). The role of such an "ambient" space will now be played by the space $\mathrm{ST}(X)$ of symbols of Toeplitz operators acting in X (precise definitions are given below). The space $\mathrm{ST}(X)_* \overset{\text{def}}{=} N(X)$ predual to $\mathrm{ST}(X)$ will be constructed from a type of projective tensor product of the spaces X_c and $\overline{(X_m)_*}$. This circumstance together with the description of $M(X)$ in Lemmas 6 and 8 in terms of membership of the derivatives in SBS's lies at the basis of our solution of the corona problem in $M(X)$.

We proceed to a detailed discussion of Toeplitz operators and their symbols. Recall that the Toeplitz operator on H^2 corresponding to a function $\varphi \in L^\infty$ is the mapping

$$T_\varphi : f \mapsto P_+(\varphi f) \quad (f \in H^2)$$

(P_+ is the Riesz projection).

The function φ is called the symbol of the operator T_φ. It is possible to give an "intrinsic" characterization of Toeplitz operators acting in H^2: a linear operator $T : H^2 \to H^2$ is a Toeplitz operator if and only if

$$T = S^* T S. \tag{13}$$

Here S is the shift operator, and S^* is the backward shift operator acting in $H^2 : Sf(z) = zf(z)$, $S^* f(z) = (f(z) - f(0))/z$. If T satisfies (13), then $T = T_\varphi$, where $\varphi \in L^\infty$ and

$$\hat{\varphi}(k) = \int_{\mathbf{T}} T(\mathbf{1})\overline{z^k}\, dm_1, \quad k = 0, 1, \dots,$$

$$\hat{\varphi}(-k) = \int_{\mathbf{T}} T(z^k)\, dm_1, \quad k = 1, 2, \dots.$$

These facts suffice for proposing a reasonable definition of a Toeplitz operator (and its symbol) for spaces of analytic functions different from H^2. Assume that the operators S and S^* act in X (we sometimes write S_X and S_X^* if we want to indicate the space where they act).

It would be natural to call a linear operator T acting from X to X a Toeplitz operator if it satisfies the relation (13); its symbol could be defined to be the function $\varphi \overset{\text{def}}{=} \sum \hat{\varphi}(n) z^n$, where $\hat{\varphi}(k-n) \overset{\text{def}}{=} \int_{\mathbf{T}} T(z^n)\overline{z^k}\, dm_1$.

However, if $X \neq X_c \overset{\text{def}}{=} \text{cl}_X(\mathscr{P}_A)$, then under this definition the symbol of a nonzero Toeplitz operator could turn out to be the zero function.

EXAMPLE. The projection of the space of Cauchy-type integrals of measures on $\mathbf{T}$ onto the space of Cauchy-type integrals of measures singular with respect to Lebesgue measure satisfies the relation (13), but its symbol is the zero function.

With each measure $\mu \in M_1$ we associate its Cauchy integral $K_\mu(z) \overset{\text{def}}{=} \frac{1}{2\pi i}\int_{\mathbf{T}} \frac{d\mu(\xi)}{\xi - z}$, and $\mathscr{K} \overset{\text{def}}{=} \{f \in \mathscr{A} : f = K_\mu\}$ for some measure $\mu \in M_1\}$. This is a Banach space with the norm $\|f\|_{\mathscr{K}} \overset{\text{def}}{=} \inf\{\|\mu\|_{M_1}$ over all measures $\mu : f = K_\mu\}$.

Let $\mathscr{K}_s = \{f : f = K_{\mu_s}$, μ_s is a measure singular with respect to Lebesgue measure on the circle$\}$.

Consider the projection $P : \mathscr{K} \to \mathscr{K}_s$ given by the formula $P(K_\mu) = K_{\mu_s}$, where μ_s is the singular component of μ. By the F. and M. Riesz theorem [14], this definition is unambiguous. Let us verify that P satisfies (13).

It is easy to see that if $\mu \in M_1$, then $K_{z\mu} = zK_\mu + \mu(\mathbf{1})$, and hence $S^*PSK_\mu = S^*P(K_{z\mu} + \mu(\mathbf{1})) = S^*K_{z\mu_s} = K_{\mu_s} = PK_\mu$. But the restriction of P to the polynomials is equal to zero; thus the symbol of P is equal to zero.

This example compels us to define a Toeplitz operator acting in X to be an operator *acting from* X_c *to* X *and satisfying condition* (13).

Let us proceed to the exact formulations. We state the requirements on X just as in §2, as axioms.

Axiom (A). General properties of a Banach space X of analytic functions.

Axiom (A1). The imbedding $X \subset \mathscr{A}$ is continuous.

Axiom (A2). $\mathscr{A}(\mathbf{D}^-) \subset X$.

Axiom (A3). If $f \in X$ and $\mu \in M_1$, then $f * \mu \in X$ and $\|f * \mu\|_X \leq \|f\|_X \cdot \|\mu\|_{\mu_1}$.

If Axiom (A) holds for the space X, then with each functional $g \in X^*$ we can associate the function $j(g) \overset{\text{def}}{=} \sum_{k=0}^{\infty} \hat{g}(k) z^k$, where $\hat{g}(k) \overset{\text{def}}{=} \overline{g(z^k)}$. Further, if $X = X_c$, then we identify the functional $g \in X^*$ with the function $j(g) \in \mathscr{A}$. If $f \in \mathscr{P}(\mathscr{A})$, then $(g, f) = \sum_{k=0}^{\infty} \overline{\hat{g}(k)} \hat{f}(k)$, and $(g, f) = \int_{\mathbf{T}} \overline{g} f\, dm_1$ for $g \in L^1 \cap X^*$. As in §2, we call this duality the "Cauchy duality," although their definitions are different. In the cases when we have to use them at the same time, we shall denote the first of them by the symbol $(f, g)_h \overset{\text{def}}{=} \int_{\mathbf{T}} fg\, dm_1$, and the second by $(f, g)_\alpha \overset{\text{def}}{=} \int_{\mathbf{T}} \overline{f} g\, dm_1$ (duality in spaces of harmonic functions and analytic functions respectively).

For a space $X \subset \mathscr{A}$ we can define the spaces X_c and X_m just as in §2:

$$X_c \overset{\text{def}}{=} \text{cl}_X \mathscr{P}_A, \qquad X_m \overset{\text{def}}{=} \left\{ f \in \mathscr{A} : \sup_{0<r<1} \|f_r\|_X < \infty \right\}.$$

The space X is dual with respect to the Cauchy duality to some space of analytic functions if and only if $X = X_m$.

To define Toeplitz operators acting in X we need the following *Axiom* (T).

Axiom (T1). The forward and backward shift operators act in X, and $r(S) = r(S^*) = 1$ ($r(L)$ is the spectral radius of an operator L).

Axiom (T2). $X = X_m$ (i.e., X is dual with respect to the Cauchy duality).

Everywhere in what follows it is assumed that Axioms (A) and (T) are satisfied.

DEFINITIONS. 1. An operator $T : X_c \to X$ is called a Toeplitz operator (written $T \in T(X)$) if $S_X^* T S_{X_c} = T$.

2. The symbol of a Toeplitz operator T is defined to be the function $\varphi \in \mathscr{H}$ with

$$P_+\varphi = T(\mathbf{1}), \qquad P_+\overline{\varphi} = T_i^*(\mathbf{1}).$$

Here $\mathbf{1}$ is the function identically equal to 1 in the disk $\mathbf{D}$, $T^* : X^* \to (X_c)^*$ is the operator adjoint to T, and $i : X_* \to (X_*)^* = X^*$ is the canonical imbedding. Note that the operator T_i^* acts from some subspace of $\mathscr{A}$, while X^* is not necessarily imbedded in $\mathscr{A}$ (we do not require the condition $X = X_c$). An equivalent definition of φ is as follows: $\hat{\varphi}(k) = (z^k, T(\mathbf{1}))$, $\hat{\varphi}(-k) = (T_i^*(\mathbf{1}), z^k)$, $k = 0, 1, \dots$. The function φ will be called the symbol of T and denoted by $\operatorname{sym} T$. Accordingly, we have constructed a mapping $\operatorname{sym} : T(X) \to \mathscr{H}$.

3. The range of the mapping sym is denoted by $\mathrm{ST}(X)$ and called the "space of symbols of Toeplitz operators acting in X."

It is easy to see that

$$(z^n, Tz^k) = \hat{\varphi}(n-k); \quad n, k = 0, 1, 2 \dots.$$

Indeed, for $n \geq k$

$$(z^n, Tz^k) = (S^k z^{n-k}, TS^k\mathbf{1}) = (z^{n-k}, S^{*k}TS^k\mathbf{1}) = (z^{n-k}, T\mathbf{1}),$$

and for $n < k$

$$(z^n, Tz^k) = (S^n\mathbf{1}, TS^n z^{k-n}) = (\mathbf{1}, Tz^{k-n}) = (T_i^*(\mathbf{1}), z^{k-n}).$$

Therefore,

$$T(p) = P_+(\varphi_p), \quad p \in \mathscr{P}_A,$$

where the product of φ by p is defined by the formula

$$\left(\widehat{\varphi \cdot \sum_{i=0}^m c_i z^i}\right)(k) \overset{\text{def}}{=} \sum_{i=0}^m c_i \hat{\varphi}(k-i).$$

Therefore, corresponding to each $\varphi \in \mathrm{ST}(X)$ there is a *unique* Toeplitz operator T_φ such that $\varphi = \operatorname{sym} T_\varphi$, and we can define the norm in the space $\mathrm{ST}(X)$ by

$$\|\varphi\|_{\mathrm{ST}(X)} \overset{\text{def}}{=} \|T_\varphi\|_{L(X_c, X)}.$$

We can now make a simple observation (which, however, plays an extremely essential role in what follows):

$$M(X) = (\mathrm{ST}(X))_A$$

(moreover, the norms in the two spaces coincide). Indeed, if $f \in (\mathrm{ST}(X))_A$, then $fX_c \subset X$, and with the use of ω-convergence (see §2) the operator of multiplication by f extends to the whole space X, so that $(\mathrm{ST}(X))_A \subset M(X)$. But if $f \in M(X)$, then the operator of multiplication by f is a Toeplitz operator acting in X.

THEOREM 9. *If* $X \in (\mathscr{A}) \cap (\mathrm{T})$, *then* $\mathrm{ST}(X) \in (\mathscr{H}) \cap (\mathrm{M})$, *and* $\mathrm{ST}(X) = (\mathrm{ST}(X))_m$.

PROOF. To verify the relations $\mathrm{ST}(X) \in (\mathscr{H})$ and $\mathrm{ST}(X) = (\mathrm{ST}(X))_m$ we describe the predual space and verify that it satisfies Axiom $(\mathscr{H})$.

Let $Y = X_*$.

For each polynomial $f \in \mathscr{P}$ we consider the expression $\|f\|_{N(X)} \stackrel{\text{def}}{=} \inf\{\sum_{i=1}^k \|x_i\|_{X_c} \cdot \|y_i\|_Y$ over all representations $f : f(\xi) = \sum_{i=1}^k x_i(\xi)\overline{y}_i(\xi)$ for all $\xi \in \mathbf{T}$; $\{x_i, y_i\} \subset \mathscr{P}_A\}$. This expression is analogous to the norm of the projective tensor product of the spaces X_c and $\overline{Y}$. It is easy to see that $\|\cdot\|_{N(X)}$ is a seminorm. We verify that it is actually a norm. If $f \in \mathscr{P}$,

$$f = \sum_{i=1}^k x_i \overline{y}_i, \qquad \{x_i, y_i\} \subset \mathscr{P}_A,$$

then $|\hat{f}(k)| = |\int_{\mathbf{T}} \sum_{i=1}^k x_i \overline{z^k y_i}\, dm_1| \le \sum_{i=1}^k \|x_i\|_X \cdot \|z^k y_i\|_Y$; therefore, $|\hat{f}(k)| \le \|S_y^k\|_{L(Y)} \|f\|_{N(X)}$ (note that if $X \in T$, then $Y = X_* \in (\mathrm{T1})$). Similarly, for negative k we have that $|\hat{f}(k)| \le \|S_X^{-k}\|_{L(X)} \cdot \|f\|_{N(X)}$; therefore, if $\|f\|_{N(X)} = 0$, then $f = 0$; i.e., $\|\cdot\|_{N(X)}$ is a norm. Denote by $N(X)$ the completion of $\mathscr{P}$ in this norm. Then we have actually verified that $N(X) \subset \mathscr{H}(\mathbf{D}^-)$, i.e., $N(X) \in (\mathscr{H}1)$. Let us verify $(\mathscr{H}2) : \mathscr{H}(\mathbf{D}^-) = \mathscr{A}(\mathbf{D}^-) + \overline{\mathscr{A}(\mathbf{D}^-)} \subset X_c + \overline{Y} \subset N(X)$. Axiom $(\mathscr{H}3)$ is valid for $N(X)$, since $N(X) = N(X)_c$.

Moreover,

$$\begin{aligned} \varphi \in N(X)^* &\Leftrightarrow |(\varphi x, \overline{y})_h| \le c\|x\|_X \cdot \|y\|_Y; \quad x, y \in \mathscr{P}_A \\ &\Leftrightarrow |(p_+(\varphi x), y)_a| \le c\|x\|_X \cdot \|y\|_Y; \quad x, y \in \mathscr{P}_A \Leftrightarrow \varphi \in \mathrm{ST}(X). \end{aligned}$$

It remains to verify Axiom (M).

If $g \in \mathscr{P}_A$, then $gX_c \subset X_c$; further, if $f \in \mathscr{P}$, then $T_{fg} = T_f \cdot T_g : X_c \to X$; therefore, $\|fg\|_{\mathrm{ST}(X)} \le \|f\|_{\mathrm{ST}(X)} \cdot \|g\|_{\mathrm{ST}(X)}$. □

REMARKS. 1. The proof of the theorem can be simplified if it is verified directly that $\mathrm{ST}(X) = \mathrm{ST}(X)_m$. But we are interested in representing the space as a "tensor product."

2. In proving the "abstract corona theorem" we used the "partial multiplication" Axiom (M) instead of the simpler Axiom $(\widetilde{\mathrm{M}})$ precisely so that we could include multiplier algebras in the general scheme.

The author does not know whether Axiom $(\widetilde{\mathrm{M}})$ holds for the space $\mathrm{ST}(X)$ if $X \in (\mathrm{A}) \cap (\mathrm{T})$.

In this section we consider mainly the case of a space X of the form $X = E_A^{(1)}$, where E is some solid Banach space of functions on $\mathbf{D}$ that are measurable with respect to Lebesgue measure m_2 (briefly, SBS($\mathbf{D}$)). The multipliers of such a space can be described with the help of the following known lemma (see, for example, [26], [27]).

Lemma 6. *If $X \in E_A^{(1)}$, where E is an* SBS *and $X \in (\mathrm{A1})$, then*

$$M(X) = \{\varphi \in H^\infty : \|\varphi' \cdot f\|_E \le c\|f\|_X,\ f \in X\}.$$

Proof. Since $X \in (\mathrm{A1})$, it follows from Landau's trick that $\|\varphi\|_\infty \le \|\varphi\|_{M(X)}$ (see the remark after Lemma 3). A function φ is in $M(X)$ if and only if $\|(\varphi \cdot f)'\|_E \le c\|f\|_X$ for all $f \in X$. It remains to use the formula $(\varphi f)' = \varphi f' + \varphi' f$. □

Corollary. *If the conditions of the lemma hold, then $M(X)$ is a full subalgebra of H^∞ (i.e., if $\varphi \in M(X)$ and $\varphi^{-1} \in H^\infty$, then $\varphi^{-1} \in M(X)$).*

To prove the corona theorem we need to make considerably more assumptions about the space X (Theorems 10 and 11). In Theorem 12 we consider examples of spaces with the L^p-norm, and in Theorem 13 the commutant of the weighted shift operator acting in l^2.

Theorem 10. *Suppose that E and F are in* SBS($\mathbf{D}$), *E is the dual to F with respect to the measure dm_2, and the space $X = E_A^{(1)}$ satisfies Axioms* (A) *and* (T), *with $X_* = F_{Ac}$. Then*

$$C(M(X), \delta) \le c\delta^{-4}, \qquad C_n(M(X), \delta) \le c \cdot n^{1/2} \cdot \delta^{-3}.$$

Proof. We have represented the algebra $M(X)$ as the analytic part of the space $\mathrm{ST}(X)$. By Theorems 1 and 9, it suffices to verify Axiom (C) for the space $\mathrm{ST}(X)$.

Suppose that $f \in D(M(X), \delta) \cap l^2(\mathscr{A}(\mathbf{D}^-))$. Let us verify (C1):

$$\|\overline{f}|f|^{-2}\|_{l^2(\mathrm{ST}(X))} = \sup\left\{\left|\int_{\mathbf{T}} \langle \overline{f}|f|^{-2}, g\rangle\, dm_1\right| : g \in l^2(\mathscr{P}),\ \|g\|_{l^2(N(X))} \le 1\right\}.$$

Note that if $g \in \mathscr{P}$, then $(g(z) - \hat{g}(0) - \hat{g}(1)z)/\overline{z} \in C^1(\mathbf{D}^-)$, and by

Lemma 1

$$\begin{aligned}\left|\int_{\mathbf{T}}\langle \overline{f}|f|^{-2}, g\rangle\, dm_1\right| &= \left|\int_{\mathbf{D}} \partial\langle \overline{f}|f|^{-2}, (g-\hat{g}(0)-\hat{g}(1)z)/\overline{z})\rangle\, dm_2\right.\\ &\qquad \left.+ \int_{\mathbf{T}}\langle \overline{f}|f|^{-2}, \hat{g}(0)+\hat{g}(1)z\rangle\, dm_1\right|\\ &\le \left|\int_{|z|\ge\frac{1}{2}}\right| + c\delta^{-4}\\ &\le c\delta^{-4} + 2\sum_{i,j}\int_{\mathbf{D}}(|f_i|\,|\langle\overline{f}, f'\rangle|\cdot|f|^{-4}|x_{ij}|\cdot|y_{ij}|\\ &\qquad + |f_i|\cdot|f|^{-2}\cdot|x'_{ij}|\cdot|y_{ij}|)\, dm_2\\ &\overset{\text{def}}{=} c\delta^{-4} + \sum\nolimits_1 + \sum\nolimits_2,\end{aligned}$$

where we substituted $g_i = \sum_j x_{ij}\overline{y}_{ij}$, $x_{ij}, y_{ij} \in \mathscr{P}_A$,

$$\sum_i\left(\sum_j \|x_{ij}\|_X\cdot\|y_{ij}\|_F\right)^2 \le 2.$$

We estimate $\sum_1$ (the sum $\sum_2$ can be estimated similarly):

$$\begin{aligned}\sum\nolimits_1 &\le c\delta^{-4}\sum_{i,j,k}\|f_i\|_\infty\cdot\|f_k\|_\infty\cdot\|f'_k x_{ij}\|_E\cdot\|y_{ij}\|_F\\ &\le c\delta^{-4}\left(\sum_k \|f_k\|^2_{M(X)}\right)\left(\sum_i \|f_i\|^2_\infty\right)^{1/2}\left(\sum_i\left(\sum_j \|x_{ij}\|_X\|y_{ij}\|_F\right)^2\right)^{1/2}\\ &\le c\delta^{-4},\end{aligned}$$

where we have used Lemma 5,

$$\int_{\mathbf{D}} |f'xy|\, dm_2 \le c\|f'x\|_E\cdot\|y\|_F \le c\|f\|_{M(X)}\cdot\|x\|_X\cdot\|y\|_F.$$

Let us verify (C2). If $g \in S_2(\mathscr{P}_A)$, $\|g\|_{S_2(N(X))} \le 1$, then we represent $g_{ij} = \sum_k x_{ijk}\overline{y_{ijk}}$; $x_{ijk}, y_{ijk} \in \mathscr{P}_A$,

$$\sum_{i,j}\left(\sum_k \|x_{ijk}\|_X \cdot \|y_{ijk}\|_F\right)^2 \le 2.$$

Then

$$\begin{aligned}\left|\int_{\mathbf{D}}\langle u_f, g\rangle\, dm_2\right| &\le c\delta^{-4}\cdot\sum_{i,j,k}(\|f_j\|_\infty\cdot\|f'_i\cdot x_{ijk}\|_E\\ &\qquad + \|f_i\|_\infty\cdot\|f'_j x_{ijk}\|_E)\cdot\|y_{ijk}\|_F \le c\delta^{-4}. \qquad \square\end{aligned}$$

THEOREM 11. *Suppose that E and F are in* SBS(**D**), *E is the dual to F with respect to the measure $(1-|z|)\,dm_2$, the space $X = E_A^{(1)}$ satisfies Axioms* (A) *and* (T), *and $X_* = F_A^{(1)}$. Moreover, assume that $M(X) \subset M(X_*)$. Then*

$$C(M(X), \delta) \leq c\delta^{-6}, \qquad C_n(M(X), \delta) \leq n \cdot c\delta^{-4}.$$

PROOF. Letting $f \in D(M(X), \delta) \cap l^2(\mathscr{A}(\mathbf{D}^-))$, we verify only (C1):

$$\left|\int_{\mathbf{T}} \langle \overline{f}|f|^{-2}, g\rangle\, dm_1\right| = \left|\langle \overline{f}|f|^{-2}, g\rangle(0) + \int_{\mathbf{D}} \Delta(\langle \overline{f}|f|^{-2}, g\rangle) \ln\frac{1}{z}\, dm_2\right|$$

$$\leq c\delta^{-4} + \left|\int_{|z|\geq\frac{1}{2}}\right|$$

$$\leq c\delta^{-4} + c\sum_{i,j}\int_{\mathbf{D}} (|f_i|\,|f|^{-2}|x'_{ij}|\,|y'_{ij}|$$

$$+ |\overline{\partial}(\overline{f}_i|f|^{-2})|\cdot|x'_{ij}|\cdot|y_{ij}| + |\partial(\overline{f}_i|f|^{-2})|\cdot|x_{ij}|\cdot|y'_{ij}|$$

$$+ |\Delta(\overline{f}|f|^{-2})|\cdot|x_{ij}|\cdot|y_{ij}|(1-|z|)\,dm_2$$

(we substituted the expansions $g_i = \sum_j x_{ij}\overline{y}_{ij}$). We differentiate and use the inequality

$$\int_{\mathbf{D}} |a|\cdot|b|\cdot(1-|z|)\,dm_2 \leq c\|a\|_E\cdot\|b\|_F, \qquad \|f'x\|_E \leq c\|f\|_{M(X)}\cdot\|x\|_X,$$

$$\|f'y\|_F \leq c\|f\|_{M(X_*)}\cdot\|y\|_{X_*} \leq c\|f\|_{M(X)}\cdot\|y\|_{X_*};$$

the rest is as in the preceding theorem. □

REMARK. If all the conditions of the theorem hold except for the condition $M(X) \subset M(X_*)$, then, using the duality

$$\mathrm{ST}(X) \cap \mathrm{ST}((X_c)^*) = (N(X_c) + N(X_*))^*$$

(recall that $X_* = ((X_c)^*)_c$, we can get the corona theorem with the same estimate in the algebra $M(X) \cap M(X_*)$.

EXAMPLE 4. Definitions. 1. ([17], [18]). A positive continuous function φ on $(0, 1)$ is said to be (ε, k)-normal if there exist an $r_0 \in (0, 1)$ and numbers ε_1 and k_1 in (ε, k) such that

$$\varphi(r)(1-r)^{-\varepsilon_1} \downarrow 0, \qquad \varphi(r)(1-r)^{-k_1} \uparrow 1, \quad r_0 \leq r, \; r \uparrow 1.$$

2. If φ is a positive function on $(0, 1)$, then

$$L^p(\varphi) \stackrel{\text{def}}{=} \left\{ f \in S(dm_2) : \int_{\mathbf{D}} |f|^p \varphi(|z|)\,dm_2 < \infty \right\}.$$

THEOREM 12. 1) *If $1 \leq p \leq \infty$ and φ is a $(-1, p-1)$-normal function, then $C(M(L_A^{p,(1)}(\varphi)), \delta) \leq c\delta^{-4}$.*

2) *If $1 \leq p \leq \infty$, φ is a $(-1, 2p-1)$-normal function, and the condition $M(X) \subset M((X_c)^*)$ holds for the space $X = L_A^{p,(1)}(\varphi)$ (this is the case if $M(X_*) = H^\infty$), then*

$$C(M(X), \delta) \leq c\delta^{-6}.$$

PROOF. To use Theorems 10 and 11 it is necessary to verify that $X_m = (Y_c)^*$, where $Y = L_A^{p/(p-1)}(\varphi^{-1/(p-1)})$ in the first part of the theorem, and $Y = L_A^{p/(p-1),(1)}(\varphi^{-1/(p-1)}(1-r)^{p/(p-1)})$ in the second.

It was shown in [17] that for this it suffices to verify that a certain projection acting in X is bounded; this was done for $p = 1$ and $p = \infty$. The general case is obtained by interpolation. □

EXAMPLE 5. The commutant of the weighted shift operator acting in l^2.

DEFINITION. Let $\beta = \{\beta_n\}_{n=0}^{\infty}$ be a sequence of positive numbers with $\overline{\lim}_{n\to\infty} \beta_n^{-1/n} = 1$. We define the following spaces:

$$H^2(\beta) \stackrel{\text{def}}{=} \left\{ f \in \mathscr{A} : \sum_{n=0}^{\infty} |\hat{f}(n)|^2 \beta_n^2 < \infty \right\}, \qquad H^\infty(\beta) \stackrel{\text{def}}{=} M(H^2(\beta)).$$

On p. 78 of [20] Shields asks whether the corona theorem holds in the algebra $H^\infty(\beta)$. Using Theorem 12, we get an answer to this question for certain weights β. The answer was known previously only for $H^\infty(\beta) = H^\infty$ or $H^\infty(\beta) = H^2(\beta)$ (then the polynomials are dense in the algebra $H^\infty(\beta)$, and its maximal ideal space can be identified with $\mathbf{D}^-$).

LEMMA 7. 1) *$H^\infty(\beta) = H^\infty$ if and only if $\beta_{n+k} \le c\beta_k$, $n, k = 0, 1, 2, \ldots$.*

2) *If φ is a convex positive function on $(0, 1)$, $\varphi(r) = O(1-r)$ as $r \uparrow 1$, and $\beta_n \stackrel{\text{def}}{=} (n+1)(\int_0^1 r^{2n}\varphi(r)\,dr)^{1/2}$, then $H^\infty(\beta) = H^\infty$.*

PROOF. 1) Let $i : H^\infty \to H^\infty(\beta)$ be an isomorphism. Then $\beta_{n+k} = \|i(z^n)z^k\|_{H^2(\beta)} \le \|i\|\cdot\beta_k$; $n, k = 0, 1, \ldots$. Conversely, if $\tilde{\beta}_k \stackrel{\text{def}}{=} \sup_{n\to k} \beta_n$, then $c\beta_k \ge \tilde{\beta}_k \ge \beta_k$ and $\tilde{\beta}_k \downarrow 0$; therefore, $H^2(\beta) = H^2(\tilde{\beta})$ and $H^\infty(\beta) = H^\infty(\tilde{\beta}) = H^\infty$ (see [20]).

$$2) \quad \int_0^1 r^{2n}\omega(r)\,dr = \frac{\omega'(1)}{(2n+1)(2n+2)} + \int_0^1 \frac{r^{2n+2}\omega''(r)}{(2n+1)(2n+2)}\,dr,$$

and we can use the criterion in the first part of the theorem. □

The next theorem follows from this lemma and Theorem 12 (with $p = 2$).

THEOREM 13. *Suppose that φ_1 is a positive function on $(0, 1)$, and let $\beta_n \stackrel{\text{def}}{=} (n+1)\cdot(\int_0^1 r^{2n}\varphi(r)\,dr)^{1/2}$, $n = 0, 1, 2, \ldots$. Then*

1) *If φ is a $(-1, 1)$-normal function, then*

$$C(H^\infty(\beta), \delta) \le c\delta^{-4}.$$

2) *If φ is a $(-1, 3)$-normal convex positive function and $\varphi(r) = O(1-r)$ as $r \uparrow 1$, then*

$$C(H^\infty(\beta), \delta) \le c\delta^{-6}.$$

In particular, substituting $\varphi(r) = (1-r)^{1-2\alpha}(\ln\frac{1}{1-r}+1)^{\gamma}$, we get the

COROLLARY. 1) *If* $\beta_n = n^{\alpha}(\ln n + 1)^{\gamma}$, $0 < \alpha < 1$, $-\infty < \gamma < \infty$, *then* $C(H^{\infty}(\beta), \delta) \le c(\alpha, \gamma)\delta^{-4}$.

2) *If* $\beta_n = (\ln n + 1)^{\gamma}$, $0 \le \gamma < \infty$, *then*

$$C(H^{\infty}(\beta), \delta) \le c(\gamma)\delta^{-6}.$$

EXAMPLE 6. Multipliers of BMOA.

DEFINITIONS. 1. A positive Borel measure μ on the disk $\mathbf{D}$ is called a Carleson measure if

$$\|\mu\|_c \stackrel{\text{def}}{=} \sup\left\{\int_{\mathbf{D}} |f|\,d\mu : \|f\|_{H'} \le 1\right\} < \infty.$$

2. The space BMOA is defined as follows:

$$\text{BMOA} \stackrel{\text{def}}{=} \{f \in \mathscr{A} : \|\,|f'|^2(1-|z|)\,dm_2\|_c < \infty\}.$$

THEOREM 14.

$$C(M(\text{BMOA}), \delta) \le c\delta^{-6}.$$

PROOF. By Lemma 6,

$$M(\text{BMOA}) = \{\varphi \in H^{\infty} : \|\,|\varphi' \cdot f|^2(1-|z|)\,dm_2\|_c \le c\|f\|^2_{\text{BMOA}}\}.$$

We use Theorem 11, but instead of describing the spaces E and F in that theorem, we verify the inequalities used in its proof:

$$\int_{\mathbf{D}} |f'|\cdot|g'|(1-|z|)\,dm_2 \le c\|f\|_{\text{BMOA}}\cdot\|g\|_{H^1},$$

$$\int_{\mathbf{D}} |\varphi'|\cdot|f'|\cdot|g|(1-|z|)\,dm_2 \le c\|\varphi\|_{\text{BMOA}}\cdot\|f\|_{\text{BMOA}}\cdot\|g\|_{H^1},$$

$$\int_{\mathbf{D}} |\varphi'|\cdot|f|\cdot|g'|(1-|z|)\,dm_2 \le c\|\varphi\|_{M(\text{BMOA})}\cdot\|f\|_{\text{BMOA}}\cdot\|g\|_{H^1}.$$

Only the last inequality will be checked. Let v be an outer function on $\mathbf{D}$ with $|v|^2 = |g|$ a.e. on $\mathbf{T}$ and $g = vu$. Then

$$|g'|^2 \le 2(|v'u|^2 + |vu'|^2) \le 2(|v'|^2 + |u'|^2)|v|^2,$$

$$\begin{aligned}\int_{\mathbf{D}} |\varphi'|\cdot|f|\cdot|g'|\cdot(1-|z|)\,dm_2 &\le \left(\int_{\mathbf{D}} |\varphi'|^2\cdot|f|^2\cdot|v|^2(1-|z|)\,dm_2\right)^{1/2} \\ &\quad\cdot\left(\int_{\mathbf{D}} 2(|v'|^2 + |u'|^2)(1-|z|)\,dm_2\right)^{1/2} \\ &\le c\cdot\|\varphi\|_{M(\text{BMOA})}\cdot\|f\|_{\text{BMOA}}\cdot\|g\|_{H^1}. \quad \square\end{aligned}$$

REMARK. It was shown in [21] and [22] that

$$M(\text{BMOA}) = H^{\infty} \cap \text{BMOA}\left(\frac{1}{\ln(1/\theta)}\right).$$

EXAMPLE 7. Besov spaces.

THEOREM 15. 1) *If* $0 < s < 1$, *then*

$$C(M(B^{p,q}_{s,A}), \delta) \leq c\delta^{-4}.$$

2) *If* $s = p = 1$, *then* $C(M(B^{p,q}_{s,A}), \delta) \leq c\delta^{-6}$.

PROOF. 1) follows from Theorem 10.

2) Recall that $f \in B^{1,q}_{1,A} \Leftrightarrow f'' \in B^{1,q}_{-1,A} = E_A$, where $E \stackrel{\text{def}}{=} \{g \in S(dm_2) : \int_0^1 \|gr\|_1^q (1-r)^{q-1}\, dr < \infty\}$ is an SBS. We describe the multipliers of this space.

LEMMA 8. $f \in M(B^{1,q}_{1,A}) \Leftrightarrow f \in H^\infty \cap B^{1,q}_{1,A}$, *and* $\|f''g\|_{B^{1,q}_{-1,A}} \leq c \cdot \|g\|_{B^{1,q}_{-1,A}}$.

PROOF OF LEMMA.

$$\begin{aligned} f \in M(B^{1,q}_{1,A}) &\Leftrightarrow \|(f, g)''\|_{B^{-1,q}_{1,A}} \leq c\|g\|_{B^{1,q}_{1,A}} \\ &\Leftrightarrow \|fg'' + 2f'g' + f''g\|_{B^{-1,q}_{1,A}} \leq c\|g\|_{B^{1,q}_{1,A}}. \end{aligned}$$

The first term is estimated if $f \in H^\infty$, and the second if $f \in B^{1,q}_{1,A}$ (see Lemma 4). □

The conclusion of the proof of the theorem is analogous to that of Theorem 11 with the help of Lemma 8 and the formula (see Lemma 1)

$$\int_{\mathbf{T}} zb\, dm_1 = \frac{1}{2} \int_{\mathbf{D}} \bar{\partial}^2 b \left(\frac{1}{z} - \bar{z}\right) dm_2. \quad \square$$

Since $M(B^{p,q}_{s,A}) = H^\infty$ for $s < 0$, Theorems 3 and 17 encompass the whole scale of Besov spaces with the exception of $s = 0$, where the answer is unknown to the author. Theorem 10 is not applicable here, because the condition $M(X) \subset M(X_*)$ fails.

§6. The corona theorem in closed subalgebras of H^∞

DEFINITIONS. 1) A Douglas algebra is defined to be a closed subalgebra of L^∞ containing H^∞.

2) Let D be a Douglas algebra. The following two symmetric algebras are called Sarason algebras: $Q_D \stackrel{\text{def}}{=} D \cap \overline{D}$; the symmetric subalgebra C_D of L^∞ generated by the functions invertible in D.

THEOREM 18. *If X is a symmetric subalgebra of L^∞, and $Q_D \supset X \supset C_D$ for some Douglas algebra D, then*

$$C(X_A, \delta) \leq 25\delta^{-6}.$$

REMARK. This theorem improves a result in [9], where the corona theorem is proved (without an estimate) for C_{DA}. The analogous theorem was proved for the algebra Q_{DA} in [24].

PROOF. Let $f \in D(H^\infty, \delta)$. Theorem 1 will be applied to the space $X = L^\infty$. Assertion 4 in Lemma 1 is used in estimating $C_2(\delta)$, and thus

$$\int_{\mathbf{T}} \langle v_0, g\rangle \, dm_1 = \frac{1}{2}\int_{\mathbf{D}} \Delta\langle v_0, g\rangle \ln\frac{1}{|z|}\, dm_2$$
$$= \int_{\mathbf{D}} 2(\langle u_f, g'\rangle + \langle \partial u_f, g\rangle) \ln\frac{1}{|z|}\, dm_2 .$$

We prepare some estimates. In part 4 of Lemma 1 put

$$a = |h|^2, \ h \in H^2 : \int_{\mathbf{D}} |h'|^2 \ln\frac{1}{|z|}\, dm_2 \le \tfrac{1}{2}\|h\|_{H^2}^2 . \tag{14}$$

If $f \in H^\infty$, $g \in H^1$, h is an outer function with $|h| = |g|^{1/2}$ a.e. on $\mathbf{T}$, and $h_1 \overset{\text{def}}{=} g/h$, then $|h_1| \le |h|$ in $\mathbf{D}$, and

$$|f'|^2 \cdot |g| \le |f'|^2 \cdot |h|^2 \le 2(|(fh)'| + |f|^2 \cdot |h'|^2).$$

Therefore, it follows from (14) that

$$\int_{\mathbf{D}} |f'|^2 \cdot |g| \cdot \ln\frac{1}{|z|}\, dm_2 \le 2\|f\|_{H^\infty}^2 \cdot \|g\|_{H^1} .$$

Since $|g'|^2 = |(hh_1)'|^2 \le 2|h|^2(|h_1'|^2 + |h'|^2)$, it follows that

$$\int_{\mathbf{D}} |f'| \cdot |g'| \ln\frac{1}{|z|}\, dm_2$$
$$\le \left(\int_{\mathbf{D}} |f'|^2 \cdot |h|^2 \ln\frac{1}{|z|}\, dm_2\right)^{1/2} \left(\int_{\mathbf{D}} |g'|^2 \cdot |h|^{-2} \ln\frac{1}{|z|}\, dm_2\right)^{1/2}$$
$$\le 2\|f\|_\infty^2 \cdot \|g\| .$$

Substituting these estimates, we get that $C_2(\delta) \le 24\delta^{-6}$. Let $v(\xi) \overset{\text{def}}{=} \int_{\mathbf{D}} \frac{u_f(z)}{\xi - z}\, dm_2(z)$, $\xi \in \mathbf{D}$, and

$$v_r(\xi) \overset{\text{def}}{=} v(r\xi), \quad \xi \in \mathbf{T}, \ 0 < r < 1 .$$

Then it follows from the definition of $C_2(\delta)$ that

$$\|p\_v_r\|_{S_2(\mathrm{BMO})} = \sup\left\{\left|\int_{\mathbf{T}} \langle v_r, g\rangle \, dm_1\right| : \|g\|_{S_2(H^1)} \le 1\right\} \le c_2(\delta),$$

and hence there exist a matrix-valued function $\mathfrak{I} \in S_2(\mathrm{BMO})$ and a sequence $r_n \uparrow 1 : p\_v_{r_n} \to p\_\mathfrak{I}$ in the weak* topology of $S_2(\mathrm{BMO})$ (i.e., the function p_v, which is defined in $\mathbf{D}$, has boundary values $\mathfrak{I}$ in a certain sense).

By adding to $\mathfrak{I}$ a suitable function in $S_2(\mathrm{BMOA})$ it is possible to satisfy the inequality $\|\mathfrak{I}\|_{S_2(L^\infty)} \le c_2(\delta)$.

Accordingly, for $f \in D(H^\infty, \delta)$ there are matrix-valued functions $v, v_{ij} \in C(\mathbf{D})$, $i, j = 1, 2, \ldots$, and $\mathfrak{I}_f$, $\|\mathfrak{I}_f\|_{S_2(L^\infty)} \le c_2(\delta)$, such that $\bar\partial v = u_f$ and $p\_v_{r_n} \to p\_\mathfrak{I}_f$ in the weak* topology of $S_2(\mathrm{BMO})$.

The following two lemmas are needed next.

LEMMA 8. *If D is a Douglas algebra, $f \in D(Q_{DA}, \delta)$, and $\{i : f_i \neq 0\}$ is finite, then $\mathfrak{I}_f \subset S_2(D)$ (the function $\mathfrak{I}_f$ was defined before the lemma).*

PROOF. If $b \in \overline{D}$ is an inner function and $0 < \delta < 1$, then $a_\delta(b) \overset{\text{def}}{=} \{z \in \mathbf{D} : |b(z)| > 1 - \delta\}$. With each function $f \in C^2(\mathbf{D})$ we associate a measure $d\mu_f \overset{\text{def}}{=} |\overline{\partial} f|^2 \ln \frac{1}{|z|}\, dm_2$, and we define the intensity of a positive measure μ on $\mathbf{D}$ by the formula $\|\mu\|_c \overset{\text{def}}{=} \sup\{\int_{\mathbf{D}} |g|\, d\mu : \|g\|_{H^1} \leq 1\}$. By Theorem 3.1 in [10], if $f \in L^\infty$, then $f \in D \Leftrightarrow$ for all $\varepsilon > 0$ there exists an inner function $b \in \overline{D}$ such that $\|\mu_f|_{a_\delta(b)}\|_c < \varepsilon$. Then for all $\varepsilon > 0$ there exists an inner function $b \in D$ such that $\|\mu_{v_{ij}}|_{a_\delta(b)}\|_c < \varepsilon$, $i, j = 1, 2, \dots$, where the properties of the matrix-valued functions are given before the lemma. Although the functions v_{ij}, $i, j = 1, 2, \dots$, are not harmonic extensions of the functions $\mathfrak{I}_{ij}$, for the arguments in [10] and [11] to be applicable it suffices that $\Delta v_{ij} = 4|\partial(u_f)_{ij}| = O(|f'|^2)$; therefore, $\mathfrak{I}_{ij} \in D$. □

LEMMA 9. *If $\mathfrak{I} \in D$ and $\varepsilon > 0$, then there exists a $w \in C_D$ such that $w - \mathfrak{I} \in H^\infty$ and $\|w\|_\infty \leq \|\mathfrak{I}\|_\infty + \varepsilon$.*

PROOF. The proof follows from the fact that the imbedding $C_D/C_{DA} \to L^\infty/H^\infty$ is an isometry and the equality $D = H^\infty + C_D$ [9]; then $C_D/C_{DA} = D/H^\infty$. □

CONCLUSION OF THE PROOF OF THE THEOREM. If $f \in D(X_A, \delta)$, then for all $\varepsilon > 0$ there exists an n such that the function

$$\tilde{f} \in l^2(X_A) : \tilde{f}_i = \begin{cases} f_i & \text{for } i \leq n, \\ 0 & \text{for } i > 0 \end{cases}$$

satisfies $\|f - \tilde{f}\|_{l^2(X_A)} \leq \varepsilon$. Then $\|\tilde{f}\|_{l^2(X_A)} \leq 1$, and $|\tilde{f}(z)| \geq \delta - \varepsilon$, $z \in \mathbf{D}$. Lemmas 8 and 9 are applicable to $\tilde{f}$; therefore, there is a function with boundary values $\mathfrak{I} : \|\mathfrak{I}\|_{S_2(C_{DA})} \leq 24(\delta - \varepsilon)^{-6} + \varepsilon$, $\overline{\partial} v = u_f$, and as a solution of the corona problem we can take $g = \overline{\tilde{f}}|\tilde{f}|^2 + v \cdot \tilde{f}$, $\|g\|_\infty \leq (\delta - \varepsilon)^{-2} + 24(\delta - \varepsilon)^{-6} + \varepsilon$. □

REMARK. Using methods in [6] and [7] and Lemmas 8 and 9, we can show that

$$C_n(X_A, \delta) \leq n \cdot c \cdot \delta^{-2} \left(\ln \frac{1}{\delta} + 1 \right)$$

(C_n was defined in the remark after Theorem 3). If we are given $h \in X_A$, $f \in l^2(X_A)$, $f_l = 0$, $l = n+1, n+2, \dots$, $\|f\|_{l^2(X_A)} \leq 1$, and $|h(z)| \leq |f(z)|^{2+\varepsilon}$, $z \in \mathbf{D}$, then $h = \langle f, g \rangle$, $g \in l^2(X_A)$, and $\|g\|_{l^2(X_A)} \leq n \cdot \varepsilon^{-1} \cdot c$ where c is an absolute constant.

We remark that the sufficient condition given for membership in an ideal has content even for the disk algebra $(X_A = C_A)$.

§7. Concluding remarks

Our results can be summed up as follows: if the "basic space X" is in duality with some other space, and this duality is estimated in terms of the moduli of derivatives, then the corona problem with "X_A-data" is solvable "with an estimate." Here X is a space of functions harmonic in the disk $\mathbf{D}$. In the case of the multiplier algebras $M(X)$ the basic space X consists of analytic functions. The requirement of invariance of X under rotations (Axioms ($\mathscr{H}$3) and (A3)) is not very essential—it is used mainly to simplify the approximation problems. This requirement is satisfied by the most interesting spaces of analytic and harmonic functions ([17], [18], [19], [27]).

At the basis of all our considerations lay the space $l^2(X)$, which can be associated with *each* space X. For many *concrete* spaces X we can also consider $X(l^2)$ along with $l^2(X)$. In these cases the problem arises of estimating the quantity

$$C(X(l^2), \delta) \overset{\text{def}}{=} \sup\{\inf\{\|g\|_{X(l^2)} : \langle f, g\rangle = 1\} : \\ \|f\|_{X(l^2)} \leq 1, |f(z)| \geq \delta, z \in \mathbf{D}\}.$$

For example, it was shown in [7] that $C(H^\infty(l^2), \delta) \leq c\delta^{-2}(\ln\frac{1}{\delta}+1)$. The derivation of estimates of such quantities is made more difficult by the fact that results about $X(l^2)$ do not follow from results about X, but require separate proofs.

We present some results.

THEOREM 4$'$. *If $B_s^{p,q}$ is an algebra, then*

$$C(B_{s,A}^{p,q}(l^2), \delta) \leq c(p, q, s)\delta^{-(3+[s])}.$$

Here the space $B_s^{p,q}(l^2)$ can be defined both in terms of harmonic extensions and in terms of the $L^p(l^2)$-modulus of continuity of the boundary values.

THEOREM 5$'$. *Under the conditions of Theorem* 5

$$C(E_A^{(k)}(l^2), \delta) \leq c(k)\delta^{-(3+k)}.$$

In the case $p = 2$ an analogue of Theorem 8 is valid.

THEOREM 8$'$. *For the space of bounded functions with finite Dirichlet integral*

$$C(H^\infty(l^2) \cap B_{1/2}^{2,2}(l^2), \delta) \leq c\delta^{-3}.$$

The proof is based on the simple proof proposed in [23] for Theorem 7 with $p = 2$, which permits Theorem 7 to be extended to the case of functions with values in l^2.

The author expresses deep gratitude to V. P. Khavin for his undivided attention to this work and for help in preparing the paper.

BIBLIOGRAPHY

1. L. Carleson, *Interpolation by bounded analytic functions and the corona theorem*, Ann. of Math. (2) **76** (1962), 547–559.

2. S. Scheinberg, *Cluster sets and corona theorems*, Lecture Notes in Math. **604** (1977), 103–106; Springer-Verlag, New York.

3. T. W. Gamelin, *Wolff's proof of the corona theorem*, Israel J. Math. **37** (1980), 113–120.

4. M. Rosenblum, *A corona theorem for countably many functions*, Integr. Eqns. Oper. Thy. **3** (1980), 125–137.

5. V. A. Tolokonnikov, *Estimates in Carleson's corona theorem and finitely generated ideals in the algebra* H^∞, Funktsional. Anal. i Prilozhen. **14** (1980), no. 4, 85–86; English transl. in Functional Anal. Appl. **14** (1980).

6. ——, *Estimates in Carleson's corona theorem, ideals in the algebra* H^∞, *and a problem of Sz.-Nagy*, Zap. Nauchn. Sem. Leningrad. Otdel. Mat. Inst. Steklov. (LOMI) **113** (1981), 178–198; English transl. in J. Soviet Math.

7. A. Uchiyama, *Corona theorems for countably many functions and estimates for their solutions*, preprint (99 pp.), UCLA, 1982.

8. K. N. Chow and D. Protas, *The maximal ideal space of bounded analytic Dirichlet finite functions*, Arch. Math. **31** (1978), 298–301.

9. S.-Y. Chang and D. Marshall, *Some algebras of bounded analytic functions containing the disk algebra*, Lecture Notes in Math. **604** (1977), 12–20, Springer-Verlag, New York.

10. S.-Y. Chang and J. B. Garnett, *Analyticity of functions and subalgebras of* L^∞ *containing* H^∞, Proc. Amer. Math. Soc. **72** (1978), 41–45.

11. S.-Y. Chang, *A characterization of Douglas subalgebras*, Acta Math. **137** (1976), 81–90.

12. L. V. Kantorovich and G. P. Akilov, *Functional analysis*, 2nd rev. ed., "Nauka", Moscow, 1977; English transl., Pergamon Press, Oxford, 1982.

13. Jöran Bergh and Jörgen Löfström, *Interpolation spaces: an introduction*, Springer-Verlag, 1976.

14. Kenneth Hoffman, *Banach spaces of analytic functions*, Prentice-Hall, Englewood Cliffs, N.J., 1962.

15. M. A. Naĭmark, *Normed algebras*, 2nd rev. ed., "Nauka", Moscow, 1968; English transl., Wolters-Noordhoff, Groningen, 1972.

16. V. V. Peller and S. V. Khrushchev, *Hankel operators, best approximations, and stationary Gaussian processes*, Uspekhi Mat. Nauk **37** (1982), no. 1 (223), 53–124; English transl. in Russian Math. Surveys **37** (1982).

17. A. L. Shields and D. L. Williams, *Bounded projections, duality, and multipliers in spaces of analytic functions*, Trans. Amer. Math. Soc. **162** (1971), 287–302.

18. ——, *Bounded projections, duality, and multipliers in spaces of harmonic functions*, J. Reine Angew. Math. **229/230** (1978), 256–279.

19. S. Jansson, *Generalizations of Lipschitz spaces and an application to Hardy spaces and* BMO, Duke Math. J. **47** (1980), 959–982.

20. A. L. Shields, *Weighted shift operators and analytic function theory, Topics in operator theory*, Math. Surveys **13** (1974), 49–128.

21. D. A. Stegenga, *Bounded Toeplitz operators on* H^1 *and applications of the duality between* H^1 *and* BMO, Amer. J. Math. **98** (1976), 573–589.

22. S. Jansson, *On functions with conditions of the mean oscillations*, Arkiv Math. **14** (1976), 189–196.

23. P. Koosis, *A theorem of Khrushchev and Peller on restrictions of analytic functions having finite Dirichlet integral to closed subsets of the unit circumference*, Proc. Conf. Harmonic Analysis (Zygmund), vol. II, ed. W. Beckner et al., Wadsworth Pub., Belmont, Calif. (1982).

24. C. Sundberg and T. Wolff, *Interpolating sequences for* QA_B, Preprint, 1983.

25. F. A. Shamoyan, *Applications of Dzhrbashyan integral representations to certain problems in analysis*, Dokl. Akad. Nauk SSSR **261** (1981), 557–561; English transl. in Soviet Math. Dokl. **24** (1981).

26. V. G. Maz′ya and T. A. Shaposhnikova, *Multipliers in spaces of differentiable functions*, Trudy Sem. S. L. Soboleva **1979**, no. 1, 37–90. (Russian)

27. D. A. Stegenga, *Multipliers of the Dirichlet space*, Illinois J. Math. **24** (1980), 113–139.

Scientific-Technical Division
Institute of Construction of Analytical Instruments
Academy of Sciences, USSR

Translated by ALLEN SHIELDS AND H. H. McFADEN

Amer. Math. Soc. Transl.
(2) Vol. **149**, 1991

The Structure of Closed Ideals in Some Algebras of Smooth Functions

L. G. KHANIN

Theorems on spectral synthesis, i.e., statements that closed ideals are representable in the form of the intersection of primary closed ideals, serve as a classical way of describing closed ideals in topological algebras. For a number of concrete algebras of functions, theorems of such type are well known [1]–[4]; however, methods of proving them are, as a rule, strictly individual.

In this work a more general approach to the problem of spectral synthesis of ideals is developed. This approach enables us to investigate from a unified viewpoint a certain class of Banach algebras of functions "of smoothness ≤ 1," defined in terms of point derivations and called below the class of D-algebras. The main result is a theorem on spectral synthesis of ideals in D-algebras (Theorem 1). This theorem enables us to simplify and unify proofs of certain theorems that are special cases (among them are theorems on spectral synthesis of ideals in Lipschitz algebras $\mathrm{Lip}(X, \rho)$ [3], [5] and in algebras of continuously differentiable functions $C^1([-1, 1]^n)$ [2], [6]), and also to obtain new statements on spectral synthesis. One such statement is a theorem on spectral synthesis of ideals in the Zygmund algebra Λ of functions of one variable (Theorem 2).

The main results of this work were announced in the note [7] and in [8]. They are expounded more thoroughly in the author's dissertation [9].

§1. Basic concepts and spaces. Known results

Let X be a compact Hausdorff space. Let $\mathscr{A}$ be a set of continuous functions on X that is a regular (in the sense of G. E. Shilov) Banach algebra

1991 *Mathematics Subject Classification.* Primary 46J15, 46J20.

Translation of Studies in the Theory of Functions of Several Real Variables, Yaroslav. Gos. Univ., Yaroslavl′, 1986, pp. 97–114; MR88d:46043.

with identity with respect to pointwise operations on the functions. ([1]) We assume that the algebra has the inversion property, i.e., for any function $f \in \mathscr{A}$ such that $f(x) \neq 0$ for all $x \in X$, $\frac{1}{f} \in \mathscr{A}$. Then [10, p. 17] its structure space coincides with X.

To each ideal I in $\mathscr{A}$ we associate the closed set $\sigma(I) = \bigcap_{f \in I} f^{-1}(0)$ and we call it the spectrum. If $\sigma(I) = \{x\}$, $x \in X$, then the ideal I is called a primary ideal at the point x. For each closed set $F \subset X$, we define M_F to be the set of functions in $\mathscr{A}$ that are equal to zero on F, and J_F to be the closure in $\mathscr{A}$ of the set of functions in $\mathscr{A}$ that are equal to zero on a neighborhood of the set F (in case $F = \{x\}$, $x \in X$, we denote these sets by M_x and J_x). It is easy to see that M_F and J_F are closed ideals in $\mathscr{A}$ with spectrum F, while for any closed ideal I in $\mathscr{A}$ with spectrum F, $J_F \subset I \subset M_F$ [10, p. 40]. The smallest closed primary ideal $I_x \supset I$ at the point $x \in \sigma(I)$ is called the primary component of I at x. It is clear that $I_x = \operatorname{clos}_{\mathscr{A}}(I + J_x)$.

We shall say that the algebra $\mathscr{A}$ admits spectral synthesis of ideals (notation: $\mathscr{A} \in$ synt), if, for any closed ideal $I \subset A$,

$$I = \bigcap_{x \in \sigma(I)} I_x. \tag{1}$$

We shall list the algebras of functions of the type described above that we consider in this paper.

The algebra of all continuous functions on X, denoted by $C(X)$, is equipped with the norm $\|f\|_X = \sup\{|f(x)| \,|\, x \in X\}$, $f \in C(X)$.

Let (X, ρ) be a compact metric space. The Lipschitz algebra $\operatorname{Lip}(X, \rho)$ [11] is defined to be the set of functions f on X such that, for some $M \geq 0$, $|f(x)| \leq M$ and $|f(x) - f(y)| \leq M\rho(x, y)$ for all $x, y \in X$; as a norm on $\operatorname{Lip}(X, \rho)$ we take $\inf M$. The closed separable subalgebra of the algebra $\operatorname{Lip}(X, \rho)$ consisting of those functions f for which $\lim_{\rho(x,y) \to 0^+} \frac{|f(x) - f(y)|}{\rho(x, y)} = 0$ is denoted by $\operatorname{lip}(X, \rho)$.

The Zygmund algebras [12] serve as the nearest analogues to the Lipschitz algebras. Before we define them, we shall introduce some notation. For $h = (h_1, \dots, h_m) \in \mathbf{R}^n$, we put $|h| = \max_{1 \leq i \leq n} |h_i|$. The cube in $\mathbf{R}^n$ with center c and edge $2d$ is the set $Q(c, d) = \{x \in \mathbf{R}^n \,|\, |x - c| \leq d\}$; a parallelepiped in $\mathbf{R}^n$ is defined analogously. If a function f is given on a cube or parallelepiped Q, then for $x \in Q$, $h \in \mathbf{R}^n$, $x \pm h \in Q$ we shall write $\Delta_h^1 f(x) = f(x+h) - f(x)$; $\Delta_h^2 f(x) = f(x+h) - 2f(x) + f(x-h)$; $\omega_2(f; Q; t) = \sup\{|\Delta_h^2 f(x)| \,|\, x, x \pm h \in Q, |h| \leq t\}$, $t \geq 0$.

The Zygmund space $\Lambda_n = \Lambda([-1, 1]^n)$ (for $n = 1$ it is denoted by Λ) is the set of all functions f on the cube $[-1, 1]^n$ such that, for some $M \geq 0$, $|f(x)| \leq M$ and $|\Delta_h^2 f(x)| \leq M|h|$ for all $x, x \pm h \in [-1, 1]^n$; the

([1]) The algebras $\mathscr{A}$ are considered real; however, all the results can be carried over almost word-for-word to the complex case.

norm on Λ_n is defined to be $\inf M$. The functions $f \in \Lambda_n$ for which $\lim_{t\to 0^+} \frac{\omega_2(f;[-1,1]^n;t)}{t} = 0$ form a closed separable subalgebra λ_n in Λ_n (for $n=1$ we write $\lambda_1 = \lambda$).

If f is a bounded function on the interval $[a, b]$, then for $x, x+h \in [a, b]$,

$$|f(x+h) - f(x)| \le A|h| \left[\int_{|h|}^{b-a} \frac{\omega_2(f;[a,b];t)}{t^2} dt + \frac{\|f\|_{[a,b]}}{b-a} \right] \tag{2}$$

which is a special case of A. Marchaud's inequality [13]. (2) From (2) it follows that for $f \in \Lambda_n$ and $x, x+h \in [-1, 1]^n$,

$$|f(x+h) - f(x)| \le A(n)\|f\|_{\Lambda_n} |h| \ln \frac{3}{|h|}. \tag{3}$$

From the estimate (3) and the identity

$$\Delta_h^2(fg)(x) = \Delta_h^2 f(x) g(x+h) + 2\Delta_h^1 f(x-h) \Delta_h^1 g(x) + f(x-h)\Delta_h^2 g(x) \tag{4}$$

it results that the space Λ_n is a Banach algebra. It is easy to see that the algebras Λ_n and λ_n satisfy all the above-mentioned requirements.

Finally, we shall denote by $C^m([-1,1]^n)$ the algebra of m times continuously differentiable functions on $[-1,1]^n$ with the natural norm.

M. H. Stone showed [1] that every closed ideal I in the algebra $C(X)$ has the form

$$I = M_F, \quad \text{where } F = \sigma(I) \tag{5}$$

(see also [10, pp. 34–35]). Equality (5) is the simplest form of the synthesis property (1). The algebra $\mathrm{lip}(X, \rho)$ also satisfies (5) [11]. On the other hand, equality (5) is not, generally speaking, satisfied for the algebra $\mathrm{Lip}(X, \rho)$. Nevertheless, $\mathrm{Lip}(X, \rho) \in \mathrm{synt}$. In the case of the algebras $\mathrm{Lip}\,\alpha(X)$, $X \subset \mathbf{R}^n$, $0 < \alpha \le 1$, this was proved by G. Glaeser [5]; the general result is due to L. Waelbroeck [3]. H. Whitney showed [2] that, for all m, $C^m([-1,1]^n) \in \mathrm{synt}$ (for $m=1$ this was established by È. È. Shnol′[6]).

§2. *D*-Algebras

Let $\mathscr{A}$ be an algebra from Section 1. A functional $D \in \mathscr{A}^*$ is called a point derivation of the algebra $\mathscr{A}$ at the point $x \in X$ if $D(fg) = f(x)Dg + g(x)Df$ for all $f, g \in \mathscr{A}$. We shall denote by $\mathscr{D}_x$ the linear space of all point derivations of the algebra $\mathscr{A}$ at the point x and we note that, by the regularity of $\mathscr{A}$, $D(J_x) = \{0\}$ for all $D \in \mathscr{D}_x$. For $x \in X$, we put $K_x = \{D \in \mathscr{D}_x | \, \|D\| \le 1\}$, where $\|\cdot\|$ is the $\mathscr{A}^*$-norm. If F is a closed subset of X, then the set $K_F = \bigcup_{x\in F} K_x$ is compact in the weak* topology on $\mathscr{A}^*$.

(2) The letter A denotes various nonnegative absolute constants (even in one chain of estimates). By the symbol $A(\ )$, where the letter A can also be applied with an index, is denoted various nonnegative constants depending on the parameter indicated in the parentheses.

To each function $f \in M_F$ we shall associate a function $\hat{f} \in C(K_F)$ by setting $\hat{f}(D) = Df$, $D \in K_F$. In this way we determine a linear mapping $d_F : M_F/J_F \to C(K_F)$, with

$$\|\hat{f}\|_{K_F} \le \|\dot{f}\|_F, \quad \dot{f} \in M_F/J_F \tag{6}$$

where $\|\cdot\|_{K_F}$ is the sup norm on $C(K_F)$ and $\|\dot{f}\|_F$ is the quotient norm of the class $\dot{f} \in \mathscr{A}/J_F$ that contains f.

DEFINITION. An algebra $\mathscr{A}$ is called a D-algebra if for any closed set $F \subset X$ there exists a constant $A(F)$ such that

$$\|\dot{f}\|_F \le A(F)\|\hat{f}\|_{K_F} \quad \text{for all } \dot{f} \in M_F/J_F. \tag{7}$$

We note that D-algebras are fully characterized by the fact that, for every closed subset F of X, d_F is a bijective mapping on a closed subspace in $C(K_F)$.

From (7) it follows that, for any closed set $F \subset X$,

$$J_F = \operatorname{clos}_{\mathscr{A}}(M_F^2). \tag{8}$$

The following algebras are D-algebras.

1. Algebras with property (5), in particular, $C(X)$, $\operatorname{lip}(X, \rho)$, and, as will be shown in Section 5, the algebra λ.
2. The algebras $\operatorname{Lip}(X, \rho)$.

For a closed set $F \subset X$ and $f \in \operatorname{Lip}(X, \rho)$, we set

$$L_F(\dot{f}) = \varlimsup_{x, y \to F} \frac{|f(x) - f(y)|}{\rho(x, y)}. \text{ (}^3\text{)}$$

It is easy to see that, if $f \in M_F$, then there exists a point derivation $D \in K_F$ such that $L_F(\dot{f}) = |Df|$. Therefore $L_F(\dot{f}) \le \|\hat{f}\|_{K_F}$. On the other hand, each function in the space $\operatorname{Lip}(Y, \rho)$, where $Y \subset X$, can be isometrically extended to a function in the space $\operatorname{Lip}(X, \rho)$ [14]. Hence it follows that $\|\dot{f}\|_F \le L_F(\dot{f})$. Thus, for the algebras $\operatorname{Lip}(X, \rho)$, inequality (7) is satisfied with $A(F) = 1$, and consequently becomes an equality.

The description and properties of point derivations in Lipschitz algebras have been presented in [11] and [8]; see also [9].

3. The algebras $C^1([-1, 1]^n)$.

In these algebras the point derivations are classical, i.e., they have the form $f \to (\nabla f(x), \eta)$, where $x \in [-1, 1]^n$, $\eta \in \mathbf{R}^n$. Relation (7) follows from Whitney's extension theorem [15] with the constant in (7) depending only on n.

4. The Zygmund algebra Λ.

In Λ, in contrast to Lipschitz algebras, bounded extension operators from arbitrary closed sets are missing. Nevertheless, the following lemma is valid.

(^{3}) The notation $x \to F$ means that $\operatorname{dist}(x, F) \to 0$.

LEMMA 1. *Let F be a closed subset of $[-1, 1]$, let $f \in \Lambda$, let $f|_F \equiv 0$, and let $M = \overline{\lim}_{\substack{x \to F \\ h \to 0^+}} \frac{|\Delta_h^2 f(x)|}{|h|}$. Then for every $\varepsilon > 0$ there exists a number $\rho > 0$ and a function $\tilde{f} \in \Lambda$ such that $\tilde{f}|_{F_\rho} = f|_{F_\rho}$ and $\|\tilde{f}\|_\Lambda \le A(M + \varepsilon)$, where $F_\rho = \{x \in [-1, 1] | \operatorname{dist}(x, F) \le \rho\}$.*

COROLLARY. *The Zygmund algebra λ has property* (5).

Proofs of Lemma 1 and its corollary are presented in Section 5.

If we set $M = 0$ in Lemma 1, we shall then find that the functional $Z_F(\dot{f}) = \overline{\lim}_{\substack{x \to F \\ h \to 0^+}} \frac{|\Delta_h^2 f(x)|}{|h|}$, $\dot{f} \in \Lambda / J_F$, determines the norm on M_F/J_F; at the same time

$$\|\dot{f}\|_F \le A Z_F(\dot{f}) \quad \text{for all } \dot{f} \in M_F/J_F. \tag{9}$$

From (3) and (4) it follows that for every function $f \in \Lambda$ there exists a point derivation $D \in K_F$ such that $Z_F(\dot{f}) = |Df|$. Consequently, (7) follows from (9) with the same constant A.

REMARK. There are justifications for assuming that, for all n, the algebras Λ_n are D-algebras and that the algebras λ_n satisfy property (5).

Statements concerning the structure and properties of point derivations in the algebras Λ_n (see [16], [8], and also [9]) are analogous to the corresponding results for Lipschitz algebras.

§3. Spectral synthesis of ideals in D-algebras

Let $\mathscr{A}$ be a D-algebra, let F be a closed subset of X. We will denote by $\mathscr{A}_F$ the algebra of restrictions to F of functions in $\mathscr{A}$, equipped with the "trace" norm $\|f\|_{\mathscr{A}_F} = \inf\{\|\tilde{f}\|_{\mathscr{A}} | \tilde{f}|_F = f\}$, $f \in \mathscr{A}_F$. It is easy to see that $\mathscr{A}_F$ is a Banach algebra with identity whose structure space coincides with F. The quotient space M_F/J_F is, in view of (8), a module over $\mathscr{A}_F$ (*a fortiori* over $\mathscr{A}$) with respect to the operation $\dot{f}g = (f\tilde{g})^{\cdot}$, $\dot{f} \in M_F/J_F$, $g \in \mathscr{A}_F$, induced by multiplication of functions; here $\tilde{g}$ is (any) function in $\mathscr{A}$ such that $\tilde{g}|_F = g$. The dual space $(M_F/J_F)^*$ is also a module over $\mathscr{A}_F$ (and over $\mathscr{A}$) with respect to the operation defined by the equality $(g\tau)(\dot{f}) = \tau(\dot{f}g)$, where $\tau \in (M_F/J_F)^*$, $g \in \mathscr{A}_F$, $\dot{f} \in M_F/J_F$.

We shall replace the quotient norm on M_F/J_F by $\|\dot{f}\| = \|\hat{f}\|_{K_F}$, $\dot{f} \in M_F/J_F$, which is equivalent by virtue of (6) and (7); the corresponding norm on $(M_F/J_F)^*$ will also be denoted by $\|\cdot\|$.

LEMMA 2. *Let $\mathscr{A}$ be a D-algebra, let I be a closed ideal in $\mathscr{A}$ with spectrum F, let σ be an extreme point on the closed unit ball B in the space $(I/J_F)^\perp$ with respect to the norm $\|\cdot\|$. Then there exists a point $x \in F$ such that*

$$g\sigma = g(x)\sigma \quad \textit{for all } g \in \mathscr{A}. \tag{10}$$

PROOF. The closed unit ball in the space M_F/J_F, with respect to the norm $\|\cdot\|$, is $\mathscr{A}_F$-convex in the sense that, if $\dot{f}_1, \dot{f}_2 \in M_F/J_F$, $\|\dot{f}_1\| \le 1$, $\|\dot{f}_2\| \le 1$, and the functions g_1 and g_2 in $\mathscr{A}_F$ are such that $g_1, g_2 \ge 0$ and $g_1 + g_2 \equiv 1$, then $\|\dot{f}_1 g_1 + \dot{f}_2 g_2\| \le 1$. In fact, for all $x \in F$ and $D \in K_x$, we have $|D(f_1\tilde{g}_1 + f_2\tilde{g}_2)| = |g_1(x)Df_1 + g_2(x)Df_2| \le g_1(x) + g_2(x) = 1$, from which, by the definition of the norm $\|\cdot\|$ and the action of $\mathscr{A}_F$ on M_F/J_F, it follows that $\|\dot{f}_1 g_1 + \dot{f}_2 g_2\| \le 1$. From the property of $\mathscr{A}_F$-convexity of the ball in M_F/J_F, it follows that, for all $\tau \in (M_F/J_F)^*$, $\|g_1\tau\| + \|g_2\tau\| = \|\tau\|$ for the above mentioned g_1 and g_2. In fact,

$$\begin{aligned}\|g_1\tau\| + \|g_2\tau\| &= \sup_{\|\dot{f}_1\|\le 1} \tau(\dot{f}_1 g_1) + \sup_{\|\dot{f}_2\|\le 1} \tau(\dot{f}_2 g_2) \\ &= \sup_{\|\dot{f}_1\|\le 1,\, \|\dot{f}_2\|\le 1} \tau(\dot{f}_1 g_1 + \dot{f}_2 g_2) \le \|\tau\|;\end{aligned}$$

the reverse inequality is obtained from the identity $g_1\tau + g_2\tau = \tau$.

Let σ be an extreme point of the set B. We shall show that for any function $g \in \mathscr{A}_F$ there exists a number $\lambda(g)$ such that $g\sigma = \lambda(g)\sigma$. It is sufficient to verify the last equality in the case when $0 \le g \le 1$ and $g\sigma \ne 0$, $g\sigma \ne \sigma$. We shall put $\sigma_1 = g\sigma/\|g\sigma\|$, $\sigma_2 = (1-g)\sigma/\|(1-g)\sigma\|$, $t_1 = \|g\sigma\|$, $t_2 = \|(1-g)\sigma\|$. Then $\sigma = t_1\sigma_1 + t_2\sigma_2$, where $\|\sigma_1\| = \|\sigma_2\| = 1$, $t_1, t_2 > 0$, and, as was shown above, $t_1 + t_2 = \|\sigma\| = 1$. We also notice that, since $(I/J_F)^{\perp}$ is a submodule of $(M_F/J_F)^*$, we have $\sigma_1, \sigma_2 \in B$. Hence $\sigma_1 = \sigma$, that is, $g\sigma = \|g\sigma\|\sigma$.

The mapping $g \to \lambda(g)$ is a nontrivial multiplicative linear functional on $\mathscr{A}_F$. Consequently, there exists a point $x \in F$ for which $\lambda(g) = g(x)$ for all $g \in \mathscr{A}_F$. This proves the lemma.

Following Waelbroeck [3], we shall call the functionals in $(M_F/J_F)^*$ that possess property (10) local functionals at the point x.

For the algebras $\mathrm{Lip}(X, \rho)$, Lemma 2 was established in [3], and the proof here is an abstract analog of the corresponding argument from [3]. The coincidence, for Lipschitz algebras, of the quotient norms on M_F/J_F and the norms generated by the transformation d_F enabled Waelbroeck to avoid the use of point derivations in establishing equality (10), which served as the result of the first stage of the proof of the theorem about spectral synthesis of ideals (at the same time the property of $\mathscr{A}_F$-convexity for the quotient norm in Lipschitz algebras follows from the fact that $\|\dot{f}\|_F = L_F(\dot{f}) = \sup_{x\in F} L_x(\dot{f})$, $\dot{f} \in M_F/J_F$). However, at the second stage of the proof he was obliged to introduce point derivatives for the sole purpose of the rather complicated construction connected with imbedding the space $\mathrm{Lip}(X, \rho)$ in $\mathrm{Lip}(X \times X, \rho \times \rho)$. In fact, this stage appears to be unnecessary, since, as will immediately be shown, local functionals in D-algebras are "traces" of point derivations.

LEMMA 3. *Let $\mathscr{A}$ be a D-algebra, let F be a closed subset of X. Then the functional $\tau \in (M_F/J_F)^*$ is local at the point $x \in F$ if and only if there exists a point derivation D of the algebra $\mathscr{A}$ at the point x such that $\tau(\dot{f}) = D(f)$ for all $f \in M_F$.*

PROOF. A functional $\tau \in (M_F/J_F)^*$ generated by a point derivation $D \in \mathscr{D}_x$ is obviously local at the point x.

Conversely, let τ be a local functional at the point $x \in F$. We can assume that $\|\tau\| = 1$. In view of our renorming, the mapping d_F is an isometry; consequently, $\tau \circ d_F^{-1}$ is a linear functional on a certain subspace of $C(K_F)$, with norm equal to 1. Extending it by the Hahn–Banach Theorem, with preservation of norm, to a linear functional on $C(K_F)$, we obtain that there exists a regular measure μ on K_F such that $|\mu|(K_F) = 1$ and

$$\tau(\dot{f}) = \int_{K_F} \hat{f}(D)\, d\mu(D) \quad \text{for all } \dot{f} \in M_F/J_F. \tag{11}$$

From (11) it follows that $\mu(\{0\}) = 0$. Consequently the formula $\nu E = |\mu|(K_E)$, where $K_E = \bigcup_{x\in E} K_x$, $E \subset F$, defines a nonnegative regular measure ν on F such that $\nu(F) = 1$.

Let $\mathscr{U}$ be an arbitrary (open in F) neighborhood of the point x. In view of the regularity of $\mathscr{A}$, we can find a function $\varphi \in \mathscr{A}$ such that $\varphi(x) = 1$ and $\varphi|_{F\setminus\mathscr{U}} \equiv 0$. Since $\mathscr{A}$ possesses the inversion property, the function $g = \frac{2\varphi}{\varphi^2+1}$ belongs to $\mathscr{A}$. We have $g(x) = 1$, $g|_{F\setminus\mathscr{U}} = 0$, and $\|g\|_F \le 1$. Since τ is local at the point x, it follows that $\tau = g\tau$. Consequently, taking advantage of (11), we get that, for all $\dot{f} \in M_F/J_F$, $|\tau(\dot{f})| = |\tau(\dot{f}g)| \le \|\hat{f}\|_{K_F}|\mu|(K_{\mathscr{U}}) = \|\dot{f}\|\nu(\mathscr{U})$, whence $\|\tau\| \le \nu(\mathscr{U})$. Then, by the regularity of the measure ν, $\|\tau\| \le \nu(\{x\})$. Because $\|\tau\| = \nu(F) = 1$, it follows that ν is concentrated at the point x. Therefore, $\tau(\dot{f}) = \int_{K_x} \hat{f}(D)\, d\mu(D)$ for all $\dot{f} \in M_F/J_F$. It remains to note that the right-hand side of the last equality defines a point derivation on $\mathscr{A}$ at the point x. This proves the lemma.

We now state the main result.

THEOREM 1. *Every D-algebra admits spectral synthesis of ideals.*

PROOF. Let $\mathscr{A}$ be a D-algebra. Let I be a closed ideal in $\mathscr{A}$ with spectrum F. We shall denote by Σ the set of all extreme points of the closed unit ball in the space $(I/J_F)^{\perp}$ with respect to the norm $\|\cdot\|$. From the Kreĭn–Mil′man Theorem it follows that $f \in I$ if and only if $f \in M_F$ and $\sigma(\dot{f}) = 0$ for all $\sigma \in \Sigma$. According to Lemmas 2 and 3, each functional $\sigma \in \Sigma$ generates a point derivation D_σ of the algebra $\mathscr{A}$ at the point $x_\sigma \in F$. We note that $D_\sigma^{-1}(0) \cap M_{x_\sigma} \supset I$ is a closed primary ideal at the point x_σ.

Therefore,

$$I = \left(\bigcap_{\sigma\in\Sigma} D_\sigma^{-1}(0)\right) \cap M_F \supset \left(\bigcap_{\sigma\in\Sigma} (D_\sigma^{-1}(0) \cap M_{x_\sigma})\right)$$

$$\cap M_F \supset \left(\bigcap_{\sigma\in\Sigma} I_{x_\sigma}\right) \cap M_F \supset \bigcap_{x\in F} I_x$$

which is equivalent to (1). This proves the theorem.

To complete the description of closed ideals in D-algebras, it remains to characterize their closed primary ideals.

The set of all closed primary ideals of the D-algebra $\mathscr{A}$ at the point x coincides with the set of closed subspaces I in $\mathscr{A}$ such that $J_x \subset I \subset M_x$. In fact, it is only necessary to check that from these inclusions it follows that I is an ideal in $\mathscr{A}$. Let $f \in I$, $g \in \mathscr{A}$; then, using (8), we obtain that $fg = f(g - g(x)) + g(x)f \in M_x^2 + I \subset J_x + I = I$. As for the ideals J_x, in the general case $J_x = \operatorname{clos}_{\mathscr{A}}(M_x^2) = M_x \cap^\perp \mathscr{D}_x$; however, for concrete D-algebras, their explicit description can be derived. So, for the algebras $C^1([-1, 1]^n)$, $J_x = \{f | f(x) = 0, \nabla f(x) = \mathbf{0}\}$; for the algebras $\operatorname{Lip}(X, \rho)$, $J_x = \{f | f(x) = 0, L_x(\dot{f}) = 0\}$ [11]; and analogously, for the algebras Λ_n, $J_x = \{f | f(x) = 0, Z_X(\dot{f}) = 0\}$ [16], [8], [9], while the same representations with the replacement of x by F are also valid for the ideals J_F. We also point out that in any D-algebra the mapping $I \to \mathscr{D}_x \cap I^\perp$ defines a bijective correspondence between the closed primary ideals I at the point x and the weak* closed subspaces in $\mathscr{D}_x$, while the reverse correspondence is established by the mapping $L \to M_x \cap^\perp L$, $L \subset \mathscr{D}_x$.

§4. Some auxiliary statements

In this section we gather various facts of an analytic character that are necessary for the proof of Lemma 1.

PROPOSITION 1. *Let f be a function defined on the interval $[0, a]$, such that $f(0) = 0$, and let $\tilde{f}$ be its extension to the interval $[-a, a]$ as an odd function. Then $\omega_2(\tilde{f}; [-a, a]; t) \le 5\omega_2(f; [0, a]; t)$ for all $t \ge 0$.*

PROOF. To estimate the quantity $\Delta_h^2 \tilde{f}(x)$, it is sufficient to assume that x, $x + h \in [0, a]$, $x - h \in [-a, 0]$. We have $\Delta_h^2 \tilde{f}(x) = f(x+h) - 2f(x) - f(h-x) = [f(x+h) - 2f(h) + f(h-x)] - 2[f(h-x) - 2f(h/2) + f(x)] + 2[f(h) - 2f(h/2) + f(0)]$, from which $\omega_2(\tilde{f}; [-a, a]; t) \le \omega_2(f; [0, a]; t) + 4\omega_2(f; [0, a]; t/2) \le 5\omega_2(f; [0, a]; t)$ for $t \ge 0$.

REMARK. To Proposition 1 we also add the following more general case. Let f be a function on the interval $l = [c-r, c+r]$, and let $\tilde{f}$ be its extension from l to $\tilde{l} = [c - 2r, c + 2r]$, defined by the equalities $\tilde{f}(x) = 2f(c + r) - f(2c + 2r - x)$ when $x \in (c + r, c + 2r]$ and $\tilde{f}(x) = 2f(c - r) - f(2c - 2r - x)$ when $x \in [c - 2r, c - r)$. Then $\omega_2(\tilde{f}; \tilde{l}; t) \le 5\omega_2(f; l; t)$, $t \ge 0$.

If f is a function on the cube (or parallelepiped) Q, then for $F \subset Q$ we shall denote by $P(f; F)$ the polynomial in $\mathscr{P}_1$ that best approximates f in $L_\infty(F)$ and by $E_1(f, F) = \|f - P(f, F)\|_F$, the smallest deviation from f in $L_\infty(F)$ by polynomials in $\mathscr{P}_1$. Here, as usual, $\mathscr{P}_1$ denotes the space of polynomials in n variables over $\mathbf{R}$ of degree ≤ 1.

PROPOSITION 2. *Let f be a bounded function on the cube $Q = Q(c, d)$. Then $E_1(f; Q) \le A(n)\omega_2(f; Q; d)$.* (4)

PROOF. We put $\widetilde{Q} = Q(c, 2d)$. We shall represent the cube $\widetilde{Q}$ in the form of a union of segments that it cuts out from all possible straight lines passing through the point c. Extending the function f from each of these segments as was done in the Remark to Proposition 1, we obtain the function $\tilde{f}$ on $\widetilde{Q}$ extending f. For x, $x \pm h \in \widetilde{Q}$ we have $\tilde{f}(x+h) - 2\tilde{f}(x) + \tilde{f}(x-h) = [\tilde{f}(x+h)-2\tilde{f}((c+x+h)/2)+\tilde{f}(c)]-2[\tilde{f}(x)-2\tilde{f}((c+x)/2)+\tilde{f}(c)]+[\tilde{f}(x-h)-2\tilde{f}((c+x-h)/2)+\tilde{f}(c)]+2[f((c+x+h)/2)-2f((c+x)/2)+f((c+x-h)/2)]$ where $(c+x)/2$, $(c+x\pm h)/2 \in Q$. From this, taking into account the Remark to Proposition 1, we find that $\omega_2(\tilde{f}; \widetilde{Q} : d) \le 20\omega_2(f; Q; d\sqrt{n}) + 2\omega_2(f; Q; d/2) \le A(n)\omega_2(f; Q; d)$. Let w be a function in $C^\infty(\mathbf{R}^n)$ with the properties $w \ge 0$; $\operatorname{supp} w \subset Q(\mathbf{0}, 1)$; $\int_{\mathbf{R}^n} w(y)\,dy = 1$; $w(-y) = w(y)$, $y \in \mathbf{R}^n$. We set $w_d(y) = \frac{1}{d^n}w(y/d)$, $y \in \mathbf{R}^n$; $g(x) = \int_{|y|\le d}\tilde{f}(x-y)w_d(y)\,dy$, $x \in Q$; $P(x) = g(c)+(\nabla g(c), x-c)$. It is clear that $g \in C^\infty(Q)$ and $P \in \mathscr{P}_1$. For $x \in Q$, $g(x) - f(x) = \frac{1}{2}\int_{|y|\le d}[\tilde{f}(x+y) - 2\tilde{f}(x) + \tilde{f}(x-y)]w_d(y)\,dy$; hence $\|g - f\|_Q \le \frac{1}{2}\omega_2(\tilde{f}; \widetilde{Q}; d) \le A(n)\omega_2(f; Q; d)$. Further, for $x \in Q$,

$$g(x) - P(x) = \sum_{i=1}^{n}\sum_{j=1}^{n}(x_i - c_i)(x_j - c_j)\int_0^t\int_0^t D^{e_i e_j} g(c + s(x-c))\,ds\,dt;$$

therefore $\|g-P\|_Q \le A(n)d^2 \max_{1\le i,j\le n}\|D^{e_i e_j} g\|_Q$. Since $\int_{|y|\le d} D^{e_i e_j} w_d(y)\,dy = 0$ and $D^{e_i e_j} w_d(-y) = D^{e_i e_j} w_d(y)$, $y \in \mathbf{R}^n$, it follows that $D^{e_i e_j} g(x) = \frac{1}{2}\int_{|y|\le d}[\tilde{f}(x+y) - 2\tilde{f}(x) + \tilde{f}(x-y)]D^{e_i e_j} w_d(y)\,dy$, $x \in Q$. Consequently, $\|D^{e_i e_j} g\|_Q \le A(n)d^{-2}\omega_2(\tilde{f}; \widetilde{Q}; d) \le A(n)d^{-2}\omega_2(f; Q; d)$. Thus, $E_1(f; Q) \le \|f - P\|_Q \le \|f - g\|_Q + \|g - P\|_Q \le A(n)\omega_2(f; Q; d)$. This proves the proposition.

REMARK. In the case when Q is a parallelepiped, the inequality from Proposition 2 takes the form $E_1(f; Q) \le A(n)\Theta^4\omega_2(f; Q; d)$, where Θ is the ratio between the greatest and least of the lengths of the edges of the parallelepiped Q. In particular, if f is a bounded function on the cube Q, and Q' is the cube with center $c' \in Q$ and edge $2d'$, then

$$E_1(f; Q' \cap Q) \le A(n)\omega_2(f; Q' \cap Q; d').$$

(4) We also note the obvious inequality $\omega_2(f; Q; d) \le 4E_1(f; Q)$.

We mention that the equivalence property holds for moduli of continuity and for best approximations of any order in the L_p norm for $1 \le p \le \infty$ [17, §2.2, Theorem 4].

PROPOSITION 3. *Let the function f be defined and bounded on the interval $[a, b]$, $a \le t_1 < t_2 \le b$, and let $P_{t_1 t_2}(f)$ be the linear interpolation polynomial for f between the points t_1 and t_2. Then*

$$\|f - P_{t_1 t_2}(f)\|_{[a,b]} \le A\omega_2\left(f; [a, b], \frac{b-a}{2}\right)\frac{b-a}{t_2 - t_1}.$$

PROOF. We write $P = P(f; [a, b])$. For all $x \in [a, b]$,

$$\begin{aligned}|f(x) - P_{t_1 t_2}(f)(x)| &= \left|f(x) - \left[\frac{t_2 - x}{t_1 - t_2} f(t_1) + \frac{x - t_1}{t_2 - t_1} f(t_2)\right]\right| \\ &= \left|[f(x) - P(x)] + \left\{\frac{t_2 - x}{t_2 - t_1}[P(t_1) - f(t_1)] + \frac{x - t_1}{t_2 - t_1}[P(t_2) - f(t_2)]\right\}\right| \\ &\le \|f - P\|_{[a,b]}\left(1 + \frac{|t_2 - x|}{t_2 - t_1}, \frac{|x - t_1|}{t_2 - t_1}\right),\end{aligned}$$

from which, taking account of Proposition 2, the required inequality follows. This proves the proposition.

An important role in proving Lemma 1 is played by

PROPOSITION 4. *Let $0 < \sigma \le 1/2$. There exists a function $\psi_\sigma \in \Lambda$ such that $\psi_\sigma(x) = x$ for $|x| \le \sigma$ and $\|\psi_\sigma\| \le A/\ln(1/\sigma)$.*

PROOF. Proposition 1 shows that it is sufficient to construct the required function on $[0, 1]$ and then to take its odd extension to $[-1, 1]$.

We denote by r the greatest natural number such that $2^{r-1}\sigma \le 1$, and we set $\nu = \frac{1}{r}\sum_{i=0}^{r-1}\delta_{2^i\sigma}$, $\mu = \delta_0 - \nu$, where δ_a is the Dirac measure at the point a. The equality $\psi_\sigma(x) = \int_{[0,x]}(x - t)\,d\mu(t)$, $x \in [0, 1]$, defines a nondecreasing piecewise-linear function such that $\psi_\sigma(x) = x$ for $0 \le x \le \sigma$ and $\psi_\sigma(x) = \psi_\sigma(2^{r-1}\sigma)$ for $2^{r-1}\sigma \le x \le 1$. We shall show that this is the desired function. First of all,

$$\|\psi_\sigma\|_{[0,1]} = \psi_\sigma(2^{r-1}\sigma) = 2^{r-1}\sigma - \frac{1}{r}\sum_{i=0}^{r-1}(2^{r-1}\sigma - 2^i\sigma) = \frac{1}{r}(2^r\sigma - \sigma) \le \frac{2^r\sigma}{r} \le \frac{2}{r}.$$

Further, we estimate $|\Delta_h^2\psi_\sigma(x)|$, where $x, x \pm h \in [0, 1]$; here it can be assumed that $h > 0$. It is easy to verify that

$$|\Delta_h^2\psi_\sigma(x)| = \int_{(x,x+h]}(x + h - t)\,d\nu(t) + \int_{(x-h,x]}[t - (x - h)]\,d\nu(t).$$

We consider the first summand. If $2^i\sigma > x$, then $2^{i+1}\sigma > 2x \ge x + h$; therefore the interval $(x, x + h]$ contains more than one point of the form $2^i\sigma$, $0 \le i \le r - 1$. Hence $\int_{(x,x+h]}(x + h - t)\,d\nu(t) \le h/r$.

We denote the second summand by $R(x, h)$. For $x \in (0, \sigma)$, $R(x, h) = 0$. Therefore we assume further that $x \geq \sigma$.

First let $x - h \geq \sigma$. We pick numbers $p \in \mathbf{N}$ and $q \in \mathbf{Z}_+$ in such a manner that $2^{p-1}\sigma \leq x - h < 2^p\sigma$ and $2^{p+q-1}\sigma \leq x < 2^{p+q}\sigma$. If $q = 0$, then $R(x, h) = 0$. If, however, $q \geq 1$, then

$$\begin{aligned} R(x, h) &= \frac{1}{r}\sum_{i=1}^{q-1}[2^{p+i}\sigma - (x-h)] = \frac{1}{r}[2^p(2^q - 1)\sigma - q(x-h)] \\ &\leq \frac{1}{r}[2x - (x-h) - q(x-h)] = \frac{1}{r}[2h - (q-1)(x-h)] \leq \frac{2h}{r}. \end{aligned}$$

Now let $x - h < \sigma$, and let the number $s \in \mathbf{N}$ be such that $2^{s-1}\sigma \leq x < 2^s\sigma$. Then

$$R(x, h) = \frac{1}{r}\sum_{i=1}^{s-1}[2^i\sigma - (x-h)] = \frac{1}{r}[(2^s - 1)\sigma - s(x-h)].$$

When $s = 1$,

$$R(x, h) = \frac{1}{r}[\sigma - (x-h)] \leq \frac{h}{r}.$$

For $s \geq 2$,

$$R(x, h) \leq \frac{1}{r}(2^s - 1)\sigma \leq \frac{3}{r}(2^{s-1}\sigma - \sigma) \leq \frac{3}{r}[x - (x-h)] \leq \frac{3h}{r}.$$

Thus, in any case $|\Delta_h^2\psi_\sigma(x)| \leq 4|h|/r$. Consequently, $\|\psi_\sigma\| \leq 4/r < A/\ln(1/\sigma)$. This proves the proposition.

REMARK. The estimate of the norm in the space Λ of extensions of the same mapping from the interval $[-\sigma, \sigma]$, as given in Proposition 4, cannot be improved in order of magnitude as $\sigma \to 0$.

A slight modification in the arguments in proving Proposition 4 leads to the following statement.

PROPOSITION 5. *Let $0 < \sigma \leq d \leq 1/2$. There exists a function $\psi_{\sigma,d} \in \Lambda$ such that $\psi_{\sigma,d}(x) = x$ when $|x| \leq \sigma$, $\psi_{\sigma,d}(x) = 0$ when $2d \leq |x| \leq 1$ and $\|\psi_{\sigma,d}\|_\Lambda \leq A/\ln(2d/\sigma)$.*

We fix a function $\omega \in C^\infty(\mathbf{R}^n)$ such that $0 \leq \omega \leq 1$, $\omega \equiv 1$ on $Q(\mathbf{0}, \frac{1}{2})$ and $\omega \equiv 0$ outside $Q(\mathbf{0}, 1)$. For $x \in \mathbf{R}^n$ and $\delta > 0$, we set $\omega_\delta^x(y) = \omega(\frac{y-x}{\delta})$, $y \in \mathbf{R}^n$.

PROPOSITION 6. *Let $x \in [-1, 1]^n$, let $\tau > 0$, let $S = Q(x, \frac{3}{2}\tau) \cap [-1, 1]^n$, and let $f \in \Lambda_n$. Suppose that for some $M \geq 0$, $\|\Delta_h^2 f(y)\| \leq M|h|$ for $y, y+h \in S$ and $|h| \leq \tau$, and also suppose that a polynomial $P \in \mathscr{P}_1$ satisfies the condition $\|f - P\|_S \leq A(n)\omega_2(f; S; \tau)$. Then $\|(f-P)\omega_\tau^x\|_{\Lambda_n} \leq A(n)M$.*

PROOF. We can assume that $\tau \leq 4$. We set $g = (f - P)\omega_\tau^x$. Then we obviously have $\|g\|_{[-1,1]^n} \leq \|f - P\|_S \leq A(n)\omega_2(f; S; \tau) \leq A(n)M\tau \leq A(n)M$.

In estimating the quantity $|\Delta_h^2 g(y)|$, where $y, y \pm h \in [-1, 1]^n$, we consider the following cases:

1. $|h| \le \tau/4$, $|y - x| \le \tau/4$. Then $|\Delta_h^2 g(y)| = |\Delta_h^2 f(y)| \le M|h|$.
2. $|h| \le \tau/4$, $\tau/4 < |y - x| \le 5\tau/4$. Taking account of (4) we obtain

$$\begin{aligned}|\Delta_h^2 g(y)| &\le |\Delta_h^2 (f - P)(y)|\omega_\tau^x(y + h) \\ &\qquad + 2|\Delta_h^1 (f - P)(y - h)|\,|\Delta_h^1 \omega_\tau^x(y)| + |(f - P)(y - h)|\,|\Delta_h^2 \omega_\tau^x(y)| \\ &\le |\Delta_h^2 f(y)| + A(n)\|f - P\|_S \frac{|h|}{\tau} \le A(n)M|h|.\end{aligned}$$

3. $|h| \le \tau/4$, $|y - x| > 5\tau/4$. In this case $\Delta_h^2 g(y) = 0$.
4. $|h| > \tau/4$. Then $|\Delta_h^2 g(y)| \le 4\|g\|_{[-1,1]^n} \le A(n)M\tau \le A(n)M|h|$.

Therefore, $\|g\|_{\Lambda_n} \le A(n)M$. This proves the proposition.

PROPOSITION 7. *Let F be a closed subset of $[-1, 1]^n$. Let $f \in \Lambda_n$ be such that $f|_F = 0$. Let $M = \overline{\lim}_{\substack{x \to F \\ h \to \mathbf{0}}} |\Delta_h^2 f(x)|/|h|$. Then*

$$\overline{\lim_{x \to F}} \frac{|f(x)|}{d(x, F)\ln(3/d(x, F))} \le A(n)M. \ (^5)$$

PROOF. For a given $\varepsilon > 0$ there exists $\delta > 0$ such that for $x, x \pm h \in [-1, 1]^n$, $d(x, F) \le \delta$ and $|h| \le \delta$. The relation $|\Delta_h^2 f(x)| \le (M + \varepsilon)|h|$ is satisfied. Let the point x be such that $0 < d(x, F) \le \delta$. We pick a point $y \in F$ in such a manner that $d(x, F) = |x - y|$. We denote by γ the straight line passing through the points x and y, as we put $l = Q(y, \delta) \cap [-1, 1]^n \cap \gamma$. The length $|l|$ of the segment l obviously satisfies the inequalities $\delta \le |l| \le 2\delta\sqrt{n}$. Employing inequality (2), we obtain

$$\begin{aligned}|f(x)| = |f(x) - f(y)| &\le A(n)|x - y| \left[\int_{|x-y|}^{|l|} \frac{\omega_2(f; l; t)}{t^2}\,dt + \frac{\|f\|}{|l|}\right] \\ &\le A(n)|x - y| \left[\int_{|x-y|}^{2\delta\sqrt{n}} \frac{(M + \varepsilon)t}{t^2}\,dt + \frac{\|f\|_{[-1,1]^n}}{\delta}\right] \\ &\le A(n)d(x, F)\ln\frac{3}{d(x, F)} \left[(M + \varepsilon)\left(1 + \frac{\ln(\delta\sqrt{n})}{\ln\frac{3}{d(x,F)}}\right) + \frac{\|f\|_{[-1,1]^n}}{\delta\ln\frac{3}{d(x,F)}}\right].\end{aligned}$$

Therefore

$$\overline{\lim_{x \to F}} \frac{|f(x)|}{d(f, F)\ln(3/d(x, F))} \le A(n)(M + \varepsilon),$$

from which the required relation follows. This proves the proposition.

We formulate the final two statements of this section for generalized Zygmund spaces, defined in the following way.

(^{5}) The distance $d(\cdot, F) = \operatorname{dist}(\cdot, F)$ is taken with respect to the norm $|\cdot|$.

Let φ be a nondecreasing function on $[0, \infty)$ such that $\varphi(0) = 0$ and $\varphi(t) > 0$ for $t > 0$. The generalized Zygmund space $\Lambda_\varphi([-1, 1]^n)$ (for $n = 1$, it is denoted simply by Λ_φ) is the set of all functions f on $[-1, 1]^n$ for which there exists $M \geq 0$ such that $|f(x)| \leq M$ and $|\Delta_h^2 f(x)| \leq M\varphi(|h|)$ for all $x, x \pm h \in [-1, 1]^n$; as the norm on $\Lambda_\varphi([-1, 1]^n)$ we take $\inf M$. In the case when $\varphi(t) = t$, $t \geq 0$, we obtain the ordinary Zygmund space.

In considering the spaces $\Lambda_\varphi([-1, 1]^n)$, it is possible to assume that the function $t \to \varphi(t)/t^2$, $t > 0$, does not increase. In fact, this property is possessed by the nondecreasing function $\tilde{\varphi}$, defined by the equalities $\tilde{\varphi}(t) = t^2 \inf_{0<s\leq t}(\varphi(s)/s^2)$ for $t > 0$, with $\tilde{\varphi}(0) = 0$, while $\Lambda_{\tilde{\varphi}}([-1, 1]^n) = \Lambda_\varphi([-1, 1]^n)$ and $\|\cdot\|_{\Lambda_\varphi([-1,1]^n)} \leq \|\cdot\|_{\Lambda_{\tilde{\varphi}}([-1,1]^n)} \leq 4\|\cdot\|_{\Lambda_\varphi([-1,1]^n)}$. (6)

Let $\Lambda_\varphi(F)$ be the space of traces of functions in $\Lambda_\varphi([-1, 1]^n)$ on the closed set $F \subset [-1, 1]^n$, equipped with the "trace" norm $\|\cdot\|_{\Lambda_\varphi(F)}$. Unlike Lipschitz spaces, an "intrinsic" description of the traces of generalized Zygmund spaces on arbitrary closed sets is known only for $n = 1, 2$ (see [18] and [19], where the Zygmund spaces are equipped with the seminorm determined by the condition $|\Delta_h^2 f(x)| \leq M\varphi(|h|)$). In the normed case, the description of the traces for $n = 1$ is as follows.

PROPOSITION 8. *Let F be a closed subset of $[-1, 1]$, let f be a function on F. In order that $f \in \Lambda_\varphi(F)$ it is necessary and sufficient that for some $M \geq 0$ the following conditions are satisfied*:

(i) $|f(x)| \leq M$ *for all* $x \in F$;

(ii) $\frac{|f(x)-f(y)|}{|x-y|} \leq M \int_{|x-y|}^3 \frac{\varphi(t)}{t^2}\,dt$ *for all* $x, y \in F$, $x \neq y$;

(iii) $\left|\frac{f(y)-f(x)}{y-x} - \frac{f(z)-f(x)}{z-x}\right| \leq M \int_{|y-x|}^{2|z-x|} \frac{\varphi(t)}{t^2}\,dt$ *for all* $x, y, z \in F$

such that $0 < |x - y| \leq |z - x|$.

At the same time, for any function $f \in \Lambda_\varphi(F)$,

$$A_1(\varphi)\|f\|_{\Lambda_\varphi(F)} \leq \inf M \leq A_2(\varphi)\|f\|_{\Lambda_\varphi(F)}. \tag{12}$$

REMARK. For a particular function f, a necessary and sufficient condition for it to belong to the trace space $\Lambda_\varphi(F)$ is condition (iii) (in the seminormed case this is equivalent to a condition found in [18]). However, without condition (i), relation (12) does not hold for all $f \in \Lambda_\varphi(F)$. If we add (i), then (12) is satisfied, but the constants A_1 and A_2 in this case depend not only on φ, but also on $\operatorname{diam} F$. This last dependence disappears under the renorming of $\Lambda_\varphi(F)$ produced by condition (ii).

The proof of Proposition 8 is given in [9, pp. 111–124].

From Proposition 8 it follows that the spaces $\Lambda_\varphi(F)$, $F \subset [-1, 1]$, have the following finiteness property.

---

(6) We suppose that $\tilde{\varphi}(t) > 0$ for $t > 0$; otherwise $\Lambda_\varphi([-1, 1]^n) = \mathscr{P}_1$.

PROPOSITION 9. *Let F be a closed subset of $[-1, 1]$, let f be a function on F. If for every subset $F \subset E$ that consists of not more than $N_1 = 3$ points there exists a function $f_E \in \Lambda_\varphi$ such that $f_E|_E = f|_E$ and $\|f_E\|_{\Lambda_\varphi} \le 1$, then there exists a function $\tilde{f} \in \Lambda_\varphi$ such that $\tilde{f}|_F = f$ and $\|\tilde{f}\|_{\Lambda_\varphi} \le A(\varphi)$.*

PROOF. Applying Proposition 8 (necessity) to the functions f_E, we obtain that f satisfies conditions (i), (ii), and (iii) for some $M \le A(\varphi)$. Then, again according to Proposition 8 (sufficiency), there exists a function $\tilde{f} \in \Lambda_\varphi$ such that $\tilde{f}|_F = f$ and $\|\tilde{f}\|_{\Lambda_\varphi} \le A(\varphi)M$, which was to be proved.

REMARK. The spaces $\Lambda_\varphi(F)$, $F \subset [-1, 1]^n$, have the finiteness property for every n, but instead of $N_1 = 3$ points it is necessary to take $N_n = 3\cdot 2^{n-1}$ points (for a seminormed Zygmund space this is established in [18] for $n = 1, 2$, and in [19] for arbitrary n).

§5. Proof of Lemma 1 and its corollary

We shall prove Lemma 1.

By the definition of the number M, in view of Proposition 7 there exists $\delta \in (0, \frac{1}{6}]$ for which

(i) $|\Delta_h^2 f(x)| \le (M + \varepsilon)|h|$ when $d(x, F) \le 9\delta$ and $|h| \le 6\delta$;
(ii) $|f(x)| \le A(M + \varepsilon)d(x, F) \ln \frac{3}{d(x, F)}$ when $d(x, F) \le 9\delta$.

Put $\rho = \delta^2$, $K = F_\rho$. We show that for any set $E \subset K$ that consists of no more than three points there exists a function $f_E \in \Lambda$ such that $f_E|_E = f|_E$ and $\|f_E\|_\Lambda \le A(M + \varepsilon)$. According to the finiteness property (Proposition 9), from this the assertion of Lemma 1 will follow.

Without loss of generality it is possible to assume that the set E consists of three points, say $E = \{x_1, x_2, x_3\}$, where $x_1 < x_2 < x_3$ and $x_2 - x_1 \le x_3 - x_2$. We select points $a_i \in F$ for which $|x_i - a_i| < \delta^2$, $i = 1, 2, 3$.

We consider the following cases:

1. $x_2 - x_1 \ge 2\delta$.

Let x be any of the points x_i and let a be the point a_i corresponding to it. To construct the function f_E it will first be necessary to construct the function $g = g_i$ for each i.

If $x \in F$, then we set $g \equiv 0$. We will assume that $x \notin F$. Of the points $a + 3\delta^2$ and $a - 3\delta^2$, at least one belongs to $[-1, 1]$; suppose it is $a + 3\delta^2$. We set $\gamma = \frac{f(a+3\delta^2) - f(a)}{3\delta^2} = \frac{f(a+3\delta^2)}{3\delta^2}$, $P(y) = \gamma(y - a)$. From condition (ii), it follows that

$$|\gamma| \le \frac{A(M + \varepsilon)}{3\delta^2} d(a + 3\delta^2, F) \ln \frac{3}{d(a + 3\delta^2, F)} \le A(M + \varepsilon) \ln \frac{1}{\delta^2}.$$

Applying Proposition 3 to the interval $S = [a - 3\delta^2, a + 3\delta^2] \cap [-1, 1]$ and points $t_1 = a$ and $t_2 = a + 3\delta^2$, we obtain that $\|f - P\|_S \le A\omega_2(f; S; 2\delta^2)$.

Now we use Proposition 6 with $\tau = 2\delta^2$. We then have $\|(f-P)\omega^a_{2\delta^2}\|_\Lambda \le A(M+\varepsilon)$. According to Proposition 5 (in which the role of σ is played by $\rho = \delta^2$, and $d = \delta/2$) there exists a function $\psi \in \Lambda$ with the following properties: $\psi(y) = y - a$ for $|y-a| \le \delta^2$, $y \in [-1, 1]$; $\psi(y) = 0$ for $|y-a| \ge \delta$, $y \in [-1-1]$; $\|\psi\|_\Lambda \le A/\ln(1/\delta^2) = A/\ln(1/\delta)$. We set $g = (f-P)\omega^a_{2\delta^2} + \gamma\psi$. Then $\|g\|_\Lambda \le A(M+\varepsilon)(1 + \frac{\ln(1/\delta^2)}{\ln(1/\delta)}) \le A(M+\varepsilon)$. Further, $g(x) = f(x) - P(x) + \gamma(x-a) = f(x)$. If, however, x' is a point in E different from x, then $|x'-a| \ge |x'-x| - |x-a| \ge 2\delta - \delta^2 > \delta > 2\delta^2$; therefore $g(x') = 0$.

It is clear that the function $f_E = g_1 + g_2 + g_3$ has all the required properties.

2. $3\delta^2 \le x_2 - x_1 < 2\delta$, $x_3 - x_2 \ge 2\delta$.

We put $x = (x_1+x_2)/2$, $\tau = x_2 - x_1$. We apply Proposition 3 to the interval $[a, b] = S = [x - \frac{3}{2}\tau, x + \frac{3}{2}\tau] \cap [-1, 1]$ and the points $t_1 = a_1$, $t_2 = a_2$, having observed beforehand that $t_1, t_2 \in [a, b]$ and $(b-a)/(t_1 - t_2) \le 3\tau/(\tau - 2\delta^2) \le 9$. We have $\|f\|_S \le A\omega_2(f; S; \tau)$. Now using Proposition 6 with $P \equiv 0$, we obtain that $\|f\omega^x_\tau\|_\Lambda \le A(M+\varepsilon)$. Let g be the function constructed in case 1 for the point x_3. Writing $f_E = f\omega^x_\tau + g$, it is easy to see that $f_E(x_1) = f(x_1)$ and $f_E(x_2) = f(x_2)$. Since $x_3 - x_2 > 2\delta > x_2 - x_1$, we have $f_E(x_3) = g(x_3) = f(x_3)$. Moreover, $\|f_E\|_\Lambda \le A(M+\varepsilon)$. Therefore, f_E is the desired function.

3. $x_2 - x_1 < 3\delta^2$, $x_3 - x_2 \ge 2\delta$.

Of the points $a_1 + 12\delta^2$, $a_1 - 12\delta^2$, at least one, say $a_1 + 12\delta^2$, belongs to $[-1, 1]$. We set

$$\gamma = \frac{f(a_1 + 12\delta^2) - f(a_1)}{12\delta^2} = \frac{f(a_1+12\delta^2)}{12\delta^2}, \qquad P(y) = \gamma(y - a_1).$$

As in case 1, it follows from (ii) that $|\gamma| \le A(M+\varepsilon)\ln(1/(4\delta^2))$. Proposition 3 applied to the interval $S = [a_1 - 12\delta^2, a_1 + 12\delta^2] \cap [-1, 1]$ and the points $t_1 = a_1$, $t_2 = a_1 + 12\delta^2$, gives $\|f - P\|_S \le A\omega_2(f; S; 8\delta^2)$; and Proposition 6, with $x = a_1$ and $\tau = 8\delta^2$, leads to the estimate $\|(f-P)\omega^{a_1}_{\gamma\delta^2}\|_\Lambda \le A(M+\varepsilon)$. Proposition 5 guarantees the existence of a function $\psi \in \Lambda$ such that $\psi(y) = y - a_1$ for $|y - a_1| \le 4\delta^2$, $y \in [-1, 1]$; $\psi(y) = 0$ for $|y - a_1| \ge \delta$, $y \in [-1, 1]$; $\|\psi\|_\Lambda \le A/\ln(1/(4\delta))$. Let g be the function constructed as in case 1 for the point x_3. We put $f_E = (f-P)\omega^{a_1}_{\gamma\delta^2} + \gamma\psi + g$. It is easy to check that the function f_E satisfies all the stated conditions.

4. $x_3 - x_2 < 2\delta$.

Assuming that, of the points $a_2 + 9\delta$ and $a_2 - 9\delta$, the first belongs to $[-1, 1]$, we put

$$\gamma = \frac{f(a_2 + 9\delta) - f(a_2)}{9\delta} = \frac{f(a_2 + 9\delta)}{9\delta}, \qquad P(y) = \gamma(y - a_2).$$

From (ii), it follows that $|\gamma| \leq A(M+\varepsilon)\ln(1/(3\delta))$. Using Proposition 3 for the interval $S = [a_2 - 9\delta, a_2 + 9\delta] \cap [-1, 1]$ and the points $t_1 = a_2$, $t_2 = a_2 + 9\delta$, we find that $\|f - P\|_S \leq A\omega_2(f; S; 6\delta)$. Using Proposition 6 with $x = a_2$ and $\tau = 6\delta$, we get the estimate $\|(f - P)\omega_{6\delta}^{a_2}\|_\Lambda \leq A(M+\varepsilon)$. According to Proposition 4, there exists a function $\psi \in \Lambda$ such that $\psi(y) = y - a_2$ for $|y - a_2| \leq 3\delta$, $y \in [-1, 1]$, and $\|\psi\|_\Lambda \leq A/\ln(1/(3\delta))$. Writing $f_E = (f - P)\omega_{6\delta}^{a_2} + \gamma\delta$, it is clear that $\|f_E\|_\Lambda \leq A(M+\varepsilon)$. For every $i = 1, 2, 3$, $|x_i - a_2| \leq |x_i - x_2| + |x_2 - a_2| < 2\delta + \delta^2 < 3\delta$; therefore $f_E(x_i) = f(x_i) - P(x_i) + \gamma(x_i - a_2) = f(x_i)$. Consequently, in case 4 as well, f_E is the required function.

This proves Lemma 1.

On the basis of Lemma 1, taking into account the discussion of Section 2 we conclude that Λ is a D-algebra. Consequently, from Theorem 1 follows

THEOREM 2. $\Lambda \in \mathrm{synt}$.

We prove the statement expressed in the Corollary to Lemma 1.

Let I be a closed ideal, with spectrum F, in the algebra λ. Since $J_F \subset I \subset M_F$, to prove the equality $I = M_F$ it is sufficient to verify the inclusion $M_F \subset J_F$. We take an arbitrary function $f \in M_F$ and fix a number $\varepsilon > 0$. We extend f to the interval $[-2, 2]$ as was done in the Remark to Proposition 1 (as previously, the function thus obtained will be denoted by f). Applying Lemma 1 to $f \in \Lambda([-2, 2])$ (noting that $M = 0$ because $f \in \lambda$), we find that there exist a function $g \in \Lambda([-2, 2])$ and a number $\delta \in (0, 1]$ such that g is equal to zero on a δ-neighborhood of the set F and $\|g - f\|_{\Lambda([-2,2])} \leq \varepsilon$.

Let w be the function introduced in the proof of Proposition 2, and let $w_\rho(x) = \frac{1}{\rho}w(\frac{x}{\rho})$, $x \in \mathbf{R}$, $\rho > 0$. For $u \in \Lambda([-2, 2])$, with $\rho \in (0, 1]$, let us consider the averaged function $u_\rho(x) = \int_{-1}^{1} u(x - t)w_\rho(t)\,dt$, $x \in [-1, 1]$. It is clear that $u_\rho \in C^\infty([-1, 1]) \subset \lambda$ and $\|u_\rho\|_\Lambda \leq \|u\|_{\Lambda([-2,2])}$.

Since $f \in \lambda([-2, 2])$, there exists $\sigma \in (0, 1]$ such that, when $|h| \leq \sigma$, $|\Delta_h^2 f(x)| \leq \varepsilon|h|$ for all $x \in [-2, 2]$. Therefore, for any $\rho \in (0, 1]$, $|h| \leq \sigma$ and $x \in [-1, 1]$, we have $|\Delta_h^2(f - f_\delta)(x)| \leq 2\varepsilon|h|$. Further, there exists $\tau \in (0, 1]$ such that $\|f - f_\rho\|_{[-1,1]} \leq \varepsilon\sigma$ for $\rho \in (0, \tau]$. Then, for $|h| > \sigma$ and $x \in [-1, 1]$, $|\Delta_h^2(f - f_\rho)(x)| \leq 4\|f - f_\rho\|_{[-1,1]} \leq 4\varepsilon\sigma < 4\varepsilon|h|$ for $\rho \in (0, \tau]$. Therefore, for $\rho \in (0, \tau]$, $\|f - f_\rho\|_\Lambda \leq 4\varepsilon$. Moreover, for all $\rho \in (0, 1]$, $\|f_\rho - g_\rho\|_\Lambda = \|(f - g)_\rho\|_\Lambda \leq \|f - g\|_{\Lambda([-2,2])} \leq \varepsilon$. Thus, for $\rho \in (0, \tau]$, $\|f - g_\rho\|_\Lambda \leq \|f - f_\rho\|_\Lambda + \|f_\rho - g_\rho\|_\Lambda \leq 5\varepsilon$, while for $\rho < \delta$ the functions g_ρ are equal to zero in a neighborhood of the set F. Therefore, by virtue of the arbitrariness of ε, $f \in J_F$, which was to be proved.

In conclusion, the author expresses sincere gratitude to Yu. A. Brudnyĭ for discussion of the results obtained and for valuable advice.

REFERENCES

1. M. H. Stone, *Applications of the theory of Boolean rings to general topology*, Trans. Amer. Math. Soc. **41** (1937), 375–481.

2. H. Whitney, *On ideals of differentiable functions*, Amer. J. Math. **70** (1948), 635–658.

3. L. Waelbroeck, *Closed ideals of Lipschitz functions*, Proc. Int. Symp. "Functional Algebras", (F. Birtel, ed.), Scott-Foresmann, Chicago, 1966, pp. 322–325.

4. L. G. Khanin, *Spectral synthesis of ideals in the algebras of functions that have generalized derivatives*, Uspekhi Mat. Nauk **39** (1984), 199–200.

5. G. Glaeser, *Synthese spèctrale des ideaux de fonctions Lipschitziennes*, C. R. Acad. Sci. Paris **260** (1965), 1539–1542.

6. È. È. Shnol′, *Closed ideals in the ring of continuously differentiable functions*, Mat. Sb. **27** (1950), 281–284.

7. L. G. Khanin, *A theorem on spectral synthesis of ideals for a certain class of Banach algebras*, Dokl. Akad. Nauk SSSR **292** (1987), no. 3, 568–571.

8. ——, *Closed ideals and point derivations in some functional algebras*, Yaroslavl′, 1985, 24 pp., Manuscript Presentations by the University of Yaroslavl′, deposed in VINITI Jan. 28, 1985, N.775-85.

9. ——, *Closed ideals in the algebras of smooth functions*, Dissertation of candidate of math.-phys. sciences, Yaroslavl′, 1984, 133 pp.

10. G. E. Shilov, *On regular normed rings*, Trudy Mat. Inst. Steklova, 1947, v. 21.

11. D. R. Sherbert, *The structure of ideals and point derivations in Banach algebras of Lipschitz functions*, Trans. Amer. Math. Soc. **111** (1964), 240–272.

12. A. Zygmund, *Smooth functions*, Duke Math. J. **12** (1945), 47–76.

13. A. Marchaud, *Sur les dérivées et sur les différences des fonctions de variables réelles*, J. Math. Pure Appl. **6** (1927), 337–425.

14. E. J. McShane, *Extension of range of functions*, Bull. Amer. Math. Soc. **40** (1934), 837–842.

15. H. Whitney, *Analytic extensions of differentiable functions defined on closed sets*, Trans. Amer. Math. Soc. **26** (1934), 63–89.

16. L. G. Khanin, *Closed primary ideals and point derivations in Zygmund algebras, Constructive Theory of Functions*, Proc. Intl. Conf. on Constructive Theory of Functions, Publishing House of the Bulgarian Academy of Sciences, Sofia, 1984, pp. 397–402.

17. Yu. A. Brudnyĭ, *Spaces defined with the use of local approximations*, Trudy Moskov. Mat. Obshch **24** (1971), 69–132.

18. P. A. Shvartsman, *On the traces of functions of two variables that satisfy Zygmund's condition. A survey of the theory of functions of several real variables*, Yaroslavl′, 1982, pp. 145–168.

19. ——, *Lipschitz sections of multiple-valued mappings and traces of functions of the Zygmund class on arbitrary compacta*, Dokl. Akad. Nauk **276** (1984), 559–562.

University of Yaroslavl′

Translated by J. AND M. MOLONEY

Amer. Math. Soc. Transl.
(2) Vol. **149**, 1991

K-Theory and Morse Theory. I

V. V. SHARKO

Introduction

The idea of studying smooth manifolds by means of the level curves of functions defined on them, whose singularities have the simplest possible structure, goes back to Poincaré, and even to Möbius (1869). But its systematic development we owe to Marston Morse. Morse observed [33] that the number of critical points of different indices of a smooth function on a manifold can be made use of in order to study the geometric properties of the manifold. He derived certain inequalities connecting the number of critical points with the ranks and torsion orders of the homology groups of the manifold. At various times, the development of this line of thought has been carried forward by Bott [21], Pitcher [34], Reeb [36], Èl′sgol′ts [27], and Chogoshvili [14].

In 1962, Smale showed [37] that on any simply-connected manifold M^n $(n > 5)$ there exists a Morse function with the minimal number of critical points (an exact Morse function). The number of critical points of index λ is

$$N_\lambda = \mu(H_\lambda(M^n, \mathbb{Z})) + \mu(\operatorname{Tors} H_{\lambda-1}(M^n, \mathbb{Z})),$$

where $\mu(H)$ is the minimal number of generators of the group H. In 1965 Barden, using the specific character of the structure of 5-dimensional manifolds, carried over Smale's result to the closed simply-connected 5-dimensional case [20].

In the mid-seventies, progress was made on the construction of special Morse functions on simply-connected 4-dimensional manifolds. For details,

1991 *Mathematics Subject Classification*. Primary 57Q10, 19B28.
Translation of Akad. Nauk Ukrain. SSR Inst. Mat. Preprint **1986**, no. 39; MR 88f: 57057a.

see the work of Akbulut, Kas, Casson, Kirby, Mandelbaum, Rudolph, and Harer, as mentioned in [4].

In the work of Volodin, Kuznetsov, and Fomenko, on the problem of characterizing the standard 3-sphere, Morse functions were made use of in an essential fashion [13].

Bogoyavlenskiĭ [2], Hajduk [28], and Wall [39] have considered various aspects of the existence of exact Morse functions on nonsimply-connected manifolds.

Using results of M. Freedman, Solopko has proved the existence of exact topological Morse functions on smooth simply-connected 5-dimensional cobordisms [11].

Starting in the mid-seventies, the author began a study of the question of constructing minimal Morse functions (definition below) on nonsimply-connected manifolds (see, e.g., [15], [16], [17], [18]). It was shown that for the existence of minimal Morse functions on a manifold with fundamental group isomorphic to π, it is necessary that π be an s-group (meaning: over the ring $\mathbb{Z}[\pi]$, all stably free modules are free). Also relevant is the Whitehead group $\mathrm{Wh}(\pi)$. Since every Morse function on a closed manifold determines a representation of the group π, the construction of minimal Morse functions is closely related to the existence of minimal representations of a group π with the requisite properties. In addition, the 2-type of the manifold also plays a part. For certain groups our information is sufficiently complete to allow a full accounting of the question of existence of a minimal Morse function.

A new impulse was given to Morse theory in 1982 by the work of Novikov [8], which dealt with Morse mappings into a circle. The situation in this case differs qualitatively from the classical one. The work of Farber [12] also goes in this direction.

The present paper is the first of a series in which will be collected the author's results concerning existence of minimal Morse functions on nonsimply-connected manifolds. We consider here the question of constructing minimal Morse functions on nonsimply-connected cobordisms. This is the simplest case; the influence of the fundamental group is in this case at a minimum.

We now briefly describe the paper's contents. Let $\mathscr{F}_\lambda(M^n)$ be the set of Morse functions on the manifold M^n with the minimal possible number of critical points of index λ. A function $f \in \bigcap_{\lambda=0}^n \mathscr{F}_\lambda(M^n)$ is called a *minimal Morse function* on M^n. A large part of the paper is devoted to developing the algebraic apparatus needed for construction of minimal Morse functions. In the first section we examine the behavior of the numerical invariant $\mu(M)$ of a module M—the minimal number of generators — under addition of a direct free summand. We also prove the existence of minimal resolutions.

We call an epimorphism $f : F \to M$ *minimal* if F is a free module of rank $\mu(M)$. If in a free resolution of a module M the epimorphisms onto the kernels are all minimal, the resolution is itself called minimal.

THEOREM 1.7. *Let Λ be an IBN, s-ring,*

$$0 \leftarrow M \overset{g_0}{\leftarrow} G_0 \overset{g_1}{\leftarrow} G_1 \leftarrow \cdots$$

an arbitrary free resolution, and

$$0 \leftarrow M \overset{f_0}{\leftarrow} F_0 \overset{f_1}{\leftarrow} F_1 \leftarrow \cdots$$

a minimal resolution. Then $\mu(F_i) \le \mu(G_i)$.

In the second section we examine the number $f-\operatorname{rank}(N, M)$ and describe its connection with minimal epimorphisms. The results of this section play an essential role in the construction of minimal chain complexes. For any submodule $N \subset M$, $f\text{-}\operatorname{rank}(N, M)$ is defined as the largest integer r such that N contains a direct summand of M isomorphic to $r\Lambda$. We say that $f\text{-}\operatorname{rank}(N, M)$ is *additive* if $f\text{-}\operatorname{rank}(N \oplus k\Lambda, M \oplus k\Lambda) = f\text{-}\operatorname{rank}(N, M) + k$. The following theorem holds.

THEOREM 2.6. *Let Λ be an IBN, s-ring, and $f : F \to M$ an epimorphism of a free module of rank k onto M. In order for the epimorphism f to be minimal, it is necessary and sufficient that $f\text{-}\operatorname{rank}(\operatorname{Ker} f, F) = 0$ and be additive. The kernel of the epimorphism need not be finitely generated.*

The third section is of an auxiliary character. In it we examine briefly the simple-homotopy type theory of chain complexes. We say that the representable dimension of a group π is less than or equal to k ($r\text{-}\dim \pi \le k$) if every element $a \in \operatorname{Wh}(\pi)$ can be realized by a matrix in $\operatorname{GL}(\mathbb{Z}[\pi], k)$. We present some results on $r\text{-}\dim \pi$ for certain groups, and we prove a lemma on matrices. At the end of the section we discuss the semilinear mappings of chain complexes associated with a given ring automorphism $\theta : \Lambda \to \Lambda$.

The fourth is a key section. In it we prove that if Λ is an IBN, s-ring, then in the homotopy type of a free chain complex there always exists a minimal chain complex. A chain complex $\{C_i, \partial_i\}$ is called *minimal in dimension λ* if for any chain complex $\{D_i, d_i\}$ homotopy equivalent to $\{C_i, \partial_i\}$, $\mu(C_\lambda) \le \mu(D_\lambda)$; it is called *minimal* if it is minimal in all dimensions.

PROPOSITION 4.8. *Let $\{C_i, \partial_i\}_\Lambda$ be the class of all chain complexes over Λ. In order that every homotopy type in $\{C_i, \partial_i\}_\Lambda$ contain a minimal chain complex, it is necessary and sufficient that Λ be an IBN, s-ring.*

PROPOSITION 4.9. *Let $\mathbb{Z}[\pi]$ be an s-ring, with $r\text{-}\dim \pi \le k$; and suppose that in the homotopy type of the chain complex $\{C_i, \partial_i\}$ the minimal complex $\{\overline{C}_i, \overline{\partial}_i\}$ is such that $\mu(\overline{C}_j) \le k$ for some j. Then $\{C_i, \partial_i\}$ is simple-homotopy equivalent to the minimal complex.*

The fifth section considers certain chain complex invariants that are "stably" isomorphic for chain complexes from the same homotopy type. These can be used to construct an algorithm for computing the ranks of the chain modules in the minimal chain complex. These invariants arise from module extensions; for that reason a large part of the section is concerned with the latter.

In the final section we construct minimal Morse functions on nonsimply-connected cobordisms. Consider a cobordism (W^n, V_0, V_1) such that $\pi = \pi_1(V_0) \hookrightarrow \pi_1(W^n) \hookleftarrow \pi_1(V_1)$ are isomorphisms, with $n \geq 6$. Bringing in the universal covering space $(\widetilde{W}^n, \widetilde{V}_0, \widetilde{V}_1)$, one constructs [31] for every Morse function f on W^n a chain complex of $\mathbb{Z}[\pi]$-modules $\mathbb{Z} \xleftarrow{\varepsilon} C_0 \xleftarrow{\partial_1} C_1 \leftarrow \cdots \xleftarrow{\partial_n} C_n$; and these are all simple-homotopy equivalent. Furthermore, any chain complex constructed via triangulation or any other method will have the same simple-homotopy type . This is a consequence of the fact that simple-homotopy type is topologically invariant. It is therefore meaningful to speak of the chain complex $C(W^n)$ of the cobordism.

PROPOSITION 6.2. *Let (W^n, V_0, V_1) be a cobordism, $n \geq 6$, and $\pi = \pi_1(V_0) \hookrightarrow \pi_1(W^n) \hookleftarrow \pi_1(V_1)$ an isomorphism. Then for the existence of a minimal Morse function on W^n it is necessary and sufficient that the chain complex $C(W^n)$ be simple-homotopy equivalent to a minimal chain complex.*

Thus, if π is an s-group and $\mathrm{Wh}(\pi) = 0$, then W^n always has a minimal Morse function. This holds in the case of free groups, free abelian groups, and certain cyclic groups $\mathbb{Z}_k$ $(k = 2, 3, 4, 6)$.

PROPOSITION 6.8. *Suppose that either π is a finite abelian group isomorphic to $(\mathbb{Z}_2)^n$; or that every Sylow p-subgroup in π is of the form $\mathbb{Z}_{p^n}$ or $\mathbb{Z}_p \oplus \mathbb{Z}_{p^n}$; or that π is a finite s-group for which $SK_1(\mathbb{Z}[\pi]) = 0$. Then W^n has a minimal Morse function, provided W^n is not an h-cobordism, i.e., provided $H_i(\widetilde{W}^n, \widetilde{V}_0) \neq 0$.*

PROPOSITION 6.9. *Suppose $\pi = \pi_1(W^n)$ is a finite s-group and*

$$\mu(H_i(W^n, V_0, \mathbb{Z})) + \mu(\mathrm{Tors}(H_{i-1}(W^n, V_0, \mathbb{Z}))) \geq 2$$

$(3 \leq i \leq n-2)$. *Then W^n has a minimal Morse function.*

Similar results hold in the case of cobordisms for which the mappings $\pi_1(V_0) \hookrightarrow \pi_1(W^n) \hookleftarrow \pi_1(V_1)$ are epimorphisms. Just as in the case of chain complexes, there exists for minimal Morse functions an algorithm for computing the number of critical points in terms of invariants connected with module extensions. At the end of the section we consider those cases where the number of critical points can be computed by using only homology.

PROPOSITION 6.11. *Let (W^n, V_0, V_1) be a cobordism, with $\pi = \pi_1(V_0) \hookrightarrow \pi_1(W^n) \hookleftarrow \pi_1(V_1)$ isomorphisms, $n \geq 6$, and π an s-group. Suppose W^n has a minimal Morse function, and the homology modules $H_i(\widetilde{W}^n, \widetilde{V}_0, \mathbb{Z})$ have free resolutions*

$$0 \leftarrow H_i(\widetilde{W}^n, \widetilde{V}_0, \mathbb{Z}) \leftarrow G_i^0 \leftarrow G_i^1 \leftarrow 0.$$

Then the number of critical points of index λ for a minimal Morse function is

$$N_\lambda = \mu(F_\lambda^0) + \mu(F_{\lambda-1}^1).$$

Here

$$0 \leftarrow H_i(\widetilde{W}^n, \widetilde{V}_0, \mathbb{Z}) \leftarrow F_i^0 \leftarrow F_i^1 \leftarrow 0$$

is a minimal free resolution of the module $H_i(\widetilde{W}^n, \widetilde{V}_0, \mathbb{Z})$, which under the given conditions always exists.

In concluding, the author has the pleasant duty to express his thanks to S. P. Novikov and A. T. Fomenko for their stimulating discussions of various questions in Morse theory. He is also sincerely grateful to A. A. Suslin for helpful advice and critical observations in connection with the algebraic part of this work.

§1. Stably free modules and minimal resolutions

Let Λ be an associative ring with unity. All Λ-modules will be taken to be left modules. Let M be a Λ-module. A family $\{m_i \in I\}$ that generates M is called a system of generators for M. If the direct sum of the mappings $m_i\Lambda \to M$ gives an isomorphism

$$f : \bigoplus \sum_i m_i\Lambda \to M,$$

then $\{m_i\}$ is called a basis for M, and the cardinality of the index set I, a basis number. The basis number depends in general on the choice of a basis, and can therefore not be taken as an invariant of the module $M = \bigoplus \sum_i m_i\Lambda$. A ring Λ such that the basis number of a free module is uniquely determined is called an IBN-ring (ring with Invariant Basis Number). It is well known that if there exists a nontrivial homomorphism of Λ into a commutative ring, then Λ is an IBN-ring. In particular, every integral group ring $\mathbb{Z}[\pi]$ is an IBN-ring, since it has the augmentation homomorphism $\varepsilon : \mathbb{Z}[\pi] \to \mathbb{Z}$; Noetherian rings are also IBN-rings.

We recall that a module M is called *stably free* if the direct sum of M with some free module is free (the modules are assumed to be finitely generated).

DEFINITION 1.1. A ring Λ is called an s-ring if every stably free module over Λ is a free module.

In our arguments, s-rings will play an important role. We note that the question of whether a given ring Λ is an s-ring is in general a complicated one. The well-known Serre conjecture, settled not so long ago in the affirmative independently by Suslin [10] and Quillen [35], amounted to the question: is every polynomial ring over a field an s-ring?

We list here some groups for which it is known that the corresponding integral group ring is an s-ring; such groups will be called s-groups. A theorem of Jacobinski asserts that if a finite group has no epimorphic mapping onto the generalized quaternion group Q^n, the binary tetrahedral group T, the octahedral $\mathscr{O}$, or the icosahedral I, it is an s-group. Consequently, among the finite groups the abelian, the simple, and those of odd order are examples of s-groups. Free groups and free abelian groups are also s-groups. (See [1], [29], [36].)

Throughout this paper we denote by $\mu(M)$ the minimal number of generators of a module M. We show that the behavior of $\mu(M)$ for projective modules is closely related with the property of the ring being an s-ring.

LEMMA 1.2. *Let Λ be an IBN, s-ring, and let M be a Λ-module. Then* $\mu(M \oplus \Lambda) = \mu(M) + 1$.

PROOF. We are asserting the truth of both of the following inequalities:

$$\mu(M \oplus \Lambda) \leq \mu(M) + 1,$$
$$\mu(M \oplus \Lambda) \geq \mu(M) + 1.$$

The first is obvious. We prove the second. Let $\gamma_1, \dots, \gamma_k$ be a minimal system of generators for $\Gamma = M \oplus \Lambda$, F a free module of rank k, $g : F \to \Gamma$ the natural epimorphism, and $p : M \oplus \Lambda \to \Lambda$ the projection onto the second summand. Consider the composition $F \xrightarrow{g} M \oplus \Lambda \xrightarrow{p} \Lambda$. We have the splitting

$$\begin{array}{ccccc} \overline{F} & \xrightarrow{\overline{g}} & M & \longrightarrow & 0 \\ \oplus & & \oplus & & \\ \Lambda & \xrightarrow{i} & \Lambda & \longrightarrow & \Lambda \end{array}$$

where $\overline{F} \oplus \Lambda = F$, $\overline{g} \oplus i = g$. Because of the assumptions on Λ, the module $\overline{F}$ is free; consequently, $\mu(\Gamma) = \mu(\overline{F} \oplus \Lambda) = \mu(\overline{F}) + 1 = k$. Since $\overline{g} : \overline{F} \to M$ is an epimorphism, we have $\mu(\overline{F}) \geq \mu(M)$. Therefore $\mu(\Gamma) = \mu(\overline{F}) + 1 \geq \mu(M) + 1$, as required. This proves the lemma.

COROLLARY 1.3. *Let $\Gamma = M \oplus F$, where F is a free module of rank k. Then $\mu(\Gamma) = \mu(M) + k$.*

LEMMA 1.4. *Let Λ be an IBN-ring. In order for Λ to be an s-ring, it is necessary and sufficient that for an arbitrary projective module P over Λ,* $\mu(P) + 1 = \mu(P \oplus \Lambda)$.

PROOF. Necessity follows from the preceding lemma. We prove sufficiency. If $\mu(P \oplus \Lambda) = \mu(P) + 1$ for a projective module P, then also $\mu(P \oplus n\Lambda) = \mu(P) + n$. Let P be a stably free module. Then there exists a k such that $P \oplus k\Lambda$ is free. Select a minimal number of generators $p_1, \dots, p_l$ for P, and let $e_1, \dots, e_k$ be a basis for $k\Lambda = \Lambda \oplus \cdots \oplus \Lambda$. Then $p_1, \dots, p_l, e_1, \dots, e_k$ are generators for the module $P \oplus k\Lambda = F$. We construct an epimorphism $f : F \to F$ by putting $f(f_i) = p_i$ $(i = 1, \dots, l)$ and $f(f_{k+j}) = e_j$ $(j = 1, \dots, k)$, where $f_1, \dots, f_{k+l}$ is a basis for F. If $R = \operatorname{Ker} f$, then since F is free we have the splitting $R \oplus F = F$. Therefore $\mu(R) + \mu(F) = \mu(F)$, so $R = 0$, and f is an isomorphism. Let $\overline{F}$ be the free submodule of F generated by the subset $f_1, \dots, f_l$ of the basis; then $f(\overline{F}) = P$ is free. This proves the lemma.

The rest of this section has to do with the study of free resolutions for modules. The behavior of the invariant $\mu(M)$ allows us to prove the existence of minimal free resolutions.

DEFINITION 1.5. An epimorphism $f : F \to M$ is called minimal if F is a free module of rank $\mu(M)$.

DEFINITION 1.6. Let Λ be an IBN-ring, and

$$0 \leftarrow M \xleftarrow{f_0} F_0 \xleftarrow{f_1} F_1 \leftarrow \cdots$$

a free resolution of the Λ-module M, where modules F_i are finitely generated. We say that the resolution is minimal if every $f_i : F_i \to \operatorname{Ker} f_{i-1}$ is a minimal epimorphism.

The following theorem indicates the reason for the term "minimal" here.

THEOREM 1.7. *Let Λ be an IBN, s-ring;*

$$0 \leftarrow M \xleftarrow{g_0} G_0 \xleftarrow{g_1} G_1 \leftarrow \cdots$$

an arbitrary free resolution of the Λ-module M, where the modules G_i are finitely generated, and

$$0 \leftarrow M \xleftarrow{f_0} F_0 \xleftarrow{f_1} F_1 \leftarrow \cdots$$

a minimal resolution of M. Then $\mu(F_i) \le \mu(G_i)$.

PROOF. Consider the segments

$$\begin{array}{ccccccc} & & G_0 & \longleftarrow & \operatorname{Ker} g_0 & \longleftarrow & 0 \\ & \swarrow^{g_0} & & & & & \\ 0 \longleftarrow & M & & & & & \\ & \nwarrow_{f_0} & & & & & \\ & & F_0 & \longleftarrow & \operatorname{Ker} f_0 & \longleftarrow & 0 \end{array}$$

of the resolutions. By Schanuel's lemma [1],

$$G_0 \oplus \operatorname{Ker} f_0 = F_0 \oplus \operatorname{Ker} g_0.$$

By Corollary 1.3,

$$\mu(G_0 \oplus \operatorname{Ker} f_0) = \mu(F_0 \oplus \operatorname{Ker} g_0) = \mu(G_0) + \mu(\operatorname{Ker} f_0) = \mu(F_0) + \mu(\operatorname{Ker} g_0).$$

By hypothesis, $\mu(G_0) \ge \mu(F_0)$. Therefore $\mu(\operatorname{Ker} f_0) \le \mu(\operatorname{Ker} g_0)$, and so $\mu(F_1) \le \mu(G_1)$. Since $\mu(\operatorname{Ker} f_0) = \mu(F_1)$, we have

$$\mu(G_0) + \mu(F_1) \le \mu(F_0) + \mu(G_1). \tag{1}$$

Now consider the segments

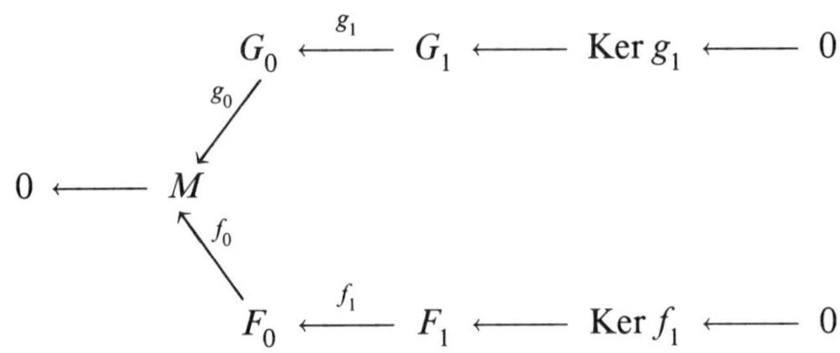

and apply again Schanuel's lemma. We have

$$\mu(G_0) + \mu(F_1) + \mu(\operatorname{Ker} g_1) = \mu(F_0) + \mu(G_1) + \mu(\operatorname{Ker} f_1).$$

Using the inequality (1), it is easily shown that $\mu(F_2) \leq \mu(G_2)$. Then since $\mu(\operatorname{Ker} f_1) = \mu(F_2)$, we have the inequality

$$\mu(G_0) + \mu(F_1) + \mu(G_2) \geq \mu(F_0) + \mu(G_1) + \mu(F_2),$$

which we now use for the next step of the proof. Repeating the argument indefinitely, we obtain the proof of the theorem.

Observe that if over a ring Λ there exist stably free modules that are not free, the theorem is false. An example is given by Swan in [38]. We reproduce it. Let $\pi = \{a, b; aba^{-1}b, b^8a^{-2}\}$ be the generalized quaternion group of order 32, and $\mathbb{Z}[\pi]$ its integral group ring. Consider the ideal generated by the element $a+4$; denote it by $\mathscr{T}$. Let $M = \mathbb{Z}[\pi]/\mathscr{T}$. Swan showed that M has a free resolution of the form

$$0 \leftarrow M \xleftarrow{p} \mathbb{Z}[\pi] \xleftarrow{i} \mathbb{Z}[\pi] \leftarrow 0,$$

which is obviously minimal. At the same time, there exists an epimorphism $f : \mathbb{Z}[\pi] \to M$ such that the ideal $K = \operatorname{Ker} f$ is not a free $\mathbb{Z}[\pi]$-module, $\mu(K) = 2$, and $K \oplus \mathbb{Z}[\pi] \approx \mathbb{Z}[\pi] \oplus \mathbb{Z}[\pi]$. It follows that the resolution

$$0 \leftarrow M \xleftarrow{f} \mathbb{Z}[\pi] \leftarrow \mathbb{Z}[\pi] \oplus \mathbb{Z}[\pi] \leftarrow \mathbb{Z}[\pi] \leftarrow 0,$$

of length 3, is minimal, but does not satisfy the conclusion of the theorem.

Definition 1.8. Let $f : F \to M$ be an epimorphism, where F is a free module. We say that from f can be extracted an epimorphism $\overline{f} : \overline{F} \to M$ if there exists a decomposition

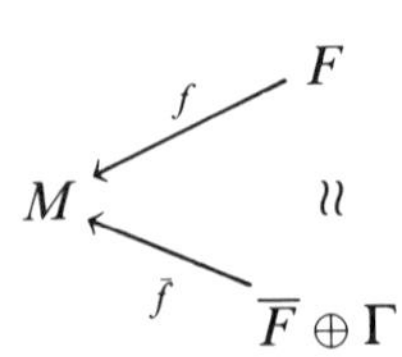

with $\overline{f}(\overline{F}) = M$, $\overline{F}(\Gamma) = 0$.

DEFINITION 1.9. By the thickening of a homomorphism $f : F \to M$ via a module G is meant the homomorphism

$$\hat{f} : F \oplus G \to M, \quad \hat{f}\Big|_{F\oplus 0} = f, \quad \hat{f}\Big|_{0\oplus G} = 0.$$

LEMMA 1.10. *Let $g : G \to M$ be an epimorphism, where G is a free module, and $f : F \to M$ an arbitrary epimorphism. Then the mapping f can be extracted from the thickening of g via F.*

The point of this lemma is that, in general, from an arbitrary epimorphism $g : G \to M$ cannot be extracted, for example, a minimal epimorphism, but by enlarging G this can be done.

PROOF. Let $g_1, \dots, g_k$ be a basis for G. Then $g(g_1), \dots, g(g_k)$ is a system of generators for M. Similarly, let $f_1, \dots, f_l$ be a basis for F; then $f(f_1), \dots, f(f_l)$ is also a system of generators for M. Obviously $g_1, \dots, g_k, f_1, \dots, f_l$ is a basis for the module $G \oplus F$. Consider the epimorphism $\hat{g} : G \oplus F \to M$, $\hat{g}\big|_{G\oplus 0} = g$, $\hat{g}(0 \oplus F) = 0$. We shall find a new basis for $G \oplus F$ in which k basis elements are contained in the kernel of $\hat{g}$. Let $\tilde{f}_i = f_i + h_i$ $(i = 1, \dots, l)$, where $h_i \in g^{-1}(f(f_i))$; then obviously $g_1, \dots, g_k, \tilde{f}_1, \dots, \tilde{f}_l$ also form a basis for $G \oplus F$. Let $\hat{F}$ be the submodule generated by $\tilde{f}_1, \dots, \tilde{f}_l$. Clearly, $\hat{g}\big|_{0\oplus\hat{F}}$ is an epimorphism onto M. Consequently, there exist elements $s_1, \dots, s_k$ in $0 \oplus \hat{F}$ such that $\hat{g}(s_j) = g(g_j)$ $(j = 1, \dots, k)$. Consider in $G \oplus F$ the basis $g_1 - s_1, \dots, g_k - s_k, \tilde{f}_1, \dots, \tilde{f}_l$. Its first k elements generate a free direct summand $\hat{G}$ in $G \oplus F$ isomorphic to G, and $\hat{g}(\hat{G}) = 0$. Thus, we have extracted f from $\hat{g}$. This proves the lemma.

From the proof it is evident that the transition matrix from the basis $\{g_i, f_j\}$ to the basis $\{\tilde{g}_i, \tilde{f}_j\}$ in the module $G \oplus F$ is representable as a product of elementary matrices; i.e., is 0 in the group $K_1(\Lambda)$.

COROLLARY 1.11. *Let $g : G \to M$ and $f : F \to M$ be epimorphisms, where G and F are free modules. Thickening g via F, and f via G, one can obtain a commutative diagram*

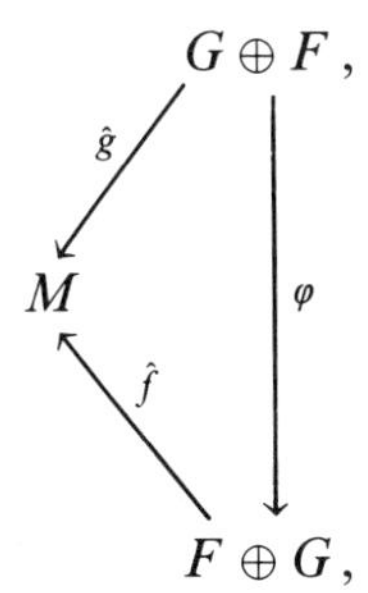

$$\hat{g}\Big|_{G\oplus 0} = g, \quad \hat{g}(0 \oplus G) = 0,$$

$$\hat{f}\Big|_{F\oplus 0} = f, \quad \hat{f}(0 \oplus G) = 0,$$

with φ an isomorphism.

PROOF. Using the lemma, extract from $\hat{g}$ the epimorphism f, and then, using bases, define the isomorphism φ.

DEFINITION 1.12. If $f : F \to G$ is a homomorphism, the *stabilization* of f via a module Γ is the homomorphism

$$f^s : F \oplus \Gamma \overset{f\oplus \mathrm{id}}{\to} G \oplus \Gamma.$$

The operation inverse to stabilization is called *contraction.*

THEOREM 1.13. *Let Λ be an IBN, s-ring, and*

$$0 \leftarrow M \overset{g_0}{\leftarrow} G_0 \overset{g_1}{\leftarrow} G_1 \leftarrow \cdots$$

an arbitrary free resolution of a Λ-module M, where the modules G_i are finitely generated. Then by stabilization of the g_i $(i \geq 1)$ one can extract from it as a direct summand an arbitrary minimal resolution.

PROOF. Consider an arbitrary minimal resolution of M:

$$0 \leftarrow M \overset{f_0}{\leftarrow} F_0 \overset{f_1}{\leftarrow} F_1 \leftarrow \cdots .$$

Stabilizing g_1 via F_0, we obtain

$$\begin{array}{ccccccccc} & & 0 & \longleftarrow & F_0 & \overset{\mathrm{id}}{\longleftarrow} & F_0 & \longleftarrow & 0 \\ & & & & \oplus & & \oplus & & \\ 0 & \longleftarrow & M & \longleftarrow & G_0 & \longleftarrow & G_1 & \longleftarrow & \cdots . \end{array} \tag{1}$$

By Lemma 1.10, from $\hat{g}_0 = g_0 \oplus 0$ we can extract f_0; after which (1) can be written as

$$\begin{array}{ccccccccccc} & & 0 & \longleftarrow & \hat{G}_0 & \longleftarrow & \hat{G}_0 & \longleftarrow & 0 & & \\ & & & & \oplus & & \oplus & & & & \\ 0 & \longleftarrow & m & \overset{f_0}{\longleftarrow} & \hat{F}_0 & \overset{\hat{g}_1}{\longleftarrow} & \hat{G}_1 & \longleftarrow & G_2 & \longleftarrow & \cdots . \end{array} \tag{2}$$

Stabilizing each g_i $(i \geq 2)$ step by step, we can successively extract from the initial resolution the given minimal resolution. This proves the theorem.

We note that minimal resolutions have been studied by a number of authors [3], [26].

§2. f-rank(N, M) and minimal epimorphisms

Let M be a Λ-module, and $N \subset M$ a submodule of M. Following Bass [1], we define f-rank(N, M) as the largest nonnegative integer r such that N contains a direct summand of M isomorphic to $r\Lambda = \Lambda \oplus \cdots \oplus \Lambda$. Let N be a submodule of M, with f-rank$(N, M) = k$. Then obviously M can be represented in the form $M = M' \oplus k\Lambda$; and therefore N can also be written as a direct sum $N = N' \oplus k\Lambda$, where $N' \subset M'$.

It is easily shown that $f\text{-rank}(N, M) = f\text{-rank}(N, M \oplus k\Lambda)$.

DEFINITION 2.1. Let $N \subset M$ be a submodule of a module M. We say that $f\text{-rank}(N, M)$ is additive if

$$f\text{-rank}(N \oplus n\Lambda, M \oplus n\Lambda) = f\text{-rank}(N, M) + n.$$

LEMMA 2.2. *Let Λ be an IBN-ring. Suppose N is a submodule, possibly infinitely generated, of a finitely generated module M. Then there exists an n_0 such that for all $n \geq n_0$, $f\text{-rank}(N \oplus n\Lambda, M \oplus n\Lambda)$ is additive.*

PROOF. Let $f\text{-rank}(N \oplus n\Lambda, M \oplus n\Lambda) = k_n$. Obviously $\mu(M \oplus n\Lambda) \leq \mu(M) + n$, and $k_n \geq n$ for all n. The module $N \oplus n\Lambda$ can be represented in the form $N \oplus n\Lambda = N_{k_n} \oplus k_n\Lambda$, where $k_n\Lambda$ is the direct summand singled out in $M \oplus n\Lambda$. If $f\text{-rank}(N \oplus n\Lambda, M \oplus n\Lambda)$ does not become additive with the growth of n, then the difference $k_n - n$ must increase unboundedly. Consequently, there exists an $\overline{n}$ such that $k_{\overline{n}} - \overline{n}$ is greater than $\mu(M)$, which is impossible. This gives a contradiction, and the lemma is proved.

LEMMA 2.3. *Let Λ be an IBN-ring, and N a submodule, possibly infinitely generated, of a module M, with $f\text{-rank}(N, M) = k$ and additive. Let N be represented in the form $N = N' \oplus k\Lambda$, where $N' \subset M'$ and $M = M' \oplus k\Lambda$. Then $f\text{-rank}(N', M') = 0$ and is likewise additive.*

PROOF. That $f\text{-rank}(N', M') = 0$ is obvious; otherwise $f\text{-rank}(N, M)$ would be greater than k, which is impossible. We show that $f\text{-rank}(N', M')$ is additive. Assume the contrary: that $f\text{-rank}(N', M')$ is not additive. Then there exists an $n > k$ such that $f\text{-rank}(N' \oplus n\Lambda, M' \oplus n\Lambda) = \overline{n} > k$. From the decompositions

$$N' \oplus n\Lambda = N' \oplus k\Lambda \oplus (n-k)\Lambda = N \oplus (n-k)\Lambda,$$
$$M' \oplus n\Lambda = M' \oplus k\Lambda \oplus (n-k)\Lambda = M \oplus (n-k)\Lambda,$$

we obtain

$$\overline{n} = f\text{-rank}(N' \oplus n\Lambda, M' \oplus n\Lambda) = f\text{-rank}(N \oplus (n-k)\Lambda, M \oplus (n-k)\Lambda).$$

But since $f\text{-rank}(N, M) = k$ and is additive, we have

$$f\text{-rank}(N \oplus (n-k)\Lambda, M \oplus (n-k)\Lambda) = k + n - k = n.$$

Consequently, $\overline{n} = n$. This gives a contradiction, and the lemma is proved.

We show now that there is a close relation between minimal epimorphisms and $f\text{-rank}(N, M)$.

LEMMA 2.4. *Let Λ be an IBN, s-ring, and N a submodule, not necessarily finitely generated, of a free module G, with $\mu(G) = k$. Suppose $p : G \to G/N$ is a minimal epimorphism. Then $f\text{-rank}(N, G) = 0$ and is additive.*

PROOF. We prove first that $f\text{-rank}(N, G) = 0$. Assume the contrary: $f\text{-rank}(N, G) = k > 0$. Consider a decomposition of the module N in the form $N = N' \oplus k\Lambda$, where $N' \subset G'$ and $G = G' \oplus k\Lambda$. The assumptions on the ring imply that G' is free. From the construction it follows that $G/N = G'/N'$. But $\mu(G) > \mu(G')$; so the epimorphism p is not minimal, and we have a contradiction.

We show now that f-rank(N, G) is additive. Assume the contrary. Then there exists an n such that $f\text{-rank}(N \oplus n\Lambda, G \oplus n\Lambda) = \overline{n} > n$. Obviously $G/N = G \oplus n\Lambda/N \oplus n\Lambda = \Gamma$. Consider a decomposition of the module $N \oplus n\Lambda$ of the form

$$N \oplus n\Lambda = \overline{N} \oplus \overline{n}\Lambda, \qquad \overline{N} \subset \overline{G}, \qquad \overline{G} \oplus \overline{n}\Lambda = G \oplus n\Lambda.$$

Since $\mu(G \oplus n\Lambda) = \mu(G) + n = \mu(\overline{G}) + \overline{n}$, we have $\mu(G) > \mu(\overline{G})$. By construction, $G \oplus n\Lambda/N \oplus n\Lambda = \Gamma = \overline{G} \oplus \overline{n}\Lambda/\overline{N} \oplus \overline{n}\Lambda = \overline{G}/\overline{N}$. But $\mu(G) = \mu(\Gamma)$, since $p : G \to G/N = \Gamma$ is minimal; and therefore $\mu(\overline{G}) \geq \mu(\Gamma) = \mu(G)$. This gives a contradiction, and the lemma is proved.

The converse assertion is also true.

LEMMA 2.5. *Let Λ be an IBN, s-ring, and N a submodule of a free module of* rank k *(N need not be finitely generated). If $f\text{-rank}(N, G) = 0$ and is additive, then $p : G \to G/N$ is a minimal epimorphism.*

PROOF. Assume the contrary: that p is not minimal. Consider the sequence

$$0 \to N \xrightarrow{i} G \xrightarrow{p} G/N \to 0,$$

and stabilize i via a free module F such that $\mu(F) = \mu(G/N)$. The assumptions on N imply that $f\text{-rank}(N \oplus F, G \oplus F) = \mu(F)$. Now consider the exact sequence

$$0 \to N \oplus F \xrightarrow{i \oplus \mathrm{id}} G \oplus F \xrightarrow{\hat{p}} G/N \to 0.$$

Using Lemma 1.10, extract from $\hat{p}$ the minimal epimorphism $f : F \to G/N$ and construct the exact sequence

$$0 \to \operatorname{Ker} f \oplus \hat{G} \xrightarrow{i \oplus \mathrm{id}} F \oplus \hat{G} \xrightarrow{f \oplus 0} G/N \to 0.$$

By construction,

$$\begin{aligned} f\text{-rank}(\operatorname{Ker} f \oplus \hat{G}, F \oplus \hat{G}) &= f\text{-rank}(\operatorname{Ker} f \oplus G, F \oplus G) \geq \mu(\hat{G}) \\ &= \mu(G) > \mu(F). \end{aligned}$$

But $\operatorname{Ker} f \oplus \hat{G} = N \oplus F$; and therefore $f\text{-rank}(N, G)$ is not additive. This gives a contradiction, and the lemma is proved.

We have now immediately, from these lemmas, the following theorem.

THEOREM 2.6. *Let Λ be an IBN, s-ring, and $f : F \to M$ an epimorphism of a free module F of rank k onto M. In order that f be a minimal epimorphism, it is necessary and sufficient that $f\text{-rank}(\operatorname{Ker} f, F) = 0$ and be additive.*

We note that the kernel of f can be either finitely or infinitely generated. The properties of f-rank(N, M) described in this section will play an essential role in the proof of the existence of minimal chain complexes.

§3. Simple-homotopy type of chain complexes

We suppose throughout that Λ is an IBN-ring. A chain complex is a sequence of free left Λ-modules and homomorphisms

$$\{C_i, \partial_i\} : 0 \leftarrow C_0 \overset{\partial_1}{\leftarrow} C_1 \leftarrow \cdots \overset{\partial_n}{\leftarrow} C_n \leftarrow 0,$$

such that $\partial_i \cdot \partial_{i+1} = 0$. If we choose a fixed basis for each module C_i, we say that we are dealing with a *based* chain complex.

We denote by $\mathrm{GL}(\Lambda)$ the full linear group over Λ, and by $E(\Lambda) \subset \mathrm{GL}(\Lambda)$ the subgroup generated by all elementary matrices. The factor-group $K_1(\Lambda) = \mathrm{GL}(\Lambda)/E(\Lambda)$ is called [7] the Whitehead group of the ring Λ. For every unit u of Λ, $[u]$ denotes the element of the group $K_1(\Lambda)$ corresponding to the 1×1 matrix (u). We eliminate the element $[-1]$ by passing to the factor-group $\overline{K}_1(\Lambda) = K_1(\Lambda)/\{0, [-1]\}$, called the reduced Whitehead group. If $\Lambda = \mathbb{Z}[\pi]$ for some multiplicative group π, then every element of π can be represented as a 1×1 matrix. Thus, we obtain a homomorphism $i : \pi \to \overline{K}_1(\mathbb{Z}[\pi])$. The factor-group $\mathrm{Wh}\,\pi = \overline{K}_1(\mathbb{Z}[\pi])/i(\pi)$ is called the Whitehead group of the group π.

DEFINITION 3.1. (see [25]). We say that the representable dimension of the group π is less than or equal to i (r-$\dim \pi \leq i$) if every element of $\mathrm{Wh}(\pi)$ can be realized as a matrix in $\mathrm{GL}(\mathbb{Z}[\pi], i)$.

We list here some known results on r-$\dim \pi$.

If π is finite, then r-$\dim \pi \leq 2$. (See [1].)

If π is abelian, then r-$\dim \pi \leq 2$. (Communicated to the author by A. A. Suslin.)

If π is finite abelian, then r-$\dim \pi \leq 1$ if and only if either $\pi \approx (\mathbb{Z}_2)^n$, or every Sylow p-subgroup of π is of the form $\mathbb{Z}_{p^n}$ or $\mathbb{Z}_p \oplus \mathbb{Z}_{p^n}$. (See [33].)

If π is finite (nonabelian), then r-$\dim \pi \leq 1$ if and only if π has no epimorphic mapping onto the generalized quaternion group, the binary tetrahedral, octahedral, or icosahedral groups, or the following groups: $\mathbb{Z}_{p^2} \oplus \mathbb{Z}_{p^2}$, $\mathbb{Z}_p \oplus \mathbb{Z}_p \oplus \mathbb{Z}_p$, $\mathbb{Z}_p \oplus \mathbb{Z}_2 \oplus \mathbb{Z}_2 \oplus \mathbb{Z}_2$, $\mathbb{Z}_4 \oplus \mathbb{Z}_2 \oplus \mathbb{Z}_2$, $\mathbb{Z}_4 \oplus \mathbb{Z}_4$ for p a prime. In particular, r-$\dim \pi \leq 1$ for all finite simple groups. The author knows of no examples of finitely representable groups for which r-$\dim \pi > 2$. On the other hand, it is unknown whether for a group π there exists an integer n such that r-$\dim \pi < n$.

We shall need the following fact.

LEMMA 3.2. *Let A be a matrix in the group $\mathrm{GL}(\Lambda, n)$, and let B be a matrix of dimension $n \times m$. Then the matrix*

$$\begin{vmatrix} A & 0 \\ 0 & B \end{vmatrix} = A \oplus B$$

can be reduced by elementary transformations to the form

$$\begin{vmatrix} E & 0 \\ 0 & \overline{B} \end{vmatrix},$$

where E *is the* $n \times n$ *identity matrix.*

PROOF. Consider the matrix

$$\begin{vmatrix} a_{11} & \cdots & a_{1n} & & & \\ \vdots & & \vdots & & 0 & \\ a_{n1} & \cdots & a_{nn} & & & \\ & & & b_{11} & \cdots & b_{1n} \\ & 0 & & \vdots & & \vdots \\ & & & b_{m1} & \cdots & b_{mn} \end{vmatrix} = A \oplus B.$$

Since the row $(a_{11}, \dots, a_{1n})$ is unimodular, there exists a linear combination of the columns of A such that $A \oplus B$ takes the form

$$\begin{vmatrix} a_{11} & \cdots & a_{1n} & 1 & & & \\ a_{21} & \cdots & a_{2n} & a_2 & & 0 & \\ \vdots & & \vdots & \vdots & & & \\ a_{n1} & \cdots & a_{nn} & a_n & & & \\ & & & & b_{11} & \cdots & b_{1n} \\ & 0 & & & \vdots & & \vdots \\ & & & & b_{m1} & \cdots & b_{mn} \end{vmatrix}.$$

Using elementary transformations on the first n rows, we can "kill" $a_2, \dots,$ a_n; i.e., obtain a matrix

$$\begin{vmatrix} a_{11} & \cdots & a_{1n} & 1 & & & \\ \tilde{a}_{21} & \cdots & \tilde{a}_{2n} & 0 & & 0 & \\ \vdots & & \vdots & \vdots & & & \\ \tilde{a}_{n1} & \cdots & \tilde{a}_{nn} & 0 & & & \\ & & & & b_{11} & \cdots & b_{1n} \\ & 0 & & & \vdots & & \vdots \\ & & & & b_{m1} & \cdots & b_{mn} \end{vmatrix}.$$

Observe that the submatrix

$$\begin{vmatrix} a_{11} & \cdots & a_{1n} \\ \tilde{a}_{21} & \cdots & \tilde{a}_{2n} \\ \vdots & & \vdots \\ \tilde{a}_{n1} & \cdots & \tilde{a}_{nn} \end{vmatrix}$$

still belongs to the group $\mathrm{GL}(\Lambda, n)$. Another sequence of column transformations brings the matrix to the form

$$\begin{vmatrix} a_{11} & \cdots & a_{1n} & 1 & \hat{a}_1 & & \\ \tilde{a}_{21} & \cdots & \tilde{a}_{2n} & 0 & 1 & & \mathbf{0} \\ \vdots & & \vdots & \vdots & & & \\ \tilde{a}_{n1} & \cdots & \tilde{a}_{nn} & 0 & \hat{a}_n & & \\ & & & b_{11} & & \cdots & b_{1n} \\ & \mathbf{0} & & \vdots & & & \vdots \\ & & & b_{m1} & & \cdots & b_{mn} \end{vmatrix}.$$

Using the second row, kill the $\hat{a}_i$ $(i = 1, 3, \ldots, n)$, obtaining

$$\begin{vmatrix} \hat{a}_{11} & \cdots & \hat{a}_{1n} & 1 & 0 & & \\ \tilde{a}_{21} & \cdots & \tilde{a}_{2n} & 0 & 1 & \mathbf{0} & \\ \vdots & & \vdots & \vdots & & & \\ \hat{a}_{n1} & \cdots & \hat{a}_{nn} & 0 & 0 & & \\ & & & b_{11} & & \cdots & b_{1n} \\ & \mathbf{0} & & \vdots & & & \vdots \\ & & & b_{m1} & & \cdots & b_{mn} \end{vmatrix}.$$

After n such steps, we arrive at a matrix

$$\begin{vmatrix} \overline{a}_{11} & \cdots & \overline{a}_{1n} & 1 & & \mathbf{0} \\ \vdots & & \vdots & & \ddots & \\ \overline{a}_{n1} & \cdots & \overline{a}_{nn} & \mathbf{0} & & 1 \\ & & & b_{11} & \cdots & b_{1n} \\ & \mathbf{0} & & \vdots & & \vdots \\ & & & b_{m1} & \cdots & b_{mn} \end{vmatrix}.$$

Now, using the 1's, kill the $\overline{a}_{ij}$ and b_{kj}, to get a matrix

$$\begin{vmatrix} 0 & E \\ \overline{B} & 0 \end{vmatrix}.$$

Finally, permuting the columns, we obtain the form desired; and the lemma is proved.

We recall the definition of the Whitehead torsion of an acyclic chain complex

$$\{C_i, \partial_i\} : 0 \leftarrow C_0 \stackrel{\partial_1}{\leftarrow} C_1 \stackrel{\partial_2}{\leftarrow} \cdots \stackrel{\partial_n}{\leftarrow} C_n \leftarrow 0.$$

Since $\{C_i, \partial_i\}$ is acyclic, there exists a deformation operator $\eta_i : C_i \to C_{i+1}$ such that $\eta_{i-1} \cdot \partial_i + \partial_{i+1} \cdot \eta_i = 1_{C_i}$. Putting $\delta_i = \eta_i \cdot \partial_{i+1} \cdot \eta_i$, form the automorphism $d_i = \partial_i + \delta_i$. Consider the modules

$$D_1 = \bigoplus \sum C_{2i}, \qquad D_2 = \bigoplus \sum C_{2i+1}.$$

There is induced an isomorphism $\sum d_i = d$, $0 \to D_1 \xrightarrow{d} D_2 \to 0$. The value $\tau(d)$ in $K_1(\Lambda)$ (or $\mathrm{Wh}\,\pi$) is called the Whitehead torsion of the chain complex $\{C_i, \partial_i\}$ and is denoted by $\tau(C_i, \partial_i)$. It is independent of the choice of the deformation operator.

Given a homotopy equivalence $f = \{f_i\}$ between two based chain complexes $\{C_i, \partial_i\}$, $\{D_i, d_i\}$, consider the mapping cylinder $M(f)$. The torsion $\tau(M(f))$ is the torsion of the homotopy equivalence $f = \{f_i\}$.

In particular, if the f_i are isomorphisms, then [22]

$$\tau(f) = \sum_{i=0}^{n} (-1)^{i+1} \tau(f_i).$$

If $\tau(f) = 0$, f is called a simple-homotopy equivalence.

We recall that a stabilization in dimension k of a based chain complex

$$0 \leftarrow C_0 \xleftarrow{\partial_1} C_1 \leftarrow \cdots \xleftarrow{\partial_n} C_n \leftarrow 0$$

is a based chain complex

$$\begin{array}{ccccccccccccccc} & & & & & & 0 & \longleftarrow & n\Lambda & \xleftarrow{\mathrm{id}} & n\Lambda & \longleftarrow & 0 & & \\ & & & & & & & & \oplus & & \oplus & & & & \\ 0 & \longleftarrow & C_0 & \xleftarrow{\partial_1} & C_1 & \longleftarrow & \cdots & \xleftarrow{\partial_{k-1}} & C_{k-1} & \xleftarrow{\partial_k} & C_k & \longleftarrow & \cdots & \xleftarrow{\partial_n} & C_n \longleftarrow 0, \end{array}$$

where the $n\Lambda$ are free modules of rank n with distinguished bases $(1, 0, \ldots, 0), \ldots, (0, \ldots, 1)$. A *contraction* of a based chain complex in dimension k is the operation inverse to a stabilization.

Two bases in a free module are said to be *equivalent* if the transition matrix from the one to the other is zero in $\overline{K}_1(\Lambda)$ (or $\mathrm{Wh}(\pi)$). We shall assume that over a based chain complex we have license to perform operations of three types: stabilization, contraction, and replacement of bases in the chain modules by equivalent ones. These operations will be called *elementary*. Obviously, under elementary operations the simple-homotopy type of a chain complex remains unchanged.

PROPOSITION 3.3 [40]. *Let $\{C_i, \partial_i\}$ and $\{D_i, d_i\}$ be based chain complexes, and $f = \{f_i\} : \{C_i, \partial_i\} \to \{D_i, d_i\}$ a simple-homotopy equivalence. Then there exists a sequence of elementary operations taking $\{C_i, \partial_i\}$ into $\{D_i, d_i\}$.*

PROPOSITION 3.4 [40]. *Let $\{C_i, \partial_i\}$ be a based chain complex with vanishing homology groups. Then it is simple-homotopy equivalent to an elementary chain complex $0 \leftarrow C_{k-1} \xleftarrow{\partial_k} C_k \leftarrow 0$, where k is arbitrary $(0 \le k-1 \le n-1)$.*

THEOREM 3.5 [22]. *Let $f = \{f_i\}$ be a homotopy equivalence between two based chain complexes $\{C_i, \partial_i\}$ and $\{D_i, d_i\}$. Then there exists a stabilization of the boundary homomorphisms ∂_i and d_i such that the new chain complexes are chain isomorphic. The torsion remains unchanged: $\tau(f^s) = \tau(f)$, where $f^s = \{f_i^s\}$ is the chain isomorphism.*

Let $\theta : \Lambda \to \Lambda$ be an automorphism of the ring Λ such that $\theta(1) = 1$. Obviously θ induces a mapping of a free module, $s_\theta : F \to F$, given by $s_\theta(\lambda_1, \dots, \lambda_n) = (\theta(\lambda_1), \dots, \theta(\lambda_n))$. From the definition of s_θ it follows immediately that $s_{\theta^{-1}} = (s_\theta)^{-1}$ and $s_\theta \cdot s_\varphi = s_{\theta\cdot\varphi}$, where φ is a second automorphism of Λ. In addition, $s_\theta(\lambda \cdot f) = \theta(\lambda) \cdot s_\theta(f)$, where $\lambda \in \Lambda$, $f \in F$; so s_θ is not a Λ-module homomorphism, although it is a homomorphism on the underlying abelian group.

We say that a mapping $g : F \to F$ is a semilinear homomorphism, associated with an automorphism $\theta : \Lambda \to \Lambda$, if g is a homomorphism of the underlying abelian group and $g(\lambda \cdot f) = \theta(\lambda) \cdot g(f)$. Obviously s_θ is a semilinear isomorphism. If $g : F_1 \to F_2$ is a semilinear homomorphism associated with $\theta : \Lambda \to \Lambda$, it can be turned into an ordinary homomorphism of Λ-modules by defining a new Λ-module structure on F_2 by the formula $\lambda \cdot f = \theta(\lambda) \cdot f$.

If $g : F \to F$ is an endomorphism of F as a Λ-module, then $g^\theta = s_\theta \cdot g \cdot s_{\theta^{-1}}$ is also an endomorphism. In particular, if $g \in \mathrm{GL}(\Lambda, n)$, then $g^\theta \in \mathrm{GL}(\Lambda, n)$; and if $g \in E(n, \Lambda)$, then $g^\theta \in E(n, \Lambda)$. We denote by Θ the group of automorphisms of Λ such that $\theta(1) = 1$. Then Θ acts on $K_1(\Lambda)$ (or $\mathrm{Wh}(\pi)$) as a group of operators according to the rule

$$\theta \cdot \tau(g) = \tau(g^\theta).$$

Let $\{C_i, \partial_i\}$: $0 \leftarrow C_0 \xleftarrow{\partial_1} C_1 \leftarrow \cdots \xleftarrow{\partial_n} C_n \leftarrow 0$ be a based chain complex of free modules over Λ. Using s_θ, construct the semilinear isomorphism s_θ: $C_i \to C_i$, and put $\partial_i^\theta = s_\theta \cdot \theta_i \cdot s_{\theta^{-1}}$: $C_i \to C_{i-1}$. Then $\partial_i^\theta \cdot \partial_{i+1}^\theta = 0$ and $\partial_i^\theta(\lambda \cdot c) = \lambda \partial_i^\theta(c)$ for $c \in C_i$; so that ∂_i^θ is a boundary operator. Let $\{C_i, \partial_i^\theta\}$ be the chain complex consisting of the modules C_i and the boundary homomorphisms ∂_i^θ. Following Whitehead, we call it the conjugate chain complex. If $g = \{g_i\} : \{C_i, \partial_i\} \to \{D_i, d_i\}$, then $g^\theta = \{g_i^\theta\}$: $\{C_i, \partial_i^\theta\} \to \{D_i, d_i^\theta\}$. Furthermore, if $\{C_i, \partial_i\}$ is acyclic and $\tau(C_i, \partial_i) = \tau_0$, then $\{C_i, \partial_i^\theta\}$ is acyclic and $\tau(C_i, \partial_i^\theta) = \theta \cdot \tau_0$.

We shall have occasion to be using semilinear chain transformations. For these, too, there exists an equivalence relation of homotopy; for details, see [40]. It is easy to show that if $f = \{f_i\} : \{C_i, \partial_i\} \to \{D_i, d_i\}$ is a homotopy equivalence associated with $\theta : \Lambda \to \Lambda$, then $f s_{\theta^{-1}} = \{f_i s_{\theta^{-1}}\}$: $\{C_i, \partial_i^\theta\} \to \{D_i, d_i\}$ is a homotopy equivalence in the ordinary sense. By definition,

$$\tau(f) = \tau(f \cdot s_{\theta^{-1}}).$$

We observe that if $f = \{f_i\} : \{C_i, \partial_i\} \to \{D_i, d_i\}$ is a chain mapping associated with $\theta : \Lambda \to \Lambda$ and inducing a semilinear isomorphism of homology modules, then a corresponding analogue of the Cockcroft-Swan theorem holds. Indeed, since $f \cdot \theta^{-1} : \{C_i, \partial_i^\theta\} \to \{D_i, d_i\}$ is an ordinary chain mapping, it is subject to the ordinary Cockcroft-Swan theorem; we have mappings

$$\begin{array}{ccccccccc} \{C_i, \partial_i\} & \xrightarrow{\theta} & \{C_i, \partial_i^\theta\} & \xrightarrow{i} & \{C_i^s, \partial_i^{\theta s}\} & \xrightarrow{h} & \{D_i^s, d_i^s\} & \xrightarrow{p} & \{D_i, d_i\}, \\ & & & & \Big\downarrow{\scriptstyle \theta^{-1}} & & & & \\ & & & & \{C_i^s, \partial_i^s\} & & & & \end{array}$$

where $\{C_i^s, \partial_i^{\theta s}\}$ and $\{D_i^s, d_i^s\}$ are stabilizations of the chain complexes $\{C_i, \partial_i^\theta\}$ and $\{D_i, d_i\}$ respectively, i is the imbedding, and p is the projection.

§4. Minimal chain complexes

This will be a key section. In it we prove that if Λ is an IBN, s-ring, then in the homotopy type of a free chain complex there always exists a minimal chain complex.

DEFINITION 4.1. A chain complex $\{C_i, \partial_i\}$ is called *minimal in dimension* λ if for any chain complex $\{D_i, d_i\}$ homotopy equivalent to $\{C_i, \partial_i\}$ we have $\mu(C_\lambda) \le \mu(D_\lambda)$.

DEFINITION 4.2. A chain complex $\{C_i, \partial_i\}$ is called *minimal* if it is minimal in all dimensions.

THEOREM 4.3. *Let* Λ *be an* IBN, *s-ring. Then for every chain complex* $\{C_i, \partial_i\}$ *there exists a homotopy equivalent chain complex* $\{\overline{C}_i, \overline{\partial}_i\}$ *such that* $f\text{-}\operatorname{rank}(\overline{\partial}_{i+1}(\overline{C}_{i+1}), \overline{C}_i) = 0$ *and is additive.*

PROOF. Consider an arbitrary based chain complex $\{C_i, \partial_i\}$: $0 \leftarrow C_0 \xleftarrow{\partial_1} C_1 \leftarrow \cdots \xleftarrow{\partial_n} C_n \leftarrow 0$. Suppose $\partial_0 : C_0 \to H_0$. Stabilize ∂_1 via the free module $n_0\Lambda$, $n_0 = \mu(H_0)$; then using Lemma 1.10, extract from ∂_0 a minimal epimorphism $\overline{\partial}_0 : \overline{C}_0 \to H_0 \to 0$, obtaining a decomposition

$$\begin{array}{ccccccccc} 0 & \longleftarrow & H_0 & \xleftarrow{\overline{\partial}_0} & \overline{C}_0 & \xleftarrow{\partial_1 \oplus \mathrm{id}} & C_1 & \xleftarrow{\partial_2} & \cdots \\ & & & & \oplus & & \oplus & & \\ & & 0 & \longleftarrow & \widetilde{C}_0 & & n_0\Lambda & & \end{array},$$

where $\widetilde{C}_0$ is a free module. We have then

$$\begin{array}{ccccccccc} 0 & \longleftarrow & H_0 & \xleftarrow{\overline{\partial}_0} & \overline{C}_0 & \longleftarrow & \widehat{C}_1 & \xleftarrow{\partial_2} & \cdots \\ & & & & \oplus & \partial_1 \oplus \mathrm{id} & \oplus & & \\ & & 0 & \longleftarrow & \widetilde{C}_0 & \longleftarrow & \widetilde{C}_0 & \longleftarrow & 0. \end{array}$$

Since Λ is an s-ring, $\widetilde{C}_1$ is a free module; and since $\partial_0 \colon \overline{C}_0 \to H_0$ is a minimal epimorphism, it follows from Lemma 2.4 that f-rank$(\partial_1 \oplus \mathrm{id}(\hat{C}_1 \oplus \widetilde{C}_0), \overline{C}_0 \oplus \widetilde{C}_0) = 0$ and is additive. Rewrite the complex in new notation:

$$\begin{array}{ccccccccc} & & \overline{C}_0 & \xleftarrow{\hat{\partial}_1} & \hat{C}_1 & \xleftarrow{\partial_2} & C_2 & \xleftarrow{\partial_3} & \cdots \\ & & \oplus & & \oplus & & & & \\ 0 & \longleftarrow & \widetilde{C}_0 & \xleftarrow{f_1} & \widetilde{C}_0 & \longleftarrow & 0 & & \end{array},$$

and consider the submodule $\partial_2(C_2) \subset \hat{C}_1$. Lemma 2.2 ensures the existence of an integer m_1, starting from which f-rank$(\partial_2(C_2) \oplus m_1\Lambda, \hat{C}_1 \oplus m_1\Lambda)$ is additive. Obviously the submodule $\widetilde{C}_0$ has no influence on the size of f-rank$(\cdot, \cdot)$. Let f-rank$(\partial_2(C_2) \oplus m_1\Lambda, \hat{C}_1 \oplus m_1\Lambda) = n_1$. Stabilizing ∂_2 via a free module of rank m_1, we obtain a chain complex

$$\begin{array}{ccccccccc} & & 0 & \longleftarrow & m_1\Lambda & \xleftarrow{\mathrm{id}} & m_1\Lambda & \longleftarrow & 0 \\ & & & & \oplus & & \oplus & & \\ & & \overline{C}_0 & \xleftarrow{\hat{\partial}_1} & \hat{C}_1 & \xleftarrow{\partial_2} & C_2 & \xleftarrow{\partial_3} & \cdots \\ & & \oplus & & \oplus & & & & \\ 0 & \longleftarrow & \widetilde{C}_0 & \xleftarrow{f_1} & \widetilde{C}_0 & \longleftarrow & 0. & & \end{array} \qquad (*)$$

Now find a decomposition of the module $\partial_2(C_2) \oplus m_1\Lambda$ such that

$$\partial_2(C_2) \oplus m_1\Lambda = B_1 \oplus \widetilde{C}_1, \qquad \mu(\widetilde{C}_1) = n_1, \qquad B_1 \subset \overline{C}_1,$$

where $\widetilde{C}_1 \oplus \overline{C}_1$ is a new decomposition of $\hat{C}_1 \oplus m_1\Lambda$. This gives a new representation of the chain complex:

$$\begin{array}{ccccccccccc} & & & & 0 & \longleftarrow & \widetilde{C}_1 & \xleftarrow{f_2} & \widetilde{C}_1 & \longleftarrow & 0 \\ & & & & & & \oplus & & \oplus & & \\ & & \overline{C}_0 & \xleftarrow{\overline{\partial}_1} & \overline{C}_1 & \xleftarrow{\hat{\partial}_2} & \hat{C}_2 & \longleftarrow & C_3 & \longleftarrow & \cdots \\ & & \oplus & & \oplus & & & & & & \\ 0 & \longleftarrow & \widetilde{C}_0 & \xleftarrow{f_1} & \widetilde{C}_0 & \longleftarrow & 0. & & & & \end{array}$$

All modules in the decomposition $(*)$ are free. Consider the submodule $B_1 = \hat{\partial}_2(\hat{C}_2)$. By the construction and by Lemma 2.3, f-rank$(B_1, \overline{C}_1) = 0$ and is additive. We note that the choice of B_1 is not unique; this fact will be examined below. Repeating the argument by stabilizing each ∂_i $(i \geq 3)$ in succession, we arrive at a diagram in which, for all i, f-rank$(\overline{\partial}_i(\overline{C}_i), \overline{C}_{i-1}) = 0$

and is additive. Thus, we obtain a representation of the chain complex in the form

$$\begin{array}{ccccccccc}
 & & 0 & \longleftarrow & \widetilde{C}_1 & \xleftarrow{f_2} & \widetilde{C}_1 & \longleftarrow & 0 \\
 & & & & \oplus & & \oplus & & \\
 & & \overline{C}_0 & \xleftarrow{\overline{\partial}_1} & \overline{C}_1 & \xleftarrow{\overline{\partial}_2} & \overline{C}_2 & \xleftarrow{\overline{\partial}_3} & \overline{C}_3 \longleftarrow \cdots \\
 & & \oplus & & \oplus & & \oplus & & \oplus \\
0 & \longleftarrow & \widetilde{C}_0 & \xleftarrow{f_1} & \widetilde{C}_0 \leftarrow 0 \leftarrow & & \widetilde{C}_2 & \xleftarrow{f_2} & \widetilde{C}_2.
\end{array}$$

Clearly, the chain complex

$$\begin{array}{ccccccccccc}
0 & \longleftarrow & \widetilde{C}_0 & \xleftarrow{f_1} & \widetilde{C}_0 \leftarrow 0 \leftarrow & \widetilde{C}_2 & \xleftarrow{f_2} & \widetilde{C}_2 & \longleftarrow & 0 & \\
 & & & & \oplus & & & \oplus & & \oplus & \\
 & & 0 & \longleftarrow & \widetilde{C}_1 & \xleftarrow{f_2} & \widetilde{C}_1 \leftarrow 0 \leftarrow & \widetilde{C}_3 & \xleftarrow{f_3} & \widetilde{C}_3 & \longleftarrow 0
\end{array}$$

is acyclic and constitutes a direct summand. Discarding it, we obtain the chain complex

$$0 \leftarrow \overline{C}_0 \overset{\overline{\partial}_1}{\leftarrow} \overline{C}_1 \overset{\overline{\partial}_2}{\leftarrow} \overline{C}_2 \overset{\overline{\partial}_3}{\leftarrow} \cdots ,$$

in which $f\text{-}\operatorname{rank}(\overline{\partial}_i(\overline{C}_i), \overline{C}_{i-1}) = 0$ and is additive. Obviously it is homotopy equivalent to the original chain complex. This proves the theorem.

It will be shown below that any chain complex for which $f\text{-}\operatorname{rank}(\partial_i(C_i), C_{i-1}) = 0$ and is additive is automatically minimal. Let us first, however, examine the construction in more detail. Our proof failed to take into consideration the bases in the chain modules. We shall now fill in this omission. When the boundary homomorphisms ∂_i are stabilized via certain free modules, we can always suppose the latter to be already furnished with canonical bases. The situation is more complicated when the chain modules are decomposed into direct sums of free modules. For example, if a module C with fixed basis $(c_1, \dots, c_k)$ is represented as a direct sum of free modules A and B, it does not at all follow that A has a basis $(a_1, \dots, a_n)$ and B a basis $(b_1, \dots, b_m)$, $n + m = k$, such that the bases $(c_1, \dots, c_k)$ and $(a_1, \dots, a_n, b_1, \dots, b_m)$ are equivalent. We show, however, that by an appropriate stabilization we can get around this difficulty.

LEMMA 4.4. *Let F be a free module, and $(f_1, \dots, f_k)$, $(\overline{f}_1, \dots, \overline{f}_k)$ any two bases for F. Then the module $F \oplus k\Lambda$ has a basis of the form $(\overline{f}_1, \dots, \overline{f}_k, g_1, \dots, g_k)$ such that the bases $(f_1, \dots, f_k, e_1, \dots, e_k)$ and $(\overline{f}_1, \dots, \overline{f}_k, g_1, \dots, g_k)$ are equivalent. Here $(e_1, \dots, e_k)$ is the canonical basis, and $(g_1, \dots, g_k)$ is a basis in the module $k\Lambda$.*

PROOF. Let M be the transition matrix from the basis $f_1, \dots, f_k$ to the basis $\overline{f}_1, \dots, \overline{f}_k$. In $k\Lambda$, consider the basis $g_1, \dots, g_k$ such that the transition matrix from the canonical basis $e_1, \ldots, e_k$ to $g_1, \ldots, g_k$ is M^{-1}. Then the transition matrix from the basis $f_1, \ldots, f_k, e_1, \ldots, e_k$ to the basis $\overline{f}_1, \ldots, \overline{f}_k, g_1, \ldots, g_k$ is of the form

$$\begin{vmatrix} M & 0 \\ 0 & M^{-1} \end{vmatrix},$$

which $= 0$ in $\overline{K}_1(\Lambda)$. Consequently, the two bases are equivalent. This proves the lemma.

It is clear now how to deal with the chain complex situation. Choose arbitrary representatives $(\overline{c}_1, \ldots, \overline{c}_k, \tilde{c}_1, \ldots, \tilde{c}_l)$ in the modules $\overline{C}_i \oplus \widetilde{C}_i$; for example, such that the homomorphisms f_i are given by identity matrices. Stabilize the boundary homomorphisms $\overline{\partial}_i \oplus f_i$ via free modules R_i^j of rank j, where

$$j = \max(\mu(\overline{C}_i) + \mu(\widetilde{C}_i) + \mu(\widetilde{C}_{i-1}),\ \mu(\overline{C}_{i+1} + \mu(\widetilde{C}_i) + \mu(\widetilde{C}_{i+1})),$$

and, selecting bases appropriately in accordance with Lemma 4.4, obtain a stabilized complex that is again a based chain complex. This is then homotopy equivalent to the original. Observe that since the homomorphisms f_i are given by identity matrices, the modules $\widetilde{C}_i$ can be discarded without changing the simple-homotopy type of the chain complex. In addition, using Proposition 3.4, the subcomplex formed by the modules R_i^j can be contracted to an elementary complex. We have arrived now at an important conclusion.

PROPOSITION 4.5. *Under the hypothesis of Theorem 4.3, the free chain complex $\{C_i, \partial_i\}$ is simple-homotopy equivalent to a chain complex of the form $\{\overline{C}_i, \overline{\partial}_i\} \oplus \{R_k, d\}$, where $\{R_k, d\}$ is an elementary chain complex in dimension k, and f-rank$(\overline{\partial}_i(\overline{C}_i), \overline{C}_{i-1}) = 0$ and is additive.*

Observe that the torsion $\tau(R_k, d)$ of the elementary complex is not uniquely determined. It depends on the choice of the submodule of boundaries constituting a direct summand in the chains.

LEMMA 4.6. *Let Λ be an IBN, s-ring, and*

$$\{C_i, \partial_i\} : 0 \leftarrow C_0 \xleftarrow{\partial_1} C_1 \leftarrow \cdots \xleftarrow{\partial_n} C_n \leftarrow 0,$$

$$\{D_i, d_i\} : 0 \leftarrow D_0 \xleftarrow{d_1} D_1 \leftarrow \cdots \xleftarrow{d_n} D_n \leftarrow 0$$

two homotopy equivalent chain complexes, in which

$$f\text{-rank}(\partial_i(C_i), C_{i-1}) = f\text{-rank}(d_i(D_i), D_{i-1}) = 0$$

and is additive for all i. Then $\mu(C_i) = \mu(D_i)$.

PROOF. Assume the contrary. Then there exists a k such that $\mu(C_k) > \mu(D_k)$ while $\mu(C_i) = \mu(D_i)$, $0 \leq i < k$. Obviously $k > 0$, since

$\partial_0 : C_0 \to H_0$ and $d_0 : D_0 \to H_0$ are minimal epimorphisms by Lemma 2.5, so that $\mu(C_0) = \mu(D_0)$. Using Theorem 3.4, we find acyclic complexes

$$\{T_i, t_i\}: 0 \leftarrow T_0 \overset{t_1}{\leftarrow} T_1 \leftarrow \cdots \overset{t_k}{\leftarrow} T_k \leftarrow 0,$$
$$\{S_i, s_i\}: 0 \leftarrow S_0 \overset{s_1}{\leftarrow} S_1 \leftarrow \cdots \overset{s_k}{\leftarrow} S_k \leftarrow 0$$

such that up to dimension $k-1$ the complexes

$$\begin{array}{ccccccccc} 0 \longleftarrow C_0 \oplus T_0 & \overset{\partial_1 \oplus t_1}{\longleftarrow} & \cdots & \overset{\partial_{k-1} \oplus t_{k-1}}{\longleftarrow} & C_{k-1} \oplus T_{k-1} & \overset{\partial_k \oplus t_k}{\longleftarrow} & C_k \oplus T_k & \overset{\partial_{k+1}}{\longleftarrow} & C_{k+1} \longleftarrow \cdots \\ \downarrow \varphi_0 & & & & \downarrow \varphi_{k-1} & & & & \\ 0 \longleftarrow D_0 \oplus S_0 & \overset{d_1 \oplus s_1}{\longleftarrow} & \cdots & \overset{d_{k-1} \oplus s_{k-1}}{\longleftarrow} & D_{k-1} \oplus S_{k-1} & \overset{d_k \oplus s_k}{\longleftarrow} & D_k \oplus S_k & \overset{d_{k+1}}{\longleftarrow} & D_{k+1} \longleftarrow \cdots \end{array}$$

are isomorphic. Observe that $\mu(T_i) = \mu(S_i)$, since $\mu(C_j) = \mu(D_j)$ for $0 \leq j < k$. Stabilize ∂_{k+1} via the module $D_k \oplus S_k$, and d_{k+1} via the module $C_k \oplus T_k$; then, as in the proof of Theorem 3.5 (see [22]), construct an isomorphism

$$\varphi_k : C_k \oplus T_k \oplus D_k \oplus S_k \to D_k \oplus S_k \oplus C_k \oplus T_k$$

such that $\varphi_{k-1} \cdot (\partial_k \oplus t_k) = (d_k \oplus s_k) \cdot \varphi_k$. Since $f\text{-rank}(\partial_{k+1}(C_{k+1}), C_k) = 0$ and is additive, we have

$$f\text{-rank}(\partial_{k+1}(C_{k+1}) \oplus D_k \oplus S_k, C_k \oplus T_k \oplus D_k \oplus S_k) = \mu(D_k) + \mu(S_k).$$

Similarly,

$$f\text{-rank}(d_{k+1}(D_{k+1}) \oplus C_k \oplus T_k, D_k \oplus S_k \oplus C_k \oplus T_k) = \mu(C_k) + \mu(T_k).$$

Since φ_k is an isomorphism and

$$\varphi_k(\partial_{k+1}(C_{k+1}) \oplus D_k \oplus S_k) = d_{k+1}(D_{k+1}) \oplus C_k \oplus T_k,$$

we have

$$\begin{aligned} & f\text{-rank}(\partial_{k+1}(C_{k+1}) \oplus D_k \oplus S_k, C_k \oplus T_k \oplus D_k \oplus S_k) \\ &= f\text{-rank}(d_{k+1}(D_{k+1}) \oplus C_k \oplus T_k, D_k \oplus S_k \oplus C_k \oplus T_k). \end{aligned}$$

But by assumption, $\mu(C_k) > \mu(D_k)$. We have therefore arrived at a contradiction, and the lemma is proved.

Observe that in this proof an essential role is played by the theorem of Cockcroft-Swan, allowing us to compare chain complexes of the same homotopy type.

We now prove an important theorem.

THEOREM 4.7. *Let Λ be an IBN, s-ring, and*

$$\{C_i, \partial_i\} : 0 \leftarrow C_0 \overset{\partial_1}{\leftarrow} C_1 \leftarrow \cdots \overset{\partial_n}{\leftarrow} C_n \leftarrow 0$$

a chain complex in which $f\text{-rank}(\partial_i(C_i), C_{i-1}) = 0$ and is additive for all i. Then $\{C_i, \partial_i\}$ is a minimal chain complex.

PROOF. Assume the contrary. Then in the class of chain complexes homotopy equivalent to $\{C_i, \partial_i\}$ there exists a complex $\{D_i, d_i\}$ such that, for some j, $\mu(C_j) > \mu(D_j)$. Obviously $j > 0$. Let k be the first value for which the inequality holds. By stabilization and contraction of the d_i, $0 < i < k-1$, and an argument similar to that in the proof of Theorem 4.3, construct a chain complex

$$0 \leftarrow D_0 \overset{\bar{d}_1}{\leftarrow} \overline{D}_1 \leftarrow \cdots \overset{\bar{d}_{k-2}}{\leftarrow} \overline{D}_{k-2} \overset{\tilde{d}_{k-1}}{\leftarrow} \widetilde{D}_{k-1} \overset{d_k}{\leftarrow} D_k \overset{d_{k+1}}{\leftarrow} D_{k+1} \leftarrow \cdots$$

in which f-$\operatorname{rank}(d_i(\overline{D}_i), \overline{D}_{i-1}) = 0$ $(0 < i < k-1)$ and is additive. If f-$\operatorname{rank}(d_k(D_k), \widetilde{D}_{k-1}) > 0$, find a submodule $\Gamma \subset d_k(D_k)$ such that the complex takes the form

$$\begin{array}{ccccccccc} 0 \leftarrow D_0 \overset{\bar{d}_1}{\leftarrow} \overline{D}_1 \leftarrow \cdots \overset{\bar{d}_{k-2}}{\leftarrow} & \overline{D}_{k-2} & \overset{\bar{d}_{k-1}}{\leftarrow} & \overline{D}_{k-1} & \overset{\tilde{d}_k}{\leftarrow} & \widetilde{D}_k & \overset{d_{k+1}}{\leftarrow} & D_{k+1} \leftarrow \cdots, \\ & & & \oplus & & \oplus & & \\ & 0 & \leftarrow & \Gamma & \leftarrow & \Gamma & \leftarrow & 0 \end{array}$$

where $\overline{D}_{k-1} \oplus \Gamma = \widetilde{D}_{k-1}$ and $\widetilde{D}_k \oplus \Gamma = D_k$. Cancelling the module Γ, we obtain a chain complex

$$0 \leftarrow D_0 \overset{\bar{d}_1}{\leftarrow} \overline{D}_1 \leftarrow \cdots \overset{\bar{d}_{k-2}}{\leftarrow} \overline{D}_{k-2} \overset{\bar{d}_{k-1}}{\leftarrow} \overline{D}_{k-1} \overset{\tilde{d}_k}{\leftarrow} \widetilde{D}_k \overset{d_{k+1}}{\leftarrow} D_{k+1} \leftarrow \cdots,$$

in which $\mu(\tilde{D}_k) < \mu(D_k)$. If f-$\operatorname{rank}(d_{k+1}(D_{k+1}), \widetilde{D}_k) > 0$, again pull out a "superfluous" module in $\widetilde{D}_k$ and D_{k+1} and cancel it. Observe that if f-$\operatorname{rank}(d_k(D_k), D_{k-1})$ is not additive, then under stabilization of d_k it can only increase, and we can therefore only decrease $\mu(D_k)$. The situation is similar in dimension $k+1$. Thus we can suppose, without loss of generality, that we have constructed a chain complex in which f-$\operatorname{rank}(\bar{d}_i(\overline{D}_i), \overline{D}_{i-1}) = 0$ and is additive for all i, and $\mu(\overline{D}_k) < \mu(D_k)$. But by Lemma 4.6 we must have $\mu(\overline{D}_k) = \mu(D_k)$. We have therefore arrived at a contradiction, and the theorem is proved.

PROPOSITION 4.8. *Let $\{C_i, \partial_i\}_\Lambda$ be the class of chain complexes over a ring Λ. In order that every homotopy type in $\{C_i, \partial_i\}_\Lambda$ contain a minimal chain complex, it is necessary and sufficient that Λ be an IBN, s-ring.*

PROOF. That the condition IBN is necessary is obvious. We show now the necessity of the condition s. Assume the contrary. Then over the ring Λ there exist stably free modules that are not free. Let M be such a module, with $M \oplus \Lambda = n\Lambda$, and construct the chain complex $\{C, \partial\}$ of the form

$$\begin{array}{ccccccccc} & \{C, \partial\}: 0 & \longleftarrow & C_i & \overset{\partial_{i+1}}{\longleftarrow} & C_{i+1} & \longleftarrow & 0 & \\ & & & \| & & \| & & & \\ 0 \longleftarrow & M & \longleftarrow & n\Lambda & \overset{i}{\longleftarrow} & \Lambda & \longleftarrow & 0, & \partial_{i+1} = i. \end{array}$$

Obviously $H_i(C, \partial) = M$. By construction, $\partial_{i+1}(C_{i+1})$ is a direct summand of C_i, and therefore f-$\operatorname{rank}(\partial_{i+1}(C_{i+1}), C_i) = 1$. But we cannot perform a contraction, since the module complementary to $\partial_{i+1}(C_{i+1})$ in C_i is only stably free.

Now stabilize $\{C, \partial\}$ in dimension i; i.e., consider the following chain complex $\{\overline{C}, \overline{\partial}\}$:

$$\begin{array}{ccccccccc}
 & & 0 & \longleftarrow & C_i & \xleftarrow{\partial_{i+1}} & C_{i+1} & \longleftarrow & 0 \\
 & & & & \oplus & & \| & & \\
0 & \longleftarrow & \Lambda & \xleftarrow{\mathrm{id}} & \Lambda & \longleftarrow 0 & \| & & \\
 & & \| & & \| & & \| & & \\
0 & \longleftarrow & \overline{C}_{i-1} & \xleftarrow{\overline{\partial}_i} & \overline{C}_i & \xleftarrow{\overline{\partial}_{i+1}} & \overline{C}_{i+1} & \longleftarrow & 0, \quad \overline{\partial}_{i+1} = \partial_{i+1}.
\end{array}$$

Here a contraction of the boundary homomorphism $\overline{\partial}_{i+1}$ is possible; and we obtain the chain complex

$$\begin{array}{ccccccccc}
0 & \longleftarrow & \widetilde{C}_{i-1} & \xleftarrow{\widetilde{\partial}_i} & \widetilde{C}_i & \longleftarrow & 0 & & \\
 & & \| & & \| & & & & \\
0 & \longleftarrow & \Lambda & \xleftarrow{p} & n\Lambda & \longleftarrow & M & \longleftarrow & 0,
\end{array}$$

which is homotopy equivalent to $\{C, \partial\}$. Obviously, the chain complex $\{\overline{C}, \overline{\partial}\}$ is minimal except for dimension $i+1$, while $\{\widetilde{C}, \tilde{\partial}\}$ is minimal except for dimension i. Consequently, the homotopy type of the complex $\{C, \partial\}$ has no minimal complex. This proves necessity. Sufficiency follows from Theorems 4.3 and 4.7; and this completes our proof.

However, the simple-homotopy type of a based chain complex $\{C_i, \partial_i\}$ does not always have a minimal chain complex in it. The obstruction, which is of the character of Whitehead torsion (see Proposition 4.5), is 0 if the mapping $\varepsilon(\{C_i, \partial_i\}) \xrightarrow{\tau} K_1(\Lambda)$ is an epimorphism. Here $\varepsilon(\{C_i, \partial_i\})$ is the group of homotopy self-equivalences of $\{C_i, \partial_i\}$, and τ means the mapping that assigns to a homotopy equivalence $f : \{C_i, \partial_i\} \to \{C_i, \partial_i\}$ the torsion $\tau(f)$. Indeed, let $\{\overline{C}_i, \overline{\partial}_i\}$ be a minimal chain complex and $g : \{C_i, \partial_i\} \to \{\overline{C}_i, \overline{\partial}_i\}$ a homotopy equivalence, with $\tau(g) = \tau_0$. Choose $f : \{C_i, \partial_i\} \to \{C_i, \partial_i\}$ such that $\tau(f) = -\tau_0$. Then $g \circ f$ is a simple-homotopy equivalence.

Again, if $\{\overline{C}_i, \overline{\partial}_i\}$ is "large" in some dimension, then $\{C_i, \partial_i\}$ is likewise simple-homotopy equivalent to a minimal chain complex. We use Lemma 3.2 to get rid of the "makeweight" arising in Proposition 4.5. For the purposes of our later arguments connected with Morse functions, the obstruction will be regarded as belonging not to $K_1(\Lambda)$, but to $\mathrm{Wh}(\pi)$.

PROPOSITION 4.9. *Suppose $\mathbb{Z}[\pi]$ is an s-ring, with r-$\dim \pi \le k$, and the chain complex $\{C_i, \partial_i\}$ has in its homotopy type a minimal chain* *complex* $\{\overline{C}_i, \overline{\partial}_i\}$ *such that* $\mu(\overline{C}_j) \geq k$ *for some* j. *Then* $\{C_i, \partial_i\}$ *is simple-homotopy equivalent to the minimal chain complex.*

PROOF. By Proposition 4.5 $\{C_i, \partial_i\}$ is simple-homotopy equivalent to a chain complex of the form

$$\begin{array}{ccccccccccccccc} 0 & \leftarrow & \overline{C}_0 & \overset{\overline{\partial}_1}{\leftarrow} & \cdots & \overset{\overline{\partial}_i}{\leftarrow} & \overline{C}_i & \overset{\overline{\partial}_{i+1}}{\leftarrow} & \overline{C}_{i+1} & \leftarrow & \cdots & \leftarrow & \overline{C}_n & \leftarrow & 0 \\ & & & & & & \oplus & & \oplus & & & & & & \\ & & & & 0 & \leftarrow & R & \overset{f}{\leftarrow} & R & \leftarrow & 0. & & & & \end{array}$$

The matrix for the isomorphism f can be reduced by elementary transformations to the form

$$\left|\begin{array}{cccc} A & & & \\ & 1 & & 0 \\ & & \ddots & \\ & 0 & & \\ & & & 1 \end{array}\right|,$$

where $A \in \mathrm{GL}(\mathbb{Z}[\pi], k)$, since $r\text{-}\dim \pi \leq k$. Apply Lemma 3.2 to the matrix of the homomorphism $\overline{\partial}_{i+1} \oplus f$, and the proposition is proved.

LEMMA 4.10. *Denote by* $\{C, \partial\}_*$ *a class of homotopy equivalent chain complexes over an* IBN, *s-ring* Λ *such that for every representative*

$$0 \leftarrow C_0 \overset{\partial_1}{\leftarrow} C_1 \leftarrow \cdots \overset{\partial_n}{\leftarrow} C_n \leftarrow 0$$

in $\{C, \partial\}_*$, $f\text{-}\mathrm{rank}(\partial_{k+1}(C_{k+1}), C_k) = 0$ *and is additive. Then the number*

$$\mu(C_0) - \mu(C_1) + \cdots + (-1)^k \mu(C_k) = \chi_k$$

is the same for all the members of $\{C, \partial\}_*$.

PROOF. Consider an arbitrary member

$$0 \leftarrow C_0 \overset{\partial_1}{\leftarrow} C_1 \leftarrow \cdots \overset{\partial_n}{\leftarrow} C_n \leftarrow 0.$$

It is well known that the Euler characteristic

$$\sum (-1)^i \mu(C_i) = \chi(C_i, \partial_i)$$

is an invariant of $\{C, \partial\}_*$. Stabilize the ∂_i, $1 \leq i \leq k$ and $i \geq k+2$, via free modules of suitable rank so that the new chain complex has as direct summand a chain complex $\{\overline{C}_i, \overline{\partial}_i\}$ in which $f\text{-}\mathrm{rank}(\overline{\partial}_i(\overline{C}_i), \overline{C}_{i-1}) = 0$ and is additive for all $1 \leq i \leq k$ and $i \geq k+2$, while the complementary chain complex is acyclic. Observe that it is not necessary to stabilize ∂_{k+1}, because of the hypothesis of the lemma. By Theorem 4.7 $\{\overline{C}_i, \overline{\partial}_i\}$ is a minimal chain complex. By construction,

$$\sum_{l \geq k+1} (-1)^l \mu(C_l) = \sum_{l \geq k+1} (-1)^l \mu(\overline{C}_l) = \kappa.$$

By Lemma 4.6 and Theorem 4.7, the ranks of the chain modules in the minimal complex are invariants of the homotopy type. Consequently, κ is constant for all $\{C_i, \partial_i\}$ in $\{C, \partial\}_*$. The conclusion of the lemma now follows from the equality $\chi_k + \kappa = \chi(C_i, \partial_i)$.

We note that although the chain complexes associated with regular Morse functions on nonsimply-connected manifolds were studied in [28], the problem of finding minimal chain complexes within the homotopy type was not considered there.

We note also that the results of this section remain valid for chain transformations associated with an automorphism $\theta : \Lambda \to \Lambda$. For any semilinear chain transformation $\{f_i\} : \{C_i, \partial_i\} \to \{D_i, d_i\}$ associated with θ, one must either consider the conjugate complex $\{C_i, \partial_i^\theta\}$ and replace $\{f_i\}$ by the transformation $\{f_i \circ \theta^{-1}\} : \{C_i, \partial_i^\theta\} \to \{D_i, d_i\}$ or define a new Λ-module structure on the complex $\{D_i, d_i\}$ given by $\lambda \cdot d = \theta(\lambda) \cdot d$, $d \in D_i$.

§5. Stable invariants of chain complexes

In this section we consider certain chain complex invariants that are "stably" isomorphic for chain complexes of the same homotopy type. These can be used to construct an algorithm for computing the ranks of the chain modules in the minimal chain complex.

We first prove several facts relating to extension of modules.

Let A and C be modules over a ring Λ. An n-fold extension of A by C is an exact sequence of Λ-modules

$$E: 0 \leftarrow C \leftarrow B_1 \leftarrow \cdots \leftarrow B_n \leftarrow A \leftarrow 0.$$

The set of classes of n-fold extensions formed by a well-known equivalence relation constitutes a group $\operatorname{Ext}^n_\Lambda(C, A)$ [3]. It is known [3] that if $E \in \operatorname{Ext}^n(C, A)$, then E can be represented by an n-fold sequence

$$\overline{E} : 0 \leftarrow C \leftarrow F_0 \leftarrow \cdots \leftarrow F_{n-2} \leftarrow B \leftarrow A \leftarrow 0$$

in which the F_i $(i = 0, \dots, n-2)$ are free modules.

DEFINITION 5.1. Let

$$E: 0 \leftarrow C \xleftarrow{f_0} F_0 \leftarrow \cdots \xleftarrow{f_{n-2}} F_{n-2} \xleftarrow{f_{n-1}} B \xleftarrow{i} A \leftarrow 0$$

and

$$\overline{E}: 0 \leftarrow C \xleftarrow{\overline{f}_0} \overline{F}_0 \leftarrow \cdots \xleftarrow{\overline{f}_{n-2}} \overline{F}_{n-2} \xleftarrow{\overline{f}_{n-1}} B \xleftarrow{i} A \leftarrow 0$$

be two congruent sequences. We say that E and $\overline{E}$ are *strictly congruent* if there exists a commutative diagram

$$\begin{array}{cccccccccccccc} E: 0 & \leftarrow & C & \leftarrow & F_0 & \leftarrow & \cdots & \leftarrow & F_{n-2} & \leftarrow & B & \leftarrow & A & \leftarrow 0 \\ & & \downarrow \mathrm{id} & & \downarrow \varphi_0 & & & & \downarrow \varphi_{n-2} & & \downarrow \varphi_{n-1} & & \downarrow \mathrm{id} & \\ \overline{E}: 0 & \leftarrow & C & \leftarrow & \overline{F}_0 & \leftarrow & \cdots & \leftarrow & \overline{F}_{n-2} & \leftarrow & B & \leftarrow & A & \leftarrow 0 \end{array}$$

where the φ_i $(i = 0, \dots, n-1)$ are isomorphisms.

Consider a sequence $E : 0 \leftarrow C \overset{f_0}{\leftarrow} F_0 \overset{f_1}{\leftarrow} \cdots \overset{f_{n-2}}{\leftarrow} F_{n-2} \overset{f_{n-1}}{\leftarrow} B \overset{i}{\leftarrow} A \leftarrow 0$. If a sequence $\overline{E}$ is obtained from E by a stabilization of the f_i $(i = 1, \dots, n-1)$, we shall denote $\overline{E}$ by $E(s) = \overline{E}$.

LEMMA 5.2. *Suppose the sequences*

$$E : 0 \leftarrow C \overset{f_0}{\leftarrow} F_0 \leftarrow \cdots \overset{f_{n-2}}{\leftarrow} F_{n-2} \overset{f_{n-1}}{\leftarrow} B \overset{i}{\leftarrow} A \leftarrow 0$$

and

$$\overline{E} : 0 \leftarrow C \overset{\overline{f}_0}{\leftarrow} \overline{F}_0 \leftarrow \cdots \overset{\overline{f}_{n-2}}{\leftarrow} \overline{F}_{n-2} \overset{\overline{f}_{n-1}}{\leftarrow} B \overset{i}{\leftarrow} A \leftarrow 0,$$

where the F_i $(\overline{F}_i)$ $(i = 0, \dots, n-2)$ are free modules, are congruent. Then there exist stabilizations of the f_i $(\overline{f}_i)$ such that $E(s)$ and $\overline{E}(s)$ are strictly congruent.

PROOF. We give a proof due to A. A. Suslin. It is easily shown, using the exact sequence for $\operatorname{Ext}^n_\Lambda(C, A)$, that if the sequences

$$E : 0 \leftarrow C \overset{f_0}{\leftarrow} F_0 \leftarrow \cdots \overset{f_{n-2}}{\leftarrow} F_{n-2} \overset{f_{n-1}}{\leftarrow} B \overset{i}{\leftarrow} A \leftarrow 0$$

and

$$\overline{E} : 0 \leftarrow C \overset{\overline{F}_0}{\leftarrow} \overline{F}_0 \leftarrow \cdots \overset{\overline{f}_{n-2}}{\leftarrow} \overline{F}_{n-2} \overset{\overline{f}_{n-1}}{\leftarrow} \overline{B} \leftarrow A \leftarrow 0$$

are congruent, so are the sequences

$$0 \leftarrow K \overset{f_1}{\leftarrow} F_1 \leftarrow \cdots \overset{f_{n-2}}{\leftarrow} F_{n-2} \overset{f_{n-1}}{\leftarrow} B \leftarrow A \leftarrow 0$$

and

$$0 \leftarrow K \overset{\overline{f}_1}{\leftarrow} \overline{F}_1 \leftarrow \cdots \overset{\overline{f}_{n-2}}{\leftarrow} \overline{F}_{n-2} \overset{\overline{f}_{n-1}}{\leftarrow} \overline{B} \leftarrow A \leftarrow 0,$$

where $K = \operatorname{Ker} f_0 = \operatorname{Ker} \overline{f}_0$. Stabilize f_1 via $\overline{F}_0$, and $\overline{f}_1$ via F_0. Using Lemma 1.10, we can get the epimorphisms f_0^s and $\overline{f}_0^s$ to be the same. Consequently, the sequences

$$\begin{array}{ccccccccccccc} 0 & \leftarrow & \operatorname{Ker} f_0^s & \leftarrow & F_1 \oplus \overline{F}_0 & \overset{f_2}{\leftarrow} & F_2 & \leftarrow & \cdots & \overset{i}{\leftarrow} & A & \leftarrow & 0 \\ & & \| & & & & & & & & & & \\ 0 & \leftarrow & \operatorname{Ker} \overline{f}_0^s & \leftarrow & \overline{F}_1 \oplus F_0 & \overset{\overline{f}_2}{\leftarrow} & \overline{F}_2 & \leftarrow & \cdots & \overset{\overline{\imath}}{\leftarrow} & A & \leftarrow & 0 \end{array}$$

are congruent. Making successive repetitions of the argument, we arrive after a finite number of steps at sequences that are strictly congruent; the lemma is proved.

COROLLARY 5.3. *Suppose the sequences*

$$E: 0 \leftarrow C \overset{f_0}{\leftarrow} F_0 \leftarrow \cdots \overset{f_{n-2}}{\leftarrow} F_{n-2} \overset{f_{n-1}}{\leftarrow} B \overset{i}{\leftarrow} A \leftarrow 0$$

and

$$\overline{E}: 0 \leftarrow C \overset{\overline{f}_0}{\leftarrow} \overline{F}_0 \leftarrow \cdots \overset{\overline{f}_{n-2}}{\leftarrow} \overline{F}_{n-2} \overset{\overline{f}_{n-1}}{\leftarrow} \overline{B} \overset{\overline{i}}{\leftarrow} A \leftarrow 0$$

are congruent, where the F_i $(\overline{F}_i)$ *are free modules. Then there exist morphisms from* E *to* $\overline{E}$ *and from* $\overline{E}$ *to* E *that are the identity on* A *and* C; *the modules* B *and* $\overline{B}$ *are stably isomorphic.*

DEFINITION 5.4. We say that a sequence

$$E: 0 \leftarrow C \overset{f_0}{\leftarrow} F_0 \leftarrow \cdots \overset{f_{n-2}}{\leftarrow} F_{n-2} \overset{f_{n-1}}{\leftarrow} B \overset{i}{\leftarrow} A \leftarrow 0$$

is *minimal* if the F_i $(i = 0, \dots, n-2)$ are free modules and the f_i : $F_i \to \operatorname{Ker} f_{i-1}$ are minimal epimorphisms.

LEMMA 5.5. *Let* Λ *be an* IBN, *s-ring, and suppose the sequences of* Λ-*modules*

$$0 \leftarrow C \overset{f_0}{\leftarrow} F_0 \leftarrow \cdots \overset{f_{n-2}}{\leftarrow} F_{n-2} \overset{f_{n-1}}{\leftarrow} B \overset{i}{\leftarrow} A \leftarrow 0$$

and

$$0 \leftarrow C \overset{\overline{f}_0}{\leftarrow} \overline{F}_0 \leftarrow \cdots \overset{\overline{f}_{n-2}}{\leftarrow} \overline{F}_{n-2} \overset{f_{n-1}}{\leftarrow} \overline{B} \overset{\overline{i}}{\leftarrow} A \leftarrow 0$$

are congruent and minimal. Then $\mu(F_i) = \mu(\overline{F}_i)$ *and* $\mu(B) = \mu(\overline{B})$.

PROOF. The sequences

$$0 \leftarrow C \overset{f_0}{\leftarrow} F_0 \leftarrow \cdots \overset{f_{n-2}}{\leftarrow} F_{n-2}$$

and

$$0 \leftarrow C \overset{\overline{f}_0}{\leftarrow} \overline{F}_0 \leftarrow \cdots \overset{\overline{f}_{n-2}}{\leftarrow} \overline{F}_{n-2}$$

are segments of minimal resolutions. By Theorem 1.7, $\mu(F_i) = \mu(\overline{F}_i)$. Lemma 5.2 allows us, by stabilizing the f_i and $\overline{f}_i$, to obtain strict congruences; i.e., to obtain a diagram

$$\begin{array}{ccccccccc}
0 & \longleftarrow & \operatorname{Ker} f^s_{n-2} \oplus F & \longleftarrow & B \oplus F & \overset{i}{\longleftarrow} & A & \longleftarrow & 0 \\
 & & \Big\downarrow{\scriptstyle \varphi_{n-2}} & & \Big\downarrow{\scriptstyle \varphi_{n-1}} & & & & \\
0 & \longleftarrow & \operatorname{Ker} \overline{f}^s_{n-2} \oplus \overline{F} & \longleftarrow & \overline{B} \oplus \overline{F} & \overset{\overline{i}}{\longleftarrow} & A & \longleftarrow & 0,
\end{array}$$

where φ_{n-2} and φ_{n-1} are isomorphisms and $\mu(F) = \mu(\overline{F})$. Then $\mu(B \oplus F) = \mu(\overline{B} \oplus \overline{F})$, and therefore $\mu(B) = \mu(\overline{B})$. This proves the lemma.

Suppose the modules A and C are isomorphic, respectively, to modules $\overline{A}$ and $\overline{C}$. Consider a pair of elements $a \in \mathrm{Ext}^2(C, A)$ and $\overline{a} \in \mathrm{Ext}^2(\overline{C}, \overline{A})$.

DEFINITION 5.6. We say that a and $\overline{a}$ are isomorphic if there exist isomorphisms $f: A \to \overline{A}$ and $g: C \to \overline{C}$ such that $g^*(\overline{a}) = f_*(a)$ in the chain of homomorphisms

$$\mathrm{Ext}^2(\overline{C}, \overline{A}) \xrightarrow{g^*} \mathrm{Ext}^2(C, \overline{A}) \xleftarrow{f_*} \mathrm{Ext}^2(C, A).$$

REMARK. Using Proposition 5.1, Chap. 3, of [3] and Corollary 5.3, one can show that this definition is equivalent to the existence of a commutative diagram

$$\begin{array}{ccccccccccc} E: 0 & \longleftarrow & C & \longleftarrow & F & \longleftarrow & B & \longleftarrow & A & \longleftarrow & 0 \\ & & \downarrow{\scriptstyle f} & & \downarrow{\scriptstyle f_1} & & \downarrow{\scriptstyle g_1} & & \downarrow{\scriptstyle g} & & \\ \overline{E}: 0 & \longleftarrow & \overline{C} & \longleftarrow & \overline{F} & \longleftarrow & \overline{B} & \longleftarrow & \overline{A} & \longleftarrow & 0 \end{array}$$

where E ($\overline{E}$) is an arbitrary sequence representing the element a ($\overline{a}$), and F ($\overline{F}$) is a free module.

Denote by $s(\Lambda)$ the abelian monoid of stable isomorphism classes of (finitely generated) modules over a ring Λ. Two modules A and B represent the same element of $s(\Lambda)$ if $A \oplus m\Lambda \approx B \oplus n\Lambda$ for some m and n; and if [A] means the class of the module A in $s(\Lambda)$, then

$$[A] + [B] = [A \oplus B].$$

LEMMA 5.7. *Suppose the sequences*

$$E: 0 \leftarrow C \leftarrow F \leftarrow B \leftarrow A \leftarrow 0$$

and

$$\overline{E}: 0 \leftarrow \overline{C} \leftarrow \overline{F} \leftarrow \overline{B} \leftarrow \overline{A} \leftarrow 0$$

represent isomorphic elements $a \in \mathrm{Ext}^2(C, A)$ *and* $\overline{a} \in \mathrm{Ext}^2(\overline{C}, \overline{A})$. *Then* $[B] = [\overline{B}]$ *in* $s(\Lambda)$.

PROOF. By definition, there exist isomorphisms $f: A \to \overline{A}$ and $g: C \to \overline{C}$ such that $g^*(a) = f_*(\overline{a})$. The sequence $g(E)$ has the form $0 \leftarrow C \leftarrow \overline{F} \leftarrow \tilde{B} \leftarrow \overline{A} \leftarrow 0$, and the sequence $(\overline{E})f$ the form $0 \leftarrow C \leftarrow F \leftarrow \hat{B} \leftarrow \overline{A} \leftarrow 0$. By construction, $\tilde{B} \approx \overline{B}$ and $\hat{B} \approx B$. Since $g(E)$ is congruent to $(\overline{E})f$, we have by Corollary 5.3 that $\tilde{B}$ is stably isomorphic to $\hat{B}$. This proves the lemma.

Observe that if Λ is an IBN, s-ring and $\mu(F) = \mu(\overline{F})$, then $\mu(\overline{B}) = \mu(B)$.

Suppose the modules A and $\overline{A}$ are isomorphic, and C and $\overline{C}$ are stably isomorphic. Consider a pair of elements $a \in \mathrm{Ext}^2(C, A)$ and $\overline{a} \in \mathrm{Ext}^2(\overline{C}, \overline{A})$.

DEFINITION 5.8. We say that a and $\overline{a}$ are *stably isomorphic* if there exist numbers m and n and isomorphisms $f: A \to \overline{A}$ and $g: C \oplus n\Lambda \to \overline{C} \oplus m\Lambda$ such that $g^*(p_1^*(\overline{a})) = f_*(p_2^*(a))$ in the chain of homomorphisms

$$\begin{array}{ccccc}
\operatorname{Ext}^2(\overline{C}, \overline{A}) & \xrightarrow{p_1^*} & \operatorname{Ext}^2(\overline{C} \oplus m\Lambda, \overline{A}) & & \\
 & & & \searrow^{g^*} & \\
 & & & & \operatorname{Ext}^2(C \oplus n\Lambda, \overline{A}), \\
 & & & \nearrow_{f_*} & \\
\operatorname{Ext}^2(C, A) & \xrightarrow{p_2^*} & \operatorname{Ext}^2(C \oplus n\Lambda, A) & &
\end{array}$$

where $p_1: \overline{C} \oplus m\Lambda \to \overline{C}$ and $p_2: C \oplus n\Lambda \to C$ are the canonical projections.

Definition 5.8 has an equivalent formulation in terms of homomorphisms of 2-fold exact sequences representing the elements a and $\overline{a}$, in the same way as for Definition 5.6. There is also an analogue of Lemma 5.7; its proof is essentially a repetition of the proof for the latter.

We consider also the action of a ring isomorphism $\theta: \Lambda \to \Lambda$. Using θ, we can define a new Λ-module structure on C by the formula $\lambda \cdot c = \theta(\lambda) \cdot c$; we denote this structure by C_θ. There is an obvious isomorphism

$$\theta: \operatorname{Ext}^n(C, A) \to \operatorname{Ext}^n(C_\theta, A_\theta).$$

We say that two elements $a \in \operatorname{Ext}^2(C, A)$ and $\overline{a} \in \operatorname{Ext}^2(\overline{C}, \overline{A})$ are θ-isomorphic (stably θ-isomorphic) if the elements a and $\theta(\overline{a})$ are isomorphic (stably isomorphic). There is an analogue of Lemma 5.7 for θ-isomorphic (stably θ-isomorphic) elements a and $\theta(\overline{a})$.

DEFINITION 5.9. Let $\{\mathbb{C}, \partial\}: 0 \leftarrow C_0 \leftarrow \cdots \xleftarrow{\partial_i} C_i \xleftarrow{\partial_{i+1}} C_{i+1} \leftarrow \cdots \xleftarrow{\partial_n} C_n \leftarrow 0$ be a free chain complex. Following Dyer [25], we shall mean by the Swan-Wall class in dimension i the class of the module $\Gamma_i = C_i/\partial_{i+1}(C_{i+1})$ in $s(\Lambda)$; and we denote it by $\Gamma_i(\mathbb{C})$.

If two chain complexes $\{\mathbb{C}, \partial\}$ and $\{\mathbb{D}, d\}$ are homotopy equivalent, then, as follows from the theorem of Cockcroft-Swan, $[\Gamma_i(\mathbb{C})] = [\Gamma_i(\mathbb{D})]$ for all i. For the chain complex $\{\mathbb{C}, \partial\}$ we have in every dimension the exact sequence

$$0 \leftarrow C_i/\partial_{i+1}(C_{i+1}) \leftarrow C_i \xleftarrow{\partial_{i+1}} C_{i+1} \hookleftarrow Z_{i+1} \leftarrow 0,$$

where $Z_{i+1} = \operatorname{Ker} \partial_{i+1}$. Factoring out $\partial_{i+2}(C_{i+2})$, we obtain the exact sequence

$$\omega_i(\mathbb{C}): 0 \leftarrow C_i/\partial_{i+1}(C_{i+1}) \leftarrow C_i \leftarrow C_{i+1}/\partial_{i+2}(C_{i+2}) \leftarrow H_{i+1}(\mathbb{C}) \leftarrow 0, \qquad (*)$$

representing an element of $\operatorname{Ext}^2(\Gamma_i(\mathbb{C}), H_{i+1}(\mathbb{C}))$ (for $i = 0$, an element of $\operatorname{Ext}^2(H_0(\mathbb{C}), H_1(\mathbb{C}))$). An obvious consequence of the Cockcroft-Swan theorem is the fact that for homotopy equivalent complexes the $\omega_i(\mathbb{C})$ are stably isomorphic. Thus, we have two stable invariants of chain complexes. Observe that specifying $\omega_i(\mathbb{C})$ determines $\Gamma_{i+1}(\mathbb{C})$. It is well known that homology and cohomology do not, in general, determine the homotopy type of a free chain complex over a ring Λ [24]. An approach has been proposed by Heller and Dold to this problem, i.e., to the description of a complete set of invariants defining the homotopy type of a chain complex. Essentially, this is the algebraic analogue of the natural Postnikov systems. Here we modify this approach, adapting it more to our purposes.

Suppose the simple-homotopy type of a chain complex $\{\mathbb{C}, \partial\}$: $0 \leftarrow C_0 \overset{\partial_1}{\leftarrow} C_1 \leftarrow \cdots \overset{\partial_n}{\leftarrow} C_n \leftarrow 0$ over a ring Λ contains a minimal chain complex.

PROPOSITION 5.10. *Let Λ be an IBN, s-ring, and assume that $H_i(C)$, $\Gamma_i(\mathbb{C})$, and $\omega_i(\mathbb{C})$ are given. Then there exists an algorithm for computing the ranks of the chain modules in the minimal chain complex.*

PROOF. Consider the sequence $\omega_0(\mathbb{C}): 0 \leftarrow H_0(\mathbb{C}) \leftarrow C_0 \leftarrow \tilde{\Gamma}_1 \leftarrow H_1(\mathbb{C}) \leftarrow 0$, and choose in its congruence class a minimal sequence $\omega_0^\mu : 0 \leftarrow H_0(\mathbb{C}) \leftarrow \overline{F}_0 \leftarrow \overline{\Gamma}_1 \leftarrow H_1(\mathbb{C}) \leftarrow 0$. By Lemma 5.5, $\mu(\overline{\Gamma}_1)$ is the same for all minimal sequences in the congruence class of $\omega_0(\mathbb{C})$. Furthermore, by the construction and by Corollary 5.3, $\tilde{\Gamma}_1$ and $\Gamma_1(\mathbb{C})$ are stably isomorphic; i.e.,

$$0 \leftarrow \Gamma_1(\mathbb{C}) \overset{p_1}{\leftarrow} \Gamma_1(\mathbb{C}) \oplus n_1\Lambda \overset{f}{\leftarrow} \tilde{\Gamma}_1 \oplus m_1\Lambda \overset{i_1}{\leftarrow} \tilde{\Gamma}_1 \leftarrow 0,$$

where f is an isomorphism and i_1 and p_1 are the canonical inclusion and projection. We have

$$0 \leftarrow \Gamma_1(\mathbb{C}) \overset{\overline{p}_1}{\leftarrow} \Gamma_1(\mathbb{C}) \oplus \overline{n}_1\Lambda \overset{\overline{f}}{\leftarrow} \overline{\Gamma}_1 \oplus \overline{m}_1\Lambda \overset{\overline{i}_1}{\leftarrow} \overline{\Gamma}_1 \leftarrow 0.$$

Using the mapping $\overline{p}_1 \cdot \overline{f}_1 \cdot \overline{i}_1$, construct from

$$\omega_1(\mathbb{C}) : 0 \leftarrow \Gamma_1(\mathbb{C}) \leftarrow F_1 \leftarrow \tilde{\Gamma}_2 \leftarrow H_2(\mathbb{C}) \leftarrow 0$$

the sequence $(\omega_1(\mathbb{C}))\overline{p}_1 \cdot \overline{f}_1 \cdot \overline{i}_1$; i.e., obtain the commutative diagram

$$\begin{array}{ccccccccccc}
(\omega_1(\mathbb{C}))\overline{p}_1 \cdot \overline{f}_1 \cdot \overline{i}_1 : 0 & \longleftarrow & \overline{\Gamma}_1 & \longleftarrow & \hat{F}_1 & \longleftarrow & \hat{\Gamma}_2 & \longleftarrow & H_2(\mathbb{C}) & \longleftarrow & 0 \\
 & & \downarrow{\scriptstyle \overline{p}_1 \cdot \overline{f}_1 \cdot \overline{i}_1} & & \downarrow & & \downarrow & & \downarrow{\scriptstyle \mathrm{id}} & & \\
\omega_1(\mathbb{C}): \quad 0 & \longleftarrow & \Gamma_1(\mathbb{C}) & \longleftarrow & F_1 & \longleftarrow & \tilde{\Gamma}_2 & \longleftarrow & H_2(\mathbb{C}) & \longleftarrow & 0.
\end{array}$$

Since $\omega_1(\mathbb{C})$ and $(\omega_1(\mathbb{C}))\overline{p}_1 \cdot \overline{f}_1 \cdot \overline{i}_1$, are by construction stably isomorphic, we have $[\hat{\Gamma}_2] = [\tilde{\Gamma}_2] = [\Gamma_2(\mathbb{C})]$. Now choose in the congruence class of $(\omega_1(\mathbb{C}))\overline{p}_1 \cdot \overline{f}_1 \cdot \overline{i}_1$ a minimal sequence

$$\omega_1^\mu : 0 \leftarrow \overline{\Gamma}_1 \leftarrow \overline{F}_1 \leftarrow \overline{\Gamma}_2 \leftarrow H_2(\mathbb{C}) \leftarrow 0.$$

Repeating the argument, we find a mapping of $\overline{\Gamma}_2$ into $\Gamma_2(\mathbb{C})$, and using it, construct an exact sequence

$$\hat{\omega}_2 : 0 \leftarrow \Gamma_2 \leftarrow F_2 \leftarrow \hat{\Gamma}_3 \leftarrow H_3(\mathbb{C}) \leftarrow 0,$$

stably isomorphic to $\omega_2(\mathbb{C})$, in whose congruence class we then choose a minimal sequence ω_2^μ. Iterating this process, we construct a set of minimal sequences

$$\omega_i^\mu : 0 \leftarrow \overline{\Gamma}_i \leftarrow \overline{F}_i \leftarrow \overline{\Gamma}_{i+1} \leftarrow H_{i+1}(\mathbb{C}) \leftarrow 0,$$

stably isomorphic to the $\omega_i(\mathbb{C})$. Consider now a minimal chain complex

$$\overline{\mathbb{C}} = \{\overline{C}_i, \overline{\partial}_i\}: 0 \leftarrow \overline{C}_0 \overset{\overline{\partial}_1}{\leftarrow} \overline{C}_1 \leftarrow \cdots \overset{\overline{\partial}_n}{\leftarrow} \overline{C}_n \leftarrow 0,$$

simple-homotopy equivalent to $\{\mathbb{C}, \partial\}$, and the corresponding sequences

$$\omega_i(\overline{\mathbb{C}}): 0 \leftarrow \widetilde{\Gamma}_i(\overline{\mathbb{C}}) \leftarrow \overline{C}_i \leftarrow \Gamma_{i+1}(\overline{\mathbb{C}}) \leftarrow H_{i+1}(\overline{\mathbb{C}}) \leftarrow 0.$$

Obviously, these $\omega_i(\mathbb{C})$ are also minimal sequences, stably isomorphic (for $i = 0$, isomorphic) to the $\omega_i(\mathbb{C})$. In view of the preceding remarks, $\mu(\Gamma_i(\overline{\mathbb{C}})) = \mu(\overline{\Gamma}_i)$. But $\mu(\Gamma_i(\overline{\mathbb{C}})) = \mu(\overline{C}_i)$; and therefore the rank of the chain module $\overline{C}_i$ of the minimal chain complex $\overline{\mathbb{C}} = \{\overline{C}_i, \overline{\partial}_i\}$ is equal to $\mu(\overline{\Gamma}_i)$. We have thus constructed an algorithm for computing the ranks of the chain modules of the minimal chain complex; the proposition is proved.

In certain cases, this computation can be done using only homology.

PROPOSITION 5.11. *Let $\{C_i, \partial_i\}$ be a chain complex over $\mathbb{Z}[\pi]$, where π is an s-group and f-$\dim \pi \le k$. Suppose the homology modules H_i of the complex $\{C_i, \partial_i\}$ have free resolutions*

$$0 \leftarrow H_i \leftarrow G_i^0 \leftarrow G_i^1 \leftarrow 0,$$

with $\mu(F_{i_0}^0 + \mu(F_{i_0-1}^1) \ge k$ for some i_0. Then $\{C_i, \partial_i\}$ is simple-homotopy equivalent to a minimal chain complex

$$\{\overline{C}_i, \overline{\partial}_i\}: 0 \leftarrow \overline{C}_0 \leftarrow \overline{C}_1 \leftarrow \cdots \leftarrow \overline{C}_n \leftarrow 0,$$

in which $\mu(\overline{C}_i) = \mu(H_i) + \mu(F_{i-1}^1)$, where

$$0 \leftarrow H_i \leftarrow F_i^0 \leftarrow F_i^1 \leftarrow 0$$

is a minimal free resolution of H_i.

COROLLARY 5.12. *Under the conditions of Proposition 5.11, if π is an s-group with* $\mathrm{Wh}(\pi) = 0$, *then the ranks of the chain modules of the minimal chain complex are given by*

$$\mu(\overline{C}_i) = \mu(H_i) + \mu(F_{i-1}^s).$$

PROOF OF PROPOSITION 5.11. In this case the modules of cycles are direct free summands in the chain modules, and we can use essentially the same argument that applies to complexes over $\mathbb{Z}$. The condition on r-$\dim \pi$ allows us to use Lemma 3.2 to get rid of any torsion block that might arise.

Let $f = \{f_i\}: \{C_i, \partial_i\} \to \{D_i, \partial_i\}$ be a homotopy equivalence associated with an automorphism $\theta: \Lambda \to \Lambda$. Then a consequence of the Cockcroft-Swan theorem (see the end of the preceding section) is the fact that $\omega_i(C_i, \partial_i)$ and $\omega_i(D_i, d_i)$ are stably θ-isomorphic (and θ-isomorphic for $i = 0$).

§6. Minimal Morse functions on nonsimply-connected cobordisms

Let W^n be a compact manifold with boundary $\partial W^n = V_0 \cup V_1$. Assume $\pi = \pi_1(V_0) \hookrightarrow \pi_1(W^n) \hookleftarrow \pi_1(V_1)$ is an isomorphism, and $n \geq 6$. We shall study in this section the question: when does there exist on W^n a minimal Morse function? With this in mind, we construct for an arbitrary regular Morse function f and regular gradient-like vector field ξ a free chain complex over the ring $\mathbb{Z}[\pi]$. The simple-homotopy type of this chain complex is determined by the manifold W^n alone, and is independent of the choice of f and ξ. Using the results of the preceding sections, we can state conditions under which W^n admits a minimal chain complex. By performing a sequence of alterations on the function f and the vector field ξ, we can then get the associated based chain complex to be minimal. This yields a minimal Morse function on W^n.

Consider a compact manifold with boundary $\partial W^n = V_0 \cup V_1$. Choose a point $x_0 \in V_0$ as base point for the fundamental group, and let $\widetilde{W}^n$ be the universal covering space of W^n. Let $p: (\widetilde{W}^n, \widetilde{V}_0, \widetilde{V}_1) \to (W^n, V_0, V_1)$ be the covering mapping; here $\widetilde{V}_i = p^{-1}(V_i)$, $i = 0, 1$. On W^n, choose a regular Morse function $f: W^n \to I$ and a regular gradient-like vector field ξ. Using the mapping p, lift f and ξ to $\widetilde{W}^n$, and denote the lifted function and vector field by $\tilde{f}$ and $\tilde{\xi}$, respectively. Using f and ξ ($\tilde{f}$ and $\tilde{\xi}$), construct chain complexes of abelian groups

$$C(W^n, f, \xi): 0 \leftarrow C_0 \xleftarrow{\partial_1} \cdots \xleftarrow{\partial_n} C_n \leftarrow 0$$

and

$$C(\widetilde{W}^n, \tilde{f}, \tilde{\xi}): 0 \leftarrow \widetilde{C}_0 \xleftarrow{\tilde{\partial}_1} \cdots \xleftarrow{\tilde{\partial}_n} \widetilde{C}_n \leftarrow 0,$$

where

$$C_i = H_i(W_i, W_{i-1}, \mathbb{Z}), \qquad \widetilde{C}_i = H_i(\widetilde{W}_i, \widetilde{W}_{i-1}, \mathbb{Z}),$$

and $W_i = f^{-1}[0, c_i]$, $\widetilde{W}_i = \tilde{f}^{-1}[0, c_i]$ are submanifolds containing all the critical points of index i. As generators of the C_i ($\widetilde{C}_i$) take the median disks of the critical points of index i, constructed for the field ξ ($\tilde{\xi}$); see [6] for details. On $\widetilde{W}^n$ there is an action of the group $\pi = \pi_1(W^n, \chi_0)$; using it, we can make $\widetilde{C}_i$ into a free finitely generated module over the group ring $\mathbb{Z}[\pi_1(W^n, \chi_0)]$. This construction gives a distinguished basis in the module $\widetilde{C}_i$. For each median disk of a critical point $x \in W^n$ of f, choose the median disk covering it for a corresponding critical point y ($x = p(y)$). The boundary homomorphism can be described by the matrix of incidence coefficients $a_{ij} \in \mathbb{Z}[\pi_1(W^n, x_0)]$, where $\tilde{\partial}_i(c_k^i) = \sum a_{ij} \tilde{c}_i^{k-1}$ and the $\tilde{c}_i^j$ are the generators of the module $\tilde{C}_j$. Since we shall be carrying out elementary operations over the complex $C(\widetilde{W}^n, \tilde{f}, \tilde{\xi})$, we must single out critical points of the function $\tilde{f}$; i.e., take fixed paths joining the critical points to the base point x_0. Without loss of generality, we can assume that the paths to critical points of index i lie in the manifold W_i. Choose distinguished points on the median and co-median spheres, and join them by paths on the median and co-median disks to the corresponding critical points. We can now alter the function f and the field ξ so as to impose on the matrix of the boundary homomorphism $\tilde{\partial}_i$ elementary operations over the ring $\mathbb{Z}[\pi_1(W^n, x_0)]$. For details, see [9], or [30], [18].

Now take a fixed isomorphism $\varphi : \pi \to \pi_1(W^n, x_0)$ of the group π with $\pi_1(W^n, x_0)$. Then the $\tilde{C}_i$ can be regarded as $\mathbb{Z}[\pi]$-modules. The construction of the chain complex $C(\widetilde{W}^n, \tilde{f}, \tilde{\xi})$ involves a certain arbitrariness, depending on the choice of the base point x_0, the isomorphism φ and the bases in the $\tilde{C}_i$. Let us examine this in detail. Suppose $x_1 \in V_0$ is a second base point, $\widetilde{W}_1^n$ the corresponding universal covering space (i.e., the set of equivalence classes of paths joining x_1 to the points of W^n), $p_1 : \widetilde{W}_1^n \to W^n$ the projection, $\tilde{f}_1$ and $\tilde{\xi}_1$ the liftings of f and ξ to $\widetilde{W}_1^n$, and $C(\widetilde{W}_1^n, \tilde{f}_1, \tilde{\xi}_1)$ the corresponding chain complex of $\mathbb{Z}[\pi_1(W^n, x_1)]$-modules. Choose a path $\gamma : I \to W^n$ joining the points x_0 and x_1. It induces a diffeomorphism $\tilde{\gamma}$ in the commutative diagram

$$\begin{array}{ccc} \widetilde{W}_1^n & \xrightarrow{\tilde{\gamma}} & \widetilde{W}_n \\ & {\scriptstyle p_1}\searrow \quad \swarrow{\scriptstyle p} & \\ & W^n & \end{array}$$

and an isomorphism $\gamma_* : \pi_1(W^n, x_1) \to \pi_1(W^n, x_0)$. Obviously $\tilde{\gamma}$ induces a chain isomorphism $\tilde{\gamma}_* : C(\widetilde{W}_1^n, \tilde{f}_1, \tilde{\xi}_1) \to C(W^n, \tilde{f}, \tilde{\xi})$ associated with the isomorphism $\overline{\gamma} : \mathbb{Z}[\pi_1(W^n, x_1)] \to \mathbb{Z}[\pi_1(W^n, x_0)]$ induced by γ_*; i.e., $\tilde{\gamma}_*(a \cdot c) = \overline{\gamma}(a) \cdot \tilde{\gamma}_*(c)$, $a \in \pi_1(W^n, x_1)$, $c \in C(\widetilde{W}_1^n, \tilde{f}_1, \tilde{\xi}_1)$. Choosing an isomorphism $\varphi_1 : \pi \to \pi_1(W^n, x_1)$, we can regard the complex $C(\widetilde{W}_1^n, \tilde{f}_1, \tilde{\xi}_1)$ as a chain complex of $\mathbb{Z}[\pi]$-modules. There now exists an isomorphism $\theta : \pi \to \pi$ that makes the diagram

$$\begin{array}{ccc} \pi_1(W^n, x_1) & \xrightarrow{\gamma_*} & \pi_1(W^n, x_0) \\ {\scriptstyle \varphi_1}\uparrow & & \uparrow{\scriptstyle \varphi} \\ \pi & \xrightarrow{\theta} & \pi \end{array}$$

commutative. Thus, the chain complexes $C(\widetilde{W}^n, \tilde{f}, \tilde{\xi})$ and $C(\widetilde{W}_1^n, \tilde{f}_1, \tilde{\xi}_1)$, taken as consisting of $\mathbb{Z}[\pi]$-modules, are θ-chain-isomorphic. If we choose a second path γ_1 joining x_0 and x_1, then γ_* is replaced by $\theta_x\gamma_*$, where θ_x is the isomorphism of conjugation in $\pi_1(W^n, x_0)$ via the element $x = \gamma\cdot\gamma_1^{-1}$. The semilinear isomorphism $\overline{\gamma}$ is replaced by $x\cdot\overline{\gamma}$. Thus, the arbitrariness in the choice of the base point determines the chain complex $C(\widetilde{W}^n, \tilde{f}, \tilde{\xi})$ up to a semilinear isomorphism. It is easily shown that the choice of the identification isomorphism $\varphi\colon \pi \to \pi_1(W^n, x_0)$ likewise determines the complex $C(\widetilde{W}^n, \tilde{f}, \tilde{\xi})$ up to a semilinear isomorphism. Finally, the bases in the modules $\widetilde{C}_i$ are determined up to multiplication by elements of the fundamental group, which allows us to choose them so that the torsion of the corresponding semilinear isomorphisms vanishes. Consequently, the simple-homotopy type of the chain complex constructed from a Morse function f and a gradient-like vector field ξ is uniquely determined.

It is known [39] that if the inclusions $V_0 \hookrightarrow W^n \hookleftarrow V_1$ induce isomorphisms $\pi_1(V_0) \hookrightarrow \pi_1(W^n) \hookleftarrow \pi_1(V_1)$, then on W^n there exist Morse functions without critical points of index $0, 1, n-1$, or n. We can therefore restrict ourselves to chain complexes of length $n-4$. All the elementary operations over the chain complexes over $\mathbb{Z}[\pi]$ can be realized by alterations on the Morse function and the vector field. We stress one feature of this process, connected with application of the Whitney lemma in codimension 2; it occurs when we are eliminating critical points of index $n-2$ and $n-3$. Let $V = f^{-1}(c)$, where $f(x^i_{n-2}) > c > f(y^j_{n-3})$, y^j_{n-3} is a critical point of index $n-3$, and x^i_{n-2} a critical point of index $n-2$. Obviously $\pi_1(V) \approx \pi_1(V_1)$, since the median spheres of critical points of index $n-3$ have codimension 3 in ∂W_{n-4} and so induce no change in the fundamental group of the level surface of the function. But since the median spheres of critical points of index $n-2$ have codimension 2 in V, we need an additional argument to allow us to eliminate critical points of index $n-2$ and $n-3$ in the case when they satisfy an algebraic annihilation condition. Suppose the intersection index for a pair of median and co-median spheres is $\varepsilon(S^2_R, S^{n-2}_L) = \pm g$, where $g \in \pi$; we prove that the corresponding critical points of index $n-2$ and $n-3$ can be suppressed. Here S^{n-2}_L is the median sphere of a critical point of index $n-2$, and S^2_R the co-median sphere of a critical point of index $n-3$. For this it is necessary to verify all the conditions for applying the Whitney lemma; i.e., to verify that $\pi_1(V\backslash S^{n-2}_L) \hookrightarrow \pi_1(V)$ is a monomorphism. But this is indeed the case, since the trajectories of the gradient-like vector field ξ of the function f determine a diffeomorphism $V_1\backslash S^1_R \approx V\backslash S^{n-2}_L$, where S^1_R is the co-median sphere of the critical point of index $n-2$. The fact that $\pi_1(V_1) \to \pi_1(W^n)$ is an isomorphism means that S^1_R realizes the trivial element of $\pi_1(V_1)$. Consequently, $\pi_1(V_1) \approx \pi_1(V_1\backslash S^1_R)$; and so $\pi_1(V\backslash S^1_R) \approx \pi_1(V\backslash S^{n-2}_L) \approx \pi_1(V)$.

PROPOSITION 6.1. *Let $C(\widetilde{W}^n, \tilde{f}, \tilde{\xi})$ and $C(\widetilde{W}^n, \tilde{g}, \tilde{\eta})$ be two free based chain complexes over $\mathbb{Z}[\pi]$, constructed from the regular Morse functions f*

and g and the regular gradient-like vector fields ξ and η, respectively. Then they are simple-homotopy equivalent. Furthermore, any free based chain complex of $\mathbb{Z}[\pi]$-modules of the form

$$0 \leftarrow C_2 \overset{\partial_3}{\leftarrow} C_3 \leftarrow \cdots \overset{\partial_{n-2}}{\leftarrow} C_{n-2} \leftarrow 0$$

simple-homotopy equivalent to $C(\widetilde{W}^n, \tilde{f}, \tilde{\xi})$ can be realized as a complex associated with some regular Morse function and regular gradient-like vector field defined on W^n.

PROOF. The first assertion is proved in [31, p. 131, Theorem 5.10]. It results from the following argument. Let $K(f)$ and $K(g)$ be the relative CW-complexes obtained by contracting the manifold W^n at the median disks of the critical points of the functions f and g, respectively. From these one can construct based chain complexes of $\mathbb{Z}[\pi]$-modules that coincide with the complexes $C(\widetilde{W}^n, \tilde{f}, \tilde{\xi})$ and $C(\widetilde{W}^n, \tilde{g}, \tilde{\eta})$. By construction, $K(f)$ and $K(g)$ are simple-homotopy equivalent to W^n, and therefore simple-homotopy equivalent to each other. It follows that $C(\widetilde{W}^n, \tilde{f}, \tilde{\xi})$ and $C(\widetilde{W}^n, \tilde{g}, \tilde{\eta})$ have the same simple-homotopy type. Here we impose no restriction on the length of the complexes $C(\widetilde{W}^n, \tilde{f}, \tilde{\xi})$ and $C(\widetilde{W}^n, \tilde{g}, \tilde{n})$; i.e., the functions f and g can have critical points of index $0, 1, n-1$, and n.

Let us prove now the second assertion. Suppose the based chain complex $C(\widetilde{W}^n, \tilde{f}, \tilde{\xi})$ associated with a regular Morse function f and gradient-like vector field ξ is simple-homotopy equivalent to $\{D_i, d_i\}$. We can assume that $\{D_i, d_i\}$ has length $n-4$, i.e., that f has no critical points of index $0, 1, n-1$, or n, since $\pi_1(V_0) \hookrightarrow \pi_1(W^n) \hookleftarrow \pi_1(V_1)$ is an isomorphism. There exists a sequence of elementary operations taking $C(\widetilde{W}^n, \tilde{f}, \tilde{\xi})$ into $\{D_i, d_i\}$ without increasing the length for any intermediate complex. Since each elementary operation can be realized by suitably altering the function f and the field ξ, we obtain our assertion. The existence of such a sequence of elementary operations is ensured by the Cockcroft-Swan theorem, applied to the complexes $C(\widetilde{W}^n, \tilde{f}, \tilde{\xi})$ and $\{D_i, d_i\}$. This completes the proof of the proposition.

We remark that the restriction on the length of the chain complex in the hypothesis of the proposition cannot be omitted. Indeed, if M^n is a closed manifold, with $\pi_1(M^n) \approx \pi$, then any element τ_0 of the group $\mathrm{Wh}\,\pi$ can be realized by an h-cobordism (W^{n+1}, M^n, N^n) such that $\tau(W^{n+1}, N^n) = \tau_0$, $n \geq 6$, and the h-cobordism admits a Morse function with critical points of index 2 and 3 only. Moreover [23], there exist examples of groups and elements $\tau_i \in \mathrm{Wh}(\pi)$ such that h-cobordisms with torsion equal to τ_i do not admit Morse functions with critical points of index 1 or 2. Consequently, not every chain complex $\{D_i, d_i\}$ simple-homotopy equivalent to a chain complex $C(\widetilde{W}^n, \tilde{f}, \tilde{\xi})$ can be realized by a Morse function, even though the latter complex is itself associated with a Morse function. In [31] it is shown that chain complexes constructed either via Morse functions or by means of a triangulation of W^n have the same simple-homotopy type. It therefore makes sense to speak of the simple-homotopy type of the chain complex $C(\widetilde{W}^n)$ of a cobordism (W^n, V_0, V_1).

PROPOSITION 6.2. *Let (W^n, V_0, V_1) be a cobordism, $n \geq 6$, and $\pi_1(V_0) \hookrightarrow \pi_1(W^n) \hookleftarrow \pi_1(V_1)$ an isomorphism. In order that there exist on W^n a minimal Morse function, it is necessary and sufficient that the chain complex $C(\widetilde{W}_n)$ of the cobordism be simple-homotopy equivalent to a minimal chain complex.*

PROOF. Necessity is obvious. We prove sufficiency. Let f be an arbitrary Morse function, ξ a gradient-like vector field, $C(\widetilde{W}^n, \tilde{f}, \tilde{\xi})$ a corresponding based chain complex, and $\{\overline{C}_i, \overline{\partial}_i\}$ a minimal chain complex simple-homotopy equivalent to $C(\widetilde{W}^n, \tilde{f}, \tilde{\xi})$. Proposition 6.1 ensures the existence of a Morse function g and gradient-like vector field that realize the complex $\{\overline{C}_i, \overline{\partial}_i\}$. Then g is a minimal Morse function. This proves the proposition.

COROLLARY 6.3. *Let (W^n, V_0, V_1) be a cobordism, $n \geq 6$, $\pi \approx \pi_1(V_0) \hookrightarrow \pi_1(W^n) \hookleftarrow \pi_1(V_1)$ an isomorphism, π an s-group, and $\mathrm{Wh}(\pi) = 0$. Then on W^n there exists a minimal Morse function.*

PROOF. Theorems 4.3 and 4.7 ensure that any chain complex corresponding to an arbitrary Morse function and gradient-like vector field on W^n will be simple-homotopy equivalent to a minimal chain complex, since $\mathrm{Wh}\,\pi = 0$.

PROPOSITION 6.4. *In the homotopy type of a cobordism (W^n, V_0, V_1), $n \geq 6$, such that $\pi \approx \pi_1(V_0) \hookrightarrow \pi_1(W^n) \hookleftarrow \pi_1(V_1)$ is an isomorphism and π is an s-group, there always exists a cobordism admitting a minimal Morse function.*

PROOF. Proposition 4.5 allows us to construct on W^n a regular Morse function f and regular gradient-like vector field ξ such that the associated chain complex is of the form $\{C_i, \partial_i\} \oplus R_{n-2}$, where R_{n-2} is an elementary chain complex generated by median disks of critical points of index $n-3$ and $n-2$. Let Γ be a cobordism, with boundary $\partial\Gamma = V \cup V_1$, containing only those critical points of index $n-3$ and $n-2$ whose median disks give the generators in R_{n-2}. Obviously Γ is an h-cobordism. Consider the manifold $\overline{W}^n = \overline{W^n \backslash \Gamma}$. Clearly, $\overline{W}^n$ and W^n have the same homotopy type. By the construction and by Theorem 4.7, there exists on $\overline{W}^n$ a minimal Morse function.

DEFINITION 6.5 [6]. The ith Morse number of a manifold M^n is the minimal number of critical points of index i, taken over all Morse functions on M^n.

PROPOSITION 6.6. *With the hypothesis of Proposition* 6.4, *suppose* $r\text{-}\dim\pi \le k$ *and the* i *th Morse number of the manifold* W^n *satisfies* $m_i(W^n) \ge k$ $(2 \le i \le n-2)$. *Then on* W^n *there exists a minimal Morse function.*

PROOF. This follows from Proposition 4.9.

From these results, and the results in §1 concerning s-groups and in §5 concerning $r\text{-}\dim\pi$, we have the following propositions.

PROPOSITION 6.7. *Let* (W^n, V_0, V_1) *be a cobordism,* $\pi \approx \pi_1(V_0) \hookrightarrow \pi_1(W^n) \hookleftarrow \pi_1(V_1)$ *an isomorphism, and* $n \ge 6$. *Suppose*

(1) π *is a free group*; *or*
(2) π *is a free abelian group*; *or*
(3) $\pi = \mathbb{Z}_k$ $(k = 2, 3, 4, 6)$.

Then on W^n *there exists a minimal Morse function.*

PROPOSITION 6.8. *Let* (W^n, V_0, V_1) *be the cobordism of the preceding proposition. Suppose*

(1) π *is a finite abelian group isomorphic to* $(\mathbb{Z}_2)^n$; *or*
(2) *every Sylow* p-*subgroup of* π *is of the form* $\mathbb{Z}_p$ *or* $\mathbb{Z}_p \oplus \mathbb{Z}_p$; *or*
(3) π *is a finite nonabelian group that has no epimorphic mapping onto* $\mathbb{Z}_{p^2} \oplus \mathbb{Z}_{p^2}$, $\mathbb{Z}_p \oplus \mathbb{Z}_p \oplus \mathbb{Z}_p$, $\mathbb{Z}_4 \oplus \mathbb{Z}_4$, $\mathbb{Z}_4 \oplus \mathbb{Z}_2 \oplus \mathbb{Z}_2$, $\mathbb{Z}_p \oplus \mathbb{Z}_2 \oplus \mathbb{Z}_2 \oplus \mathbb{Z}_2$ *for* p *prime, and* $H_i(W^n, V_0, \mathbb{Z}) \neq 0$ *for at least one value of* i.

Then on W^n *there exists a minimal Morse function.*

PROPOSITION 6.9. *Let* (W^n, V_0, V_1) *be a cobordism,* $n \ge 6$, $\pi \approx \pi_1(V_0) \hookrightarrow \pi_1(W^n) \hookleftarrow \pi_1(V_1)$ *an isomorphism,* π *a finite* s-*group, and*

$$\mu(H_{i_0}(W^n, V_0, \mathbb{Z})) + \mu(\operatorname{Tors} H_{i_0-1}(W^n, V_0, \mathbb{Z})) \ge 2$$

$(3 \le i_0 \le n-2)$. *Then on* W^n *there exists a minimal Morse function.*

Finally, we note that Proposition 4.9 guarantees that on a cobordism (W^n, V_0, V_1) for which $\pi \approx \pi_1(V_0) \hookrightarrow \pi_1(W^n) \hookleftarrow \pi_1(V_1)$ is an isomorphism and π an s-group, with $n \ge 6$, there exists a Morse function for which, with the possible exception of two adjacent indices, the number of critical points of index i $(2 \le i \le n-3)$ is equal to the i th Morse number of the cobordism.

We turn now to the question of computing the number of critical points for a minimal Morse function on those cobordisms that have been considered in this paper.

For chain complexes either associated with regular Morse functions or constructed by means of a triangulation of the cobordism (W^n, V_0, V_1), the Swan-Wall classes are well defined for all dimensions. We shall denote them by $\Gamma_\lambda(\widetilde{W}^n)$. Similarly to what was done in the preceding section, we can define for these complexes certain elements of $\operatorname{Ext}^2(\Gamma_\lambda(\widetilde{W}^n), H_{\lambda+1}(\widetilde{W}^n, \widetilde{V}_0, \mathbb{Z}))$, which we denote by $\omega_\lambda(\widetilde{W}^n)$.

PROPOSITION 6.10. *Let (W^n, V_0, V_1) be a cobordism, $\pi \approx \pi_1(V_0) \hookrightarrow \pi_1(W^n) \hookleftarrow \pi_1(V_1)$ isomorphisms, $n \geq 6$, and π an s-group. Suppose there exists on W^n a minimal Morse function. Assume given $H_i(\widetilde{W}^n, \widetilde{V}_0, \mathbb{Z})$, $\Gamma_i(\widetilde{W}^n)$, and $\omega_i(\widetilde{W}^n)$. Then there exists an algorithm for finding the number of critical points of a minimal Morse function.*

PROOF. See the proof of Proposition 5.10.

Similarly, again, to the preceding section, we have

PROPOSITION 6.11. *Let (W^n, V_0, V_1) be a cobordism, $\pi \approx \pi_1(V_0) \hookrightarrow \pi_1(W^n) \hookleftarrow \pi_1(V_1)$ isomorphisms, $n \geq 6$, π an s-group, and r-$\dim \pi \leq k$. Suppose the homology modules $H_i(\widetilde{W}^n, \widetilde{V}_0, \mathbb{Z})$ have free resolutions*

$$0 \leftarrow H_i(\widetilde{W}^n, \widetilde{V}_0, \mathbb{Z}) \leftarrow G_i^0 \leftarrow G_i^1 \leftarrow 0,$$

with $\mu(H_{i_0}(\widetilde{W}^n, \widetilde{V}_0, \mathbb{Z})) + \mu(F^1_{i_0-1}) \geq k$ for some i_0. Then the number of critical points of index i for a minimal Morse function is equal to $\mu(H_i(\widetilde{W}^n, \widetilde{V}_0, \mathbb{Z})) + \mu(F^i_{i-1})$. Here

$$0 \leftarrow H_i(\widetilde{W}^n, \widetilde{V}_0, \mathbb{Z}) \leftarrow F_i^0 \leftarrow F_i^1 \leftarrow 0$$

is a minimal free resolution of the module $H_i(\widetilde{W}^n, \widetilde{V}_0, \mathbb{Z})$, which under the hypothesis always exists.

COROLLARY 6.12. *Let (W^n, V_0, V_1) be the cobordism of the preceding proposition, π an s-group, and* $\mathrm{Wh}(\pi) = 0$. *Suppose the homology modules $H_i(\widetilde{W}^n, \widetilde{V}_0, \mathbb{Z})$ have free resolutions of the form*

$$0 \leftarrow H_i(\widetilde{W}^n, \widetilde{V}_0, \mathbb{Z}) \leftarrow G_i^0 \leftarrow G_i^1 \leftarrow 0.$$

Then the number of critical points of index i for a minimal Morse function is

$$N_i = \mu(H_i(\widetilde{W}^n, \widetilde{V}_0, \mathbb{Z})) + \mu(F^1_{i-1}),$$

where

$$0 \leftarrow H_i(\widetilde{W}^n, \widetilde{V}_0, \mathbb{Z}) \leftarrow F_i^0 \leftarrow F_i^1 \leftarrow 0$$

is a minimal free resolution of the module $H_i(\widetilde{W}^n, \widetilde{V}_0, \mathbb{Z})$.

THEOREM 6.13. *Let (W^n, V_0, V_1) be a cobordism, $\pi_1(V_0) \hookrightarrow \pi_1(W^n) \hookleftarrow \pi_1(V_1)$ isomorphisms, and $n \geq 6$. Suppose $\pi_1(W^n) \approx \pi$ is an s-group, with r-$\dim \pi \leq k$. Suppose also that the ith Morse number satisfies $m_i(W^n) \geq k$ $(3 \leq i \leq n-3)$. Then there exists on W^n a minimal Morse function.*

PROOF. Consider $\pi_2(W^n, V_0) = G_1$ and $\pi_2(W^n, V_1) = G_2$, and choose minimal systems $\{a_1, \dots, a_s\}$ of generators in G_1 and $\{b_1, \dots, b_t\}$ in G_2 as crossed modules, realizing them by disjoint imbeddings

$$\alpha_i \colon (D_i^2, S_i^1) \hookrightarrow (W^n, V_0),$$
$$\beta_j \colon (D_j^2, S_j^1) \hookrightarrow (W^n, V_1).$$

Tubular neighborhoods of $A_i = \alpha_i(D_i^2, S_i^1)$ and $B_j = \beta_j(D_j^2, S_j^1)$ can be taken as handles of index 2, attached to collars of V_0 and V_1, respectively. Denote these by Γ_1 and Γ_2. Let $f_1: \Gamma_1 \to [0, 1/3]$, $f_1^{-1}(0) = V_0$, and $f_3: \Gamma_2 \to [2/3, 1]$, $f_3^{-1}(1) = V_1$, be Morse functions on Γ_1 and Γ_2, constructed by handlebody decomposition. It is easily shown that the cobordism

$$\overline{W}^n = \overline{W^n \backslash (\Gamma_1 \cup \Gamma_2)}, \qquad \partial \overline{W}^n = \overline{V}_1 \cup \overline{V}_2,$$

has the property that $\pi \approx \pi_1(\overline{V}_1) \hookrightarrow \pi_1(\overline{W}^n) \hookleftarrow \pi_1(\overline{V}_2)$ is an isomorphism, and consequently it satisfies the hypothesis of Proposition 6.6. Let* $f_2: \overline{W}^n \to [1/2, 2/3]$, $f_2^{-1}(2/3)$, be a minimal Morse function on $\overline{W}^n$. Consider the function $g = f_1 \cup f_2 \cup f_3$. We assert that* it is a minimal Morse function on W^n. Indeed, let f be an arbitrary regular Morse function on W^n, and ξ an associated gradient-like vector field. Using f and ξ, construct a free chain complex

$$0 \leftarrow C_2 \overset{\partial_3}{\leftarrow} C_3 \leftarrow \cdots \overset{\partial_{n-2}}{\leftarrow} C_{n-2} \leftarrow 0.$$

We may assume that f has no critical points of index $0, 1, n-1$, or n since $\pi_1(W^n, V_0) = \pi_1(W^n, V_1) = 0$. Furthermore, the critical points of index i, $3 \le i \le n-3$, have no effect on the fundamental group of the level surface of the Morse function; so we can also assume that f is chosen so that

$$f\text{-rank}(\partial_4(C_4), C_3) = 0 \tag{$*$}$$

and is additive, since $\pi_1(V) \approx \pi_1(W^n)$, where $V = f^{-1}(c)$, $f(x_i) < c < f(y_j)$, with the x_i critical points of index 2 and the y_j critical points of index 3. By Lemma 4.10, the quantity $\chi = \mu(C_2) - \mu(C_3)$ is constant for all Morse functions on W^n for which the associated chain complexes satisfy condition $(*)$. Consequently, the number of critical points of index 3 reaches its minimum whenever the number of index 2 does. Since critical points of index 2 give generators of $\pi_2(W^n, V_0)$, there cannot be less than s of them. Similarly, using the function f, one shows that g is a minimal function for index $n-3$. This proves the theorem.

Suppose the boundary of the cobordism $(W^n, \partial W^n)$ consists of several components $V_0, \ldots, V_k$, $\overline{V}_0, \ldots, \overline{V}_l$. Group them as: $A = \bigcup_i V_i$, $B = \bigcup_j \overline{V}_j$. Using the preceding theorem, it is easy to state conditions for W^n to have a minimal Morse function $f: W^n \to I$ such that $f^{-1}(0) = A$ and $f^{-1}(1) = B$. By the van Kampen-Seifert theorem,

$$\pi_1(V_0 \# V_1 \# \cdots \# V_k) \approx \pi_1(V_0) * \pi_1(V_1) * \cdots * \pi_1(V_k),$$
$$\pi_1(\overline{V}_0 \# \overline{V}_1 \# \cdots \# \overline{V}_l) \approx \pi_1(\overline{V}_0) * \pi_1(\overline{V}_1) * \cdots * \pi_1(\overline{V}_l).$$

Using handles of index 1, join the components $V_0, \ldots, V_k$ and $\overline{V}_0, \ldots, \overline{V}_l$, obtaining manifolds V and $\overline{V}$.

* Translator's note. This sentence seems to be especially garbled in the original.

PROPOSITION 6.14. *Suppose the inclusions $\pi_1(V) \hookrightarrow \pi_1(W^n) \hookleftarrow \pi_1(\overline{V})$ induce epimorphisms, $n \geq 6$, $\pi_1(W^n) \approx \pi$ is an s-group, r-$\dim \pi \leq k$, and the i th Morse number satisfies $m_i(W^n) \geq k$ $(3 \leq i \leq n-3)$. Then on W^n there exists a minimal Morse function.*

We remark that similar results hold for noncompact cobordisms admitting adjunction of a boundary. Furthermore, the condition on r-$\dim \pi$ can be omitted. We do not prove these results here.

REFERENCES

1. Hyman Bass, *Algebraic K-theory*, Benjamin, New York, 1968.

2. O. I. Bogoyavlenskiı, *On an exact function on manifolds*, Mat. Zametki **8** (1970), 77–83; English transl. in Math. Notes **8** (1970).

3. Saunders Mac Lane, *Homology*, Springer-Verlag, Berlin, and Academic Press, New York, 1963.

4. Richard Mandelbaum, *Four-dimensional topology: an introduction*, Bull. Amer. Math. Soc. (N. S.) **2** (1980), 1–159.

5. J. W. Milnor, *Morse Theory*, Ann. of Math. Stud., No. 51, Princeton Univ. Press, Princeton, N. J., 1963.

6. J. W. Milnor, *Lectures on the h-Cobordism Theorem*, Princeton Univ. Press, Princeton, N. J., 1965.

7. J. W. Milnor, *Whitehead torsion*, Bull. Amer. Math. Soc. **72** (1966), 358–426.

8. S. P. Novikov, *The Hamiltonian formalism and a multivalued analogue of Morse theory*, Uspekhi Mat. Nauk **37** (1982), no. 5 (227), 3–49; English transl. in Russian Math. Surveys **37** (1982).

9. C. P. Rourke and B. J. Sanderson, *Introduction to Piecewise-Linear Topology*, Springer-Verlag, New York, 1972.

10. A. A. Suslin, *Projective modules over polynomial rings are free*, Dokl. Akad. Nauk SSSR **229** (1976), 1063–1066; English transl. in Soviet Math. Dokl. **17** (1976).

11. I. O. Solopko, *Exact Morse functions in the category TOP (dimension > 5)*, Ukrain. Mat. Zh. **35** (1983), 792–796; English transl. in Ukrainian Math. J. **35** (1983).

12. M. Sh. Farber, *Sharpness of the Novikov inequalities*, Funktsional. Anal. i Prilozhen. **19** (1985), no. 1, 49–59, 96; English transl. in Functional Anal. Appl. **19** (1985).

13. A. T. Fomenko, *Topological Variational Problems*, Moskov. Gos. Univ., Moscow, 1964. (Russian)

14. G. S. Chogoshvili, *On level surfaces and regions of smaller values*, Akad. Nauk Gruzin. SSR. Trudy Tbiliss. Mat. Inst. Razmadze **17** (1949), 203–243. (Russian)

15. V. V. Sharko, *Minimal resolutions and Morse functions*, Dokl. Akad. Nauk Ukrain. SSR. Ser. A **1980**, no. 11, 31–33. (Russian)

16. V. V. Sharko, *Minimal chain complexes, monogenic functions and mappings*, Akad. Nauk Ukrain. SSR, Inst. Mat., Kiev, 1982, pp. 20–38. (Russian)

17. V. V. Sharko, *Minimal resolutions and Morse functions*, Topology (Moscow, 1979), Trudy Mat. Inst. Steklov. **154** (1983), 306–311; English transl. in Proc. Steklov Inst. Math. **1984**.

18. V. V. Sharko, *Minimal Morse functions*, Geometry and topology in global nonlinear problems, Novoe Global. Anal., Voronezh. Gos. Univ., Voronezh, 1984, pp. 123–141. (Russian)

19. R. C. Alperin, R. K. Dennis, and M. R. Stein, *The non-triviality of $SK_1(\mathbb{Z}\pi)$*, Lecture Notes in Math., Vol. 353, Springer, Berlin, 1973, pp. 1–7.

20. D. Barden, *Simply connected five-manifolds*, Ann. of Math. (2) **82** (1965), 365–385.

21. Raoul Bott and Hans Samelson, *Applications of the theory of Morse to symmetric spaces*, Amer. J. Math. **80** (1958), 964–1029.

22. W. H. Cockcroft and R. G. Swan, *On the homotopy type of certain two-dimensional complexes*, Proc. London Math. Soc. (3) **11** (1961), 194–202.

23. M. M. Cohen, *Whitehead torsion, group extensions, and Zeeman's conjecture in high dimensions*, Topology **16** (1977), no. 1, 79–88.

24. Albrecht Dold, *Zur Homotopietheorie der Kettencomplexe*, Math. Ann. **140** (1960), 278–298.

25. M. N. Dyer, *On the essential height of homotopy trees with finite fundamental group*, Compositio Math. **36** (1978), no. 2, 209–224.

26. Samuel Eilenberg, *Homological dimension and syzygies*, Ann. of Math. (2) **64** (1956), 328–336.

27. L. Èl′sgol′ts, *Die Änderung der Bettischen Zahlen der Niveauflächen einer stetigen Funktion, die auf einer Mannigfaltigkeit definiert ist*, Mat. Sb. (N. S.) **5** (47) (1939), 559–564. (Russian; German summary)

28. B. Hajduk, *Comparing handle decompositions of homotopy equivalent manifolds*, Fund. Math. **95** (1977), no. 1, 35–47.

29. H. Jacobinski, *Genera and decompositions of lattices over orders*, Acta Math. **121** (1968), 1–29.

30. M. A. Kervaire, *Le théorème de Barden-Mazur-Stallings*, Comment. Math. Helv. **40** (1965), 31–42.

31. R. C. Kirby and L. C. Siebenmann, *Foundational essays on topological manifolds, smoothings, and triangulations*, Ann. of Math. Stud. **88**, Princeton Univ. Press, Princeton, N. J., 1977.

32. B. A. Magurn, R. M. Oliver, and Leonid Vaserstein, *Units in Whitehead groups of finite groups*, J. Algebra **84** (1983), no. 2, 324–360.

33. Marston Morse, *The calculus of variations in the large*, Amer. Math. Soc., New York, 1934.

34. Everett Pitcher, *Inequalities of critical point theory*, Bull. Amer. Math. Soc. **64** (1958), 1–30.

35. Daniel Quillen, *Projective modules over polynomial rings*, Invent. Math. **36** (1976), 167–171.

36. Georges Reeb, *Sur les variétés de niveau d'une fonction numérique*, C. R. Acad. Sci. Paris **224** (1947), 1324–1326.

37. Stephen Smale, *On the structure of manifolds*, Amer. J. Math. **84** (1962), 387–399.

38. R. G. Swan, *Projective modules over group rings and maximal orders*, Ann. of Math. (2) **76** (1962), 55–61.

39. C. T. C. Wall, *Geometrical connectivity*. I, J. London Math. Soc. (2) **3** (1971), 597–604.

40. J. H. C. Whitehead, *Simple homotopy types*, Amer. J. Math. **72** (1950), 1–57.

Translated by J. A. ZILBER

Amer. Math. Soc. Transl.
(2) Vol. **149**, 1991

K-Theory and Morse Theory. II

V. V. SHARKO

Introduction

We continue in this paper our study of Morse functions on nonsimply-connected manifolds. Let M^n be a closed manifold or a manifold with one boundary component. The procedures developed in the preceding paper for the construction of minimal functions run into difficulties connected with the application of the Whitney lemma in codimension 2. These difficulties are bypassed by drawing upon results in the homotopy classification of 2-dimensional cell complexes. As a consequence, additional restrictions arise on the fundamental group and the first k-invariant of the manifold. Furthermore, an adequate algebraic model for the description of Morse functions is furnished, not by a chain complex of $\mathbf{Z}[\pi_1(M^n)]$-modules, but by a homotopy system in the sense of Whitehead [28].

Let us briefly describe the content of the paper. In the first section we present the facts we shall be needing concerning homotopy systems. The basic element of a homotopy system is a crossed module, i.e., a triple (C, G, d), where C and G are groups, $d: C \to G$ is a homomorphism, G acts on C on the left, and d satisfies the conditions $c_1 + c_2 - c_1 = d(c_1) \cdot c_2$ and $d(g \cdot c) = g \cdot d(c) \cdot g^{-1}$. A homotopy system is a sequence of groups and homomorphisms $e \leftarrow \pi \xleftarrow{d_1} G \xleftarrow{d_2} C_2 \xleftarrow{d_3} C_3 \leftarrow \cdots \xleftarrow{d_n} C_n \leftarrow 0$ such that (C_2, G, d_2) is a free crossed module; for $i \geq 3$ the C_i are free $\mathbf{Z}[\pi]$-modules and the d_i π-module homomorphisms, and $d_i \circ d_{i+1} = 0$. If the system is made commutative, it yields a chain complex of $\mathbf{Z}[\pi]$-modules, called the chain complex *associated* with the homotopy system. We also review Whitehead's theory of simple-homotopy type for homotopy systems.

In the second section we study the first k-invariant of homotopy systems and its connection with group presentation. Also discussed are certain facts

1991 *Mathematics Subject Classification*. Primary 57R70; Secondary 18G55, 57Q10, 19B28.
Translation of Akad. Nauk Ukrain. SSR Inst. Mat. Preprint **1986**, no. 40; MR 88f:57057.

concerning 2-dimensional cell complexes.

Any crossed module (C, G, d) determines an exact sequence $e \leftarrow G/d(C) \leftarrow G \leftarrow C \leftarrow \operatorname{Ker} d \leftarrow 0$. We put $G/d(C) = B$ and $\operatorname{Ker} d = A$. Following Mac Lane and Whitehead, we call this a crossed sequence. The set of crossed sequences is isomorphic, under a natural equivalence relation, to the cohomology group $H^3(B, A)$. It is known that every element of $H^3(B, A)$ can be represented by a crossed sequence $s\colon e \leftarrow B \leftarrow \mathcal{F} \leftarrow D \leftarrow A \leftarrow 0$, where $\mathcal{F}$ is a free group. By the Euler characteristic of the sequence we mean the number $\chi(s) = -\mu(\mathcal{F}) + \mu(D)$, where $\mu(D)$ is the minimal number of generators of D as an $\mathcal{F}$-module. The Euler characteristic of an element $a \in H^3(B, A)$ is the number $\chi(a) = \min_s(\chi(s))$, where s runs over all crossed sequences representing a. Every homotopy system determines the crossed sequence

$$e \leftarrow G/d_2(C_2) \leftarrow G \leftarrow C_2/d_3(C_3) \leftarrow \operatorname{Ker} d_2/d_3(C_3) \leftarrow 0,$$

and consequently, if we put $G/d_2(C_2) = B$ and $\operatorname{Ker} d_2/d_3(C_3) = A$, an element of $H^3(B, A)$. This was how Mac Lane and Whitehead defined the first k-invariant of the homotopy system. Now let (ρ_i, d_i) be the homotopy system associated with a CW-complex K. Here $\rho_i = \pi_i(K^i, K^{i-1})$ for $i \geq 2$ and $\rho_1 = \pi_1(K^1)$, where K^i is the i-skeleton of K. The first k-invariant of the cell complex is the first k-invariant of the corresponding homotopy system; consequently, the Euler characteristic of the complex is well defined. Every Morse function on a closed manifold or a manifold with one boundary component, that has one critical point of index 0, determines a cell complex, and therefore a homotopy system.

Given a group G, we denote by $d(G)$ its defect. A presentation $G = \{a_1, \dots, a_{\mu(G)}, r_1, \dots, r_l\}$, where $\mu(G)$ is the minimal number of generators, is called a minimal presentation. It is known that for any presentation of G there exists a free resolution

$$0 \leftarrow \mathbf{Z} \leftarrow \mathbf{Z}[G] \xleftarrow{d_1} k\mathbf{Z}[G] \xleftarrow{d_2} l\mathbf{Z}[G] \leftarrow \cdots$$

(the Lyndon–Fox resolution). The first three terms of this resolution are what interest us, since they have a natural geometric interpretation. We say that G satisfies the condition (h, d) if for any presentation of G with defect d there exists a minimal presentation such that the segments of the free resolutions constructed via the two presentations are homotopy equivalent. The group G satisfies condition (h) if it satisfies condition (h, d) for all d, where d is the defect of a presentation. It is known that condition (h) is satisfied by free groups, the group $\mathbf{Z} \oplus \mathbf{Z}$, finite abelian groups, and the dihedral group D_{2k+1}. For arbitrary finite groups the condition (h, d) is satisfied at least for $d \geq d(G)+1$. We say that a group G has essential height $\leq k$ if any two CW-complexes K and L of dimension 2, with $\pi_1(K) \approx \pi_1(L) \approx G$ and Euler characteristics $\chi(K) = \chi(L) \geq d(G) + k = 1$, are homotopy equivalent. It is known that for any finite abelian group the essential height is ≤ 1; and for any finite group, ≤ 2.

In the third section we prove the existence of minimal homotopy systems.

THEOREM 3.2. *In order that there exist, for every homotopy system* (ρ_i, d_i) *associated with a CW-complex* K, *a minimal homotopy system* $(\overline{\rho}_i, \overline{d}_i) \sim (\rho_i, d_i)$, *it is necessary and sufficient that* $\pi_1(K)$ *satisfy condition* (h) *and be an* s*-group.*

THEOREM 3.3. *Suppose* (ρ_i, d_i) *is associated with a CW-complex* K, *where* $\pi_1(K)$ *satisfies condition* $(h, \chi(k^3(K)))$ *and is an* s*-group. Then there exists a minimal homotopy system* $(\overline{\rho}_i, \overline{d}_i)$ *homotopy equivalent to* (ρ_i, d_i).

In the final two sections we apply the results on homotopy systems to the construction of minimal Morse functions.

THEOREM 4.4. *Let* W^n *be a manifold with boundary* $\partial W^n = V$, $n \geq 6$, *and* $\pi_1(V) \hookrightarrow \pi_1(W^n) \approx \pi$ *an isomorphism. Suppose* π *is an* s*-group and satisfies condition* (h, χ_0) $(\chi_0 = \chi(k^3(W^n)))$, *and* $r\text{-}\dim \pi \leq k$. *Suppose also that the Morse number* $\mu_\lambda(W^n)$ *is* $\geq k$ $(\lambda \geq 3)$. *Then* W^n *has a minimal Morse function.* (*For example, if* π *is a free group or is* $\mathbf{Z} \oplus \mathbf{Z}$.)

THEOREM 4.12. *Suppose* $\pi_1(W^n)$ *is isomorphic to one of the groups* $\mathbf{Z}_k$, $\bigoplus \sum \mathbf{Z}_{p_j} \oplus \mathbf{Z}_{p_j^i}$ (p_j *a prime*), *or the dihedral group* D_{2k+1}. *Then* W^n *has a minimal Morse function.*

Similar theorems hold for closed manifolds.

In conclusion, the author wishes to express his thanks to S. P. Novikov, A. T. Fomenko, and E. Mikhaĭlyuk for their stimulating discussions.

The author is sincerely grateful to the participants of the seminars directed by M. M. Postnikov, A. V. Chernavskiĭ and Yu. Yu. Trokhimchuk for their critical observations in connection with various aspects of this work.

Finally, the author wishes to thank W. Metzler for forwarding preprints of W. Browning, and also H. Zieschang for his assistance and interest in the publication of this paper.

§1. Homotopy systems

The basic element of a homotopy system is a crossed module. A crossed module is a triple (C, G, d), where C and G are groups, $d: C \to G$ is a homomorphism, and G acts on C on the left (the action will be denoted by $g \cdot c$). Furthermore, the homomorphism d is to satisfy the conditions

$$c_1 + c_2 - c_1 = d(c_1) \cdot c_2,$$
$$d(g \cdot c) = g \cdot d(c) \cdot g^{-1}.$$

A morphism of crossed modules is a pair of group homomorphisms

$$(\alpha, \beta): (C, G, d) \to (C', G', d'), \quad \alpha: C \to C', \quad \beta: G \to G',$$

such that $\beta \circ d' \circ \alpha$ and $\alpha(g \cdot c) = \beta(g) \cdot \alpha(c)$.

The following are immediate consequences of the definition.

1) $d(C)$ is a normal subgroup of G.
2) $\operatorname{Ker} d$ belongs to the center of C.
3) The action of G on C induces a natural structure of $G/d(C)$-module on the center of C, of which $\operatorname{Ker} d$ is a submodule.
4) The action of G on C induces a structure of $G/d(C)$-module on $C^{\mathrm{ab}} = C/[C, C]$.

An important case of a crossed module is the so-called free crossed module.

DEFINITION 1.1 [28]. Let (C, G, d) be a crossed module and $\{c_{i\in I}\}$ a fixed set of elements of C. Then (C, G, d) is called a free crossed module with basis $\{c_{i\in I}\}$ if for every crossed module (C', G', d') and arbitrary set of elements $\{c'_{i\in I}\}$ of C' and homomorphism $\alpha: G \to G'$ such that $\alpha \cdot d(c_i) = d'(c'_i)$, there exists a unique homomorphism $\beta: C \to C'$ such that $\beta(c_i) = c'_i$ and (α, β) is a crossed-module homomorphism.

A constructive description of free crossed modules was given by Whitehead [28], to whom is also due the following fundamental proposition.

THEOREM 1.2 [28]. *Let A be an arcwise connected space, and X a space obtained from A by attaching 2-cells. Then $\pi_2(X, A)$ is a free crossed $\pi_1(A)$-module with basis corresponding to the cells so attached.*

We list here several important properties of free crossed modules that we need; their proofs can be found in [12], [13], [16].

If (C, G, d) is a free crossed module with basis $\{c_{i\in I}\}$, then $C^{\mathrm{ab}} = C/[C, C]$ is a free $G/d(C)$-module with basis the elements $c_i[C, C]$.

A set of elements $\{c'_{i\in I}\}$ of C is a basis if and only if

1) the set of elements $c'_i[C, C]$ is a basis for C^{ab},
2) the normal closure of $d(c'_i)$ in G is $\operatorname{Im} d$.

The cardinal number of the basis elements of a crossed module is the same for all bases. The free module C^{ab} will be called the module *associated* with the crossed module (C, G, d).

We give now the definition of a homotopy system.

DEFINITION 1.3 [13], [28]. A homotopy system (C_i, G, d_i) is a sequence of groups and homomorphisms

$$e \leftarrow \pi \xleftarrow{d_1} G \xleftarrow{d_2} C_2 \leftarrow \cdots \xleftarrow{d_n} C_n \leftarrow 0$$

with the following properties:

1) (C_2, G, d_2) is a free crossed module, and $\pi = G/d_2(C_2)$.
2) For $i \geq 3$, C_i is a free $\mathbf{Z}[\pi]$-module, and d_i a π-homomorphism of modules; d_2 commutes with the action of G; and $d_3(C_3)$ is a $\mathbf{Z}[\pi]$-module.
3) $d_i \circ d_{i+1} = 0$.

REMARK. Obviously, G acts on each C_i, $i \geq 3$. A morphism $f:(C_i, G, d_i) \to (C_i', G', d_i')$ of homotopy systems is a set of homomorphisms $f_1: G \to G'$, $f_i: C_i \to C_i'$, $i \geq 2$, that preserves the structures on the C_i and under which the corresponding homomorphism diagram is commutative.

One may define a relation of homotopy between morphisms of homotopy systems, and a relation of homotopy equivalence between homotopy systems; for details, see [13], [28]. We say that (C_i, G, d_i) has dimension n if $C_i = 0$, $i > n$, but $C_n \neq 0$. For any homotopy system (C_i, G, d_i) there exists the free chain complex of $\mathbf{Z}[\pi]$-modules

$$(C_i^{\text{ab}}, d_i): 0 \leftarrow \mathbf{Z} \leftarrow \mathbf{Z}[\pi] \overset{d_1}{\longleftarrow} n\mathbf{Z}[\pi] \overset{d_2}{\longleftarrow} C_2^{\text{ab}} \leftarrow C_3 \leftarrow \cdots \leftarrow C_n \leftarrow 0,$$

where $n = \mu(G)$. We shall say that the complex (C_i^{ab}, d_i) is *associated* with the system (C_i, G, d_i). To any morphism of homotopy systems corresponds a morphism of the associated chain complexes.

THEOREM 1.4 [28]. *A morphism $f: (C_i, G, d_i) \to (C_i', G', d_i')$ of homotopy systems is a homotopy equivalence if and only if the induced chain mapping on the associated chain complexes is a homotopy equivalence.*

We recall what is meant by a free crossed submodule of a free crossed module. Let $(C, \mathscr{F}, d)$ be a free crossed module with basis $(c_1, \dots, c_k)$, and $d(c_i) = f_i$, where $\mathscr{F}$ is a free group with basis $(f_1, \dots, f_s)$. Suppose $\overline{\mathscr{F}}$ is the subgroup generated by part of the basis $(f_{i_1}, \dots, f_{i_r})$, $\overline{C}$ is the subgroup of C generated by $(c_{j_1}, \dots, c_{j_t})$ with the action of $\overline{\mathscr{F}}$, and let $d(\overline{C}) \subset \overline{\mathscr{F}}$; if $\mathscr{F}/d(C) \approx \overline{\mathscr{F}}/d(\overline{C})$, then, as shown by Whitehead, $(\overline{C}, \overline{\mathscr{F}}, d)$ is a free crossed module with basis $(c_{j_1}, \dots, c_{j_t})$.

Let $\widehat{\mathscr{F}}$ be a free group with basis $(\hat{f}_1, \dots, \hat{f}_{\hat{k}})$. We can then construct a free crossed module $(\widehat{C}, \mathscr{F} \cdot \widehat{\mathscr{F}}, \hat{d})$ with basis $(c_1, \dots, c_k, \hat{c}_1, \dots, \hat{c}_{\hat{k}})$, where $\hat{d}(c_i) = f_i$ and $\hat{d}(\hat{c}_j) = \hat{f}_j$. Obviously there exists a canonical imbedding $(C, \mathscr{F}, d) \to (\widehat{C}, \mathscr{F} \cdot \widehat{\mathscr{F}}, \hat{d})$; i.e., $(C, \mathscr{F}, d)$ is a submodule of $(\widehat{C}, \mathscr{F} \cdot \widehat{\mathscr{F}}, \hat{d})$.

DEFINITION 1.5. The module $(\widehat{C}, \mathscr{F} \cdot \widehat{\mathscr{F}}, \hat{d})$ is called a *stabilization* of the module $(C, \mathscr{F}, d)$, and $(C, \mathscr{F}, d)$ a *contraction* of $(\widehat{C}, \mathscr{F} \cdot \widehat{\mathscr{F}}, \hat{d})$.

Let M be a free module with basis $(m_1, \dots, m_n)$ over the ring $\mathbf{Z}[\mathscr{F}/d(C)]$. Consider the group $C \times M$, and put $\hat{d}|_{C \times 0} = d$, $\hat{d}|_{0 \times M} = 0$. Obviously $(C \times M, \mathscr{F}, \hat{d})$ is a free crossed module with basis $(c_1, \dots, c_k, m_1, \dots, m_n)$.

DEFINITION 1.6. The module $(C \times M, \mathscr{F}, \hat{d})$ is called a *thickening* of the module $(C, \mathscr{F}, d)$.

Let $(C, \mathscr{F}, d)$ be a free crossed module with basis $(c_1, \dots, c_k)$. We say that a basis $(c_1', \dots, c_k')$ is obtained from the basis $(c_1, \dots, c_k)$ by an elementary transformation if

1) $c_{i_0}' = f \cdot c_{i_0}$, $c_i = c_i'$, $i \neq i_0$; or

2) $c'_{i_0} = -c_{i_0}$, $c_i = c'_i$, $i \neq i_0$; or

3) $c'_{i_0} = c_{i_0} + c_j$ $(c'_{i_0} = c_j + c_{i_0})$, $i \neq i_0$, $j \neq i_0$.

DEFINITION 1.7. Let $(C, \mathcal{F}, d)$ be a free crossed module with bases $(c_1, \dots, c_k)$ and $(c'_1, \dots, c'_k)$. The bases are said to be *equivalent* if one can be obtained from the other by a sequence of elementary transformations.

Let the bases in the homotopy systems $(C_i, \mathcal{F}, d_i)$ and $(\overline{C}_i, \overline{\mathcal{F}}, \overline{d}_i)$ be fixed, as well as the bases in the groups $\mathcal{F}$ and $\overline{\mathcal{F}}$. A homomorphism $f: (C_i, \mathcal{F}, d_i) \to (\overline{C}_i, \overline{\mathcal{F}}, \overline{d}_i)$ is called a *simple isomorphism* if the image of a basis element is, up to sign, a basis element. In the homotopy system $(C_i, \mathcal{F}, d_i)$, let c_i be a basis element in the crossed module C_2, and f_i a basis element in the group $\mathcal{F}$. Suppose $d_2(c_2) = x \cdot f_j \cdot y$, where f_j does not enter into either of the words x or y. We construct a new homotopy system $(\overline{C}_i, \overline{\mathcal{F}}, \overline{d}_i)$ by removing the elements c_i and f_j and adjusting in the obvious fashion the boundary homomorphisms in dimensions 2 and 3. This operation will be called an *elementary collapse* in dimension 2.

DEFINITION 1.8. We say that a homotopy system $(C_i, \mathcal{F}, d_i)$ is simple-homotopy equivalent to a homotopy system $(\overline{C}_i, \overline{\mathcal{F}}, \overline{d}_i)$, and write $(C_i, \mathcal{F}, d_i) \sim (\overline{C}_i, \overline{\mathcal{F}}, \overline{d}_i)$ (Σ), if we can pass from the one system to the other by a sequence of the following transformations: 1) an elementary collapse or expansion in dimension 2; 2) an elementary transformation on the basis of the crossed module C_i or on the basis of C_i, $i \geq 3$; 3) contraction or stabilization of a d_i.

THEOREM 1.9 [29]. *A mapping* $f: (C_i, \mathcal{F}, d_i) \to (\overline{C}_i, \overline{\mathcal{F}}, \overline{d}_i)$ *is a simple-homotopy equivalence if the associated chain mapping of chain complexes* $(C_i^{\mathrm{ab}}, d_i) \xrightarrow{f^{\mathrm{ab}}} (\overline{C}_i^{\mathrm{ab}}, \overline{d}_i)$ *is a simple-homotopy equivalence.*

For homotopy systems there is also an analogue of the Cockcroft-Swan theorem. Since it will not be needed, we omit the proof. Because of the noncommutativity in the first two terms, the proof itself is rather awkward, but in concept it is quite transparent.

§2. The k-invariant of a homotopy system

Any crossed module (C, G, d) determines an exact sequence

$$e \leftarrow G/d(C) \leftarrow G \leftarrow C \leftarrow \operatorname{Ker} d \leftarrow 0,$$

called by Mac Lane and Whitehead in [17] a *crossed sequence.* Put

$$G/d(C) = B, \qquad \operatorname{Ker} d = A.$$

The set of crossed sequences starting with a B-module A and ending with the group B we denote by $\Omega(B, A)$. On $\Omega(B, A)$ we define an equivalence relation. We say that a sequence s is congruent to a sequence s' if there exists a crossed-module homomorphism (φ, ψ) such that the following diagram commutes:

$$\begin{array}{cccccccccccc} s: & e & \longleftarrow & B & \longleftarrow & G & \longleftarrow & C & \longleftarrow & A & \longleftarrow & 0 \\ & & & \big\downarrow{\scriptstyle \mathrm{id}} & & \big\downarrow{\scriptstyle \varphi} & & \big\downarrow{\scriptstyle \psi} & & \big\downarrow{\scriptstyle \mathrm{id}} & & \\ s': & e & \longleftarrow & B & \longleftarrow & G' & \longleftarrow & C' & \longleftarrow & A & \longleftarrow & 0. \end{array}$$

This congruence relation generates an equivalence relation on $\Omega(B, A)$. We denote by $\mathrm{Xseq}(B, A)$ the corresponding set of equivalence classes. Using composition, $\mathrm{Xseq}(B, A)$ can be turned into a group, which is isomorphic to $H^3(B, A)$ (see [12], [13], [16]). We shall need a specific description of this isomorphism η: $\mathrm{Xseq}(B, A) \to H^3(B, A)$. It is known that in each equivalence class of crossed sequences $e \leftarrow B \leftarrow G \leftarrow C \leftarrow A \leftarrow 0$ there exists a sequence for which G is a free group. It is obtained by choosing an arbitrary free presentation of B, $e \leftarrow B \leftarrow \mathscr{F} \leftarrow R \leftarrow e$, and constructing the diagram

$$\begin{array}{ccccccccccc} e & \longleftarrow & B & \longleftarrow & \mathscr{F} & \longleftarrow & D & \longleftarrow & A & \longleftarrow & 0 \\ & & \big\downarrow{\scriptstyle \mathrm{id}} & & \big\downarrow{\scriptstyle \varphi} & & \big\downarrow{\scriptstyle \psi} & & \big\downarrow{\scriptstyle \mathrm{id}} & & \\ e & \longleftarrow & B & \overset{p}{\longleftarrow} & G & \longleftarrow & C & \longleftarrow & A & \longleftarrow & 0, \end{array}$$

where

$$\begin{array}{ccccccc} R & \longleftarrow & D & \longleftarrow & A & \longleftarrow & 0 \\ \big\downarrow{\scriptstyle \overline{\varphi}} & & \big\downarrow{\scriptstyle \psi} & & \big\downarrow{\scriptstyle \mathrm{id}} & & \\ \operatorname{Ker} p & \longleftarrow & C & \longleftarrow & A & \longleftarrow & 0 \end{array}$$

is a pull-back with $\overline{\varphi} = \varphi|_R$. The action of $\mathscr{F}$ on D is given by $f(r \times c) = (frf^{-1} \times \varphi(f) \cdot c)$, $f \in \mathscr{F}$, $r \in R$, $c \in C$. Now let

$$e \leftarrow B \overset{p}{\longleftarrow} \mathscr{F} \overset{d}{\longleftarrow} C \overset{i}{\longleftarrow} A \leftarrow 0 \qquad (*)$$

be a crossed sequence with $\mathscr{F}$ a free group. Consider the subsequence $e \leftarrow R \overset{d}{\longleftarrow} C \leftarrow A \leftarrow 0$, $R = \ker p$, and make it commutative: $0 \leftarrow R^{\mathrm{ab}} \leftarrow C^{\mathrm{ab}} \leftarrow A \leftarrow 0$. Since $\mathscr{F}$ acts on R by conjugation, R^{ab} can be turned into a $\mathbf{Z}[B]$-module. It is easily seen that also C^{ab} is a $\mathbf{Z}[B]$-module. Since A is in the center of C, C can be represented as a product $C = R \times A$. Let $s: R \to C$ be a homomorphism such that $d \cdot s = \mathrm{id}$. Then every element $c \in C$ has a unique representation $c = a_c + s \cdot d(c)$. If $p(f) = b \in B$, define the action of b on C^{ab} by $b \cdot c = b \cdot a_c + s \cdot d(f \cdot c) \bmod [C, C]$. From the sequence of groups $e \leftarrow B \leftarrow \mathscr{F} \leftarrow R \leftarrow e$ we can construct a sequence of $\mathbf{Z}[B]$-modules $0 \leftarrow I[B] \leftarrow I[\mathscr{F}] \otimes Z[B] \leftarrow R^{\mathrm{ab}} \leftarrow 0$, where $I[B]$ and $I[\mathscr{F}]$ are the augmentation ideals; see [11]. Combining this with the sequence $0 \leftarrow R^{\mathrm{ab}} \leftarrow C^{\mathrm{ab}} \leftarrow A \leftarrow 0$, we obtain an exact sequence of $\mathbf{Z}[B]$-modules

$$0 \leftarrow I[B] \leftarrow I[\mathscr{F}] \otimes \mathbf{Z}[B] \leftarrow C^{\mathrm{ab}} \leftarrow A \leftarrow 0. \qquad (**)$$

The correspondence $(*) \mapsto (**)$ determines the isomorphism $\mathrm{Xseq}(B, A) \to H^3(B, A)$. Observe that although in $(*)$ C is representable as a direct product of groups $A \times R$, the crossed sequence $(*)$ determines the zero element of $H^3(B, A)$ only when C is representable as a direct product, not just of groups, but of operator $\mathscr{F}$-groups.

We define now a numerical invariant of crossed sequences, which will play an important role in the construction of minimal homotopy systems.

DEFINITION 2.1. Let $e \leftarrow B \leftarrow \mathscr{F} \leftarrow C \leftarrow A \leftarrow 0$ be a crossed sequence, in which $\mathscr{F}$ is a free group of rank n. By the Euler characteristic of this sequence is meant the number $\chi(s) = -n + \mu(C)$, where $\mu(C)$ is the minimal number of generators of C as a crossed $\mathscr{F}$-module. By the Euler characteristic of an element $a \in H^3(B, A)$ one means the number $\chi(a) = \min_s(\chi(s))$, where s runs over all crossed sequences representing a.

We note that, as shown by examples, the Euler characteristics of two crossed sequences defining the same element of $H^3(B, A)$ can be different; e.g., consider a non-Hopf finitely representable group G and construct $s \in H^3(G, 0)$.

Whitehead showed in [28] that every CW-complex K determines a homotopy system $\rho(K)$. Without loss of generality we can suppose the 0-skeleton of K to consist of a single point x. Put $\rho_n = \pi_n(K^n, K^{n-1})$, $\rho_1 = \pi_1(K^1, x)$ (where K^i is the i-skeleton of K). Using the exact homotopy sequences of pairs, one easily obtains homomorphisms $d_n \colon \rho_n \to \rho_{n-1}$ and thereby a homotopy system

$$e \leftarrow \pi_1(K, x) = \rho_1/d_2(\rho_2) \xleftarrow{p} \rho_1 \xleftarrow{d_2} \rho_2 \leftarrow \cdots .$$

The system depends, however, on the decomposition of K into cells.

Let M^n be a manifold. A homotopy system can be constructed by using either a cellular structure on M^n or a Morse function f. Let f be a regular Morse function on M^n with one critical point x of index 0, and ξ a regular gradient-like vector field. Let $M_\lambda = f^{-1}[0, c_\lambda]$, where $f(x_\lambda) < c_\lambda < f(y_{\lambda+1})$ for x_λ (resp. $y_{\lambda+1}$) is a critical point of index λ (resp. $\lambda + 1$). Put $\rho_1 = \pi_1(M_1, x)$, $\rho_i = \pi_i(M_i, M_{i-1})$. As in the case of CW-complexes, we obtain a homotopy system

$$\rho \colon e \leftarrow \pi_1(M^n, x) = \rho_1/d_2(\rho_2) \xleftarrow{p} \rho_1 \xleftarrow{d_2} \rho_2 \leftarrow \cdots .$$

If on the median spheres of critical points of index λ belonging to the manifold $\partial M_{\lambda-1}$ we choose distinguished points and join them by paths to x, then every median disk of a critical point of index λ determines a generator of ρ_λ.

Given a homotopy system

$$(\rho_i, d_i): e \leftarrow \rho_1/d_2(\rho_2) \xleftarrow{d_1} \rho_1 \xleftarrow{d_2} \rho_2 \leftarrow \cdots,$$

construct the crossed sequence

$$e \leftarrow \rho_1/d_2(\rho_2) \leftarrow \rho_1 \leftarrow \rho_2/d_3(\rho_3) \leftarrow \operatorname{Ker} d_2/d_3(\rho_3) \leftarrow 0. \qquad (*)$$

Put $\rho_1/d_2(\rho_2) = B$, $\operatorname{Ker} d_2/d_3(\rho_3) = A$. From what has already been said, $(*)$ determines an element of $H^3(B, A)$. This was how Mac Lane and Whitehead defined the first k-invariant of a homotopy system. If (ρ_i, d_i) is the system constructed for a CW-complex K, then $(*)$ takes the form

$$e \leftarrow \pi_1(K, x) \leftarrow \pi_1(K^1, x) \leftarrow \pi_2(K, K^1, x) \leftarrow \pi_2(K, x) \leftarrow 0. \qquad (**)$$

They called the triple $(\pi_1(K, x), \pi_2(K, x), k(K))$ corresponding to the sequence $(**)$, where $k(K) \in H^3(\pi_1(K, x), \pi_2(K, x))$, the algebraic 3-type of the complex K. Later the term has settled down to "2-type", and we shall stick to that in the sequel.

An abstract 2-type is by definition a triple (π, π_2, k), consisting of a group π, a $\mathbf{Z}[\pi]$-module π_2, and an element $k \in H^3(\pi, \pi_2)$. Two 2-types $T = (\pi, \pi_2, k)$ and $T' = (\pi, \pi_2', k')$ with the same π are equivalent (notation; $T \approx T'$) if there exist isomorphisms $f: \pi \to \pi$, $f': \pi_2 \to \pi_2'$, such that $f'(x \cdot a) = f(x) \cdot f'(a)$ $(x \in \pi, a \in \pi_2)$ and $f'_*(k) = f^*(k')$ in the chain of mappings

$$f'_*: H^3(\pi, \pi_2) \to H^3(\pi, (\pi_2')_f) \leftarrow H^3(\pi, \pi_2'): f^*,$$

where $(\pi_2')_f$ means the $\mathbf{Z}[\pi]$-module with the module structure induced by the isomorphism f.

Denote by $A(\pi)$ the set of equivalence classes of 2-types (π, π_2, k) with the same group π, and by $[T]$ the equivalence class of the 2-type T. It is known that every abstract 2-type can be realized by a 3-dimensional CW-complex K such that $T \approx T(K)$ [17].

We say that two CW-complexes K and L have the same homotopy 2-type if there exist mappings $f: K^3 \to L^3$, $g: L^3 \to K^3$ such that

$$g \circ f|_{K^2} \sim i: K^2 \to K^3, \qquad f \circ g|_{L^2} \sim i: L^2 \to L^3.$$

Theorem 1 of [17] asserts that K and L have the same homotopy 2-type if and only if $T(K) \approx T(L)$.

Every regular Morse function f with one critical point x of index 0 and regular gradient-like vector field ξ on a manifold M^n defines a crossed sequence

$$e \leftarrow \pi_1(M^n, x) \leftarrow \pi_1(M_1, x) \leftarrow \pi_2(M^n, M_1, x) \leftarrow \pi_2(M^n, x) \leftarrow 0,$$

which determines an element $k^3(M^n)$ in $H^3(\pi_1(M^n, x), \pi_2(M^n, x))$. Here M_1 is a level submanifold containing all critical points of index 0 or 1. The element $k^3(M^n)$ is called the first k-invariant of M^n. Using a general-position argument, it is easily shown that any two regular Morse functions f and g on M^n have the same homotopy 2-type. This follows from the existence of a diffeomorphism, isotopic to the identity, that takes M_λ^f into M_λ^g and conversely (M_2^f and M_2^g are level submanifolds of f and g containing the critical points of index 0, 1, or 2 only). Consequently, by the Mac Lane–Whitehead theorem, f and g determine equivalent algebraic 2-types.

We show now that $\chi(k) = \chi(k')$ for equivalent 2-types $T = (\pi_1, \pi_2, k)$ and $T' = (\pi_1', \pi_2', k')$. Suppose the crossed sequence

$$s: e \leftarrow \pi_1 \xleftarrow{g_1} \mathscr{F} \xleftarrow{g_2} \Gamma_2 \xleftarrow{g_3} \pi_2 \leftarrow 0$$

that realizes $k \in H^3(\pi_1, \pi_2)$ has Euler characteristic χ_0. Since (π_1, π_2, k) and (π_1', π_2', k') are equivalent, there exist isomorphisms $f: \pi_1 \to \pi_1'$ and $f': \pi_2 \to \pi_2'$ with $f_*'(k) = f^*(k')$ in

$$f_*': H^3(\pi_1, \pi_2) \to H^3(\pi_1, (\pi_2')_f) \leftarrow H^3(\pi_1', \pi_2'): f^*.$$

Using f, construct the pull-back induced from s,

$$\begin{array}{ccccccccc}
sf: e & \longleftarrow & \pi_1' & \xleftarrow{e_1} & \mathscr{F}' & \xleftarrow{e_2} & \Gamma_2 & \xleftarrow{e_3} & \pi_2 & \longleftarrow & 0 \\
 & & \downarrow f & & \downarrow \varphi & & \downarrow \mathrm{id} & & \downarrow \mathrm{id} & & \\
e & \longleftarrow & \pi_1 & \xleftarrow{g_1} & \mathscr{F} & \xleftarrow{g_2} & \Gamma_2 & \xleftarrow{g_3} & \pi_2 & \longleftarrow & 0,
\end{array}$$

where $\mathscr{F}' = \{(a, b) \in \mathscr{F} \times \pi_1', g_1(a) = f(b)\}$. The group $\mathscr{F}'$ acts on Γ_2 by $(a, b) \cdot c = a \cdot c$, $c \in \Gamma_2$. Obviously φ is an isomorphism, and Γ_2, whether as a crossed $\mathscr{F}$- or a crossed $\mathscr{F}'$-module, has the same minimal number of generators. Now using f', construct the push-out induced from sf:

$$\begin{array}{ccccccccccc}
e & \longleftarrow & \pi_1' & \longleftarrow & \mathscr{F}' & \longleftarrow & \Gamma_2 & \longleftarrow & \pi_2 & \longleftarrow & 0 \\
 & & \downarrow \mathrm{id} & & \downarrow \mathrm{id} & & \downarrow \psi & & \downarrow f' & & \\
e & \longleftarrow & \pi_1' & \longleftarrow & \mathscr{F}' & \longleftarrow & \Gamma_2' & \longleftarrow & \pi_2' & \longleftarrow & 0,
\end{array}$$

where $\Gamma_2' = (\pi_2' \times \Gamma_2)/N$, $N = \{(-f'(x), e_3(x)), x \in \pi_2\}$; the action of $\mathscr{F}'$ on Γ_2' is induced by the diagonal action of $\mathscr{F}'$ on $\pi_2' \times \Gamma_2$. Clearly ψ is an isomorphism, and the crossed $\mathscr{F}$-module Γ_2' has the same minimal number of generators as the crossed $\mathscr{F}$-module Γ_2. By the construction, therefore, the crossed sequence

$$0 \leftarrow \pi_1' \leftarrow \mathscr{F}' \leftarrow \Gamma_2' \leftarrow \pi_2' \leftarrow 0$$

has also the Euler characteristic χ_0. Thus, Definition 2.1 can be used to define unambiguously the Euler characteristic of the first k-invariant of a manifold.

We recall that by the defect of a presentation of a group $G = \{a_1, \dots, a_k, r_1, \dots, r_l\}$ one means the number $d(a_i, r_j) = l - k$. The defect of G is by definition the number $d(G) = \min d(a_i, r_j)$, where (a_i, r_j) runs over all presentations of G. A minimal presentation of G is one of the form $G = \{a_1, \dots, a_{\mu(G)}, r_1, \dots, r_l\}$, where $\mu(G)$ is the minimal number of generators of G. Let $d^{\mu}(G) = \min d(a_i, r_j)$, where (a_i, r_j) runs over all minimal presentations. Obviously, $d^{\mu}(G) \geq d(G)$; but the author knows of no example of a finitely presentable group for which $d^{\mu}(G) > d(G)$.

For any presentation $G = \{a_1, \dots, a_k, r_1, \dots, r_l\}$, there exists a two-dimensional CW-complex K with $\pi(K) \approx G$, that has one 0-cell, k 1-cells, and l 2-cells. Starting with a given presentation of a group G, there is a canonical construction of a free resolution (sometimes called the Lyndon–Fox resolution [20])

$$0 \leftarrow \mathbf{Z} \xleftarrow{\varepsilon} \mathbf{Z}[G] \xleftarrow{d_1} k\mathbf{Z}[G] \xleftarrow{d_2} l\mathbf{Z}[G] \leftarrow \cdots, \tag{$*$}$$

where $\varepsilon\colon \mathbf{Z}[G] \to \mathbf{Z}$ is the augmentation, with kernel the ideal $I[G]$ generated by the elements $a_i - 1$. The free module $k\mathbf{Z}[G]$ of rank k, with basis $[a_1], \dots, [a_k]$, is mapped by the rule $d_1[a_i] = a_i - 1$. The free module $l\mathbf{Z}[G]$, with basis $[r_1], \dots, [r_l]$, is mapped by the rule $d_2[r_j] = \sum \frac{\partial r_j}{\partial a_i}[a_i]$; i.e., d_2 is given by a Fox matrix [20]. This resolution has a simple geometric interpretation: if K^2 is the 2-dimensional CW-complex constructed from the given group presentation, then the chain complex on the universal covering $\widehat{K}_2$ is precisely the segment $(*)$ of the resolution.

DEFINITION 2.2. We say that a group G satisfies the condition (h, d) if for any presentation of G with defect d, there exists a minimal presentation of G such that the segments of the two free resolutions constructed from these presentations are homotopy equivalent.

DEFINITION 2.3. We say that a group G satisfies condition (h) if it satisfies (h, d) for all d, where d is the defect of the presentation.

Definition 2.2 can be given an equivalent formulation in terms of 2-dimensional CW-complexes: a group G satisfies condition (h, d) if for any 2-dimensional CW-complex K^2 constructed from a presentation of G with defect d there exists a homotopy equivalent CW-complex L^2 constructed from a minimal presentation of G. There is also an equivalent formulation for Definition 2.3. The equivalence of the definitions follows from Theorems 1 and 4 of Dyer and Sieradski [7].

PROPOSITION 2.4. *The following groups satisfy condition* (h):

1) *free groups*;
2) $\mathbf{Z} \oplus \mathbf{Z}$;
3) *finite abelian groups*;
4) *the dihedral group* D_n *(for odd* n*).*

PROOF. 1) follows from a theorem of Wall (see [7], and [26]); 2) from results of Jajodia [14, Theorem 2]; 3) from results of Metzler [20], Sieradski [25, Theorems 1 and 2], and Browning [4, Theorem 1.7]; 4) from results of Jajodia and Magurn [15, Theorem 12].

PROPOSITION 2.5. *All finite groups satisfy condition* (h, d) *for* $d \geq d^{\mu}(G)+1$; *in particular, nilpotent groups for* $d \geq d(G)+1$.

PROOF. This follows from Theorem 1.1 of Browning [5]. For nilpotent groups we must use a theorem of Rapaport [23]. We note that the results in Propositions 2.4 and 2.5 are highly nontrivial, obtained by specialists concerned with classifying not simply-connected 2-dimensional cell complexes.

It is conjectured that every s-group satisfies condition (h). It is unknown whether there exists an $n = n(G)$ such that for all $N > n$ the group G satisfies condition (h, N).

We shall need certain other results concerning two-dimensional CW-complexes. Let G be a finitely presentable group. Denote by $\{K_i^2, G\}$ the class of 2-dimensional CW-complexes with fundamental group isomorphic to G, and by $[K^2]$ the class of all members of $\{K_i^2, G\}$ having the homotopy type of K^2. Dyer and Sieradski constructed [7] an oriented graph $H(G)$, whose vertices are the classes $[K^2] \subset \{K_i^2, G\}$. From a vertex $[K_{i_0}^2]$ an arrow exits to a vertex $[K_{i_1}^2]$ whenever $[K_{i_1}^2]$ is homotopy equivalent to $K_{i_0}^2 \vee S^2$. Denote by $\chi(K^2)$ the Euler characteristic of the complex K^2, and let $\chi_{\min} = \min \chi(K_i^2)$. Obviously $\chi_{\min} = d(G)+1$. By the *level* of the vertex $[K_{i_0}^2]$ is meant the number $\chi(K_{i_0}^2) - \chi_{\min}$. A complex $K_{i_0}^2$ is called a *root* if the vertex $[K_{i_0}^2]$ has no *entering* arrow, and a *minimal root* if the vertex $[K_{i_0}^2]$ has level 0.

DEFINITION 2.6 [8]. A group G has essential height $\leq k$ if any two 2-dimensional CW-complexes K^2 and L^2 with fundamental group isomorphic to G and Euler characteristics $\chi(K^2) = \chi(L^2) \geq d(G)+1+k$ are homotopy equivalent.

For example, if $d(G) = d^{\mu}(G)$, then any group with essential height k satisfies the condition $(h, d(G)+N)$, where $N \geq k$.

The following theorem is due to Browning [5].

THEOREM 2.7. *Any finite group has essential height* ≤ 1.

For finite abelian groups this was proved by Dyer in [8].

§3. Minimal homotopy systems

DEFINITION 3.1. A homotopy system

$$(\rho_i, d_i)\colon e \leftarrow \pi_1 \xleftarrow{d_1} \rho_1 \xleftarrow{d_2} \rho_2 \leftarrow \cdots$$

is called *minimal* if for any other homotopy system

$$(\overline{\rho}_i, \overline{d}_i): e \leftarrow \pi_1 \xleftarrow{\overline{d}_1} \overline{\rho}_1 \xleftarrow{\overline{d}_2} \overline{\rho}_2 \leftarrow \cdots$$

homotopy equivalent to (ρ_i, d_i) we have

$$\mu(\rho_i) \leq \mu(\overline{\rho}_i).$$

Take a fixed group π, and consider the class of all finite CW-complexes for which the fundamental group is isomorphic to π. We can suppose without loss of generality that the 0-skeleton of any such complex consists of a single point. Denote by $\rho(K)$ the class of homotopy systems associated with these CW-complexes. (As shown by Whitehead [28] not every homotopy system can be represented as associated with a CW-complex.) The following theorems give an indication of when, in the homotopy type of a system $(\rho_i, d_i) \in \rho(K)$, there exists a minimal system. The question is well posed, since by a theorem of Whitehead, if $(\rho_i, d_i) \in \rho(K)$ and a homotopy system $(\overline{\rho}_i, \overline{d}_i)$ is homotopy equivalent to (ρ_i, d_i), then $(\overline{\rho}_i, \overline{d}_i) \in \rho(K)$.

THEOREM 3.2. *In order that there exist for every homotopy system* $(\rho_i, d_i) \in \rho(K)$ *a homotopy equivalent minimal homotopy system* $(\overline{\rho}_i, \overline{d}_i)$, *it is necessary and sufficient that* $\pi = \rho_1/d_2(\rho_2)$ *satisfy condition* (h) *and be an* s*-group.*

PROOF. Necessity. Necessity of the condition s follows from Proposition 4.8 of [3]. We show now necessity of (h). Assume the contrary. Then there exists a presentation $\pi = \{a_1, \dots, a_k, r_1, \dots, r_l\}$ such that $k > \mu(\pi)$ but for which there is no minimal presentation of π such that the chain complexes constructed from the two presentations are homotopy equivalent. For this presentation of π, construct an n-dimensional manifold W^n, $n \geq 6$, having one handle of index 0, k handles of index 1, and l of index 2. At the same time, this manifold W^n can also be given a handle decomposition for which there is one handle of index 0 and $\mu(\pi)$ handles of index 1. Choose any such decomposition P. Let W_2 be the submanifold of W^n formed by the handles of index 0, 1, and 2 in the decomposition P. Obviously $\pi_1(W_2, x) \to \pi_1(W^n, x)$ is an isomorphism. Consider the exact homology sequence for the pair $(\widetilde{W}^n, \widetilde{W}_2)$ that constitutes the universal covering of the pair (W^n, W_2). It is of the form

$$0 \to H_3(\widetilde{W}^n, \widetilde{W}_2, \mathbf{Z}) \to H_2(\widetilde{W}_2, \mathbf{Z}) \to H_2(\widetilde{W}^n, \mathbf{Z}) \to 0.$$

If $H_3(\widetilde{W}^n, \widetilde{W}_2, \mathbf{Z}) = 0$, then $\overline{W^n \backslash W_2} = H$ is an h-cobordism; and consequently W_2 has the same homotopy type as W^n. But this would mean that the condition $(h, \mu(\pi))$ is satisfied for the given presentation of π. Therefore $H_3(\widetilde{W}^r, \widetilde{W}_2, \mathbf{Z}) \neq 0$, and the decomposition P of W^n contains handles of index at least 3. This implies that the homotopy system associated with the decomposition P cannot be minimal (it has length at least 3). This proves necessity.

Sufficiency. Suppose (ρ_i, d_i) is associated with an n-dimensional CW-complex K^n, $(\overline{\rho}_i, \overline{d}_i) \sim (\rho_i, d_i)$ is minimal in dimensional 2, and $(\hat{\rho}_i, \hat{d}_i) \sim (\rho_i, d_i)$ minimal in dimension 3, with $(\overline{\rho}_i, \overline{d}_i) = \rho(K_1^m)$, $(\hat{\rho}_i, \hat{d}_i) = \rho(K_2^l)$. Imbed the complexes K_1^m and K_2^l in a Euclidean space of sufficiently high dimension, and construct tubular neighborhoods for them, obtaining manifolds W_1^N and W_2^N. The cellular structures on K_1^m and K_2^l induce on W_1^N and W_2^N handle decompositions such that corresponding to every cell of dimension i is a handle of index i. It is easily shown that $\pi_1(\partial W_1^N) \to \pi_1(W_1^N) \approx \pi$, $\pi_1(\partial W_2^N) \to \pi_1(W_2^N) \approx \pi$ are isomorphisms. Let $\overline{W}_2$ be the submanifold in W_2^N formed by the handles of index 0, 1, and 2. The handle decomposition of $\overline{W}_2$ gives a presentation of the fundamental group $\pi_1(\overline{W}_2) \approx \pi$. Since π satisfies condition (h), there exists a minimal presentation of π such that the corresponding segments of the two free resolutions are homotopy equivalent. Realize this presentation by a 2-dimensional CW-complex K^2 and, using (h), imbed the latter in $\overline{W}_2$. A tubular neighborhood of the image is a submanifold $\overline{\overline{W}}_2 \subset \overline{W}_2$, containing one handle of index 0, $\mu(\pi)$ handles of index 1, and a certain number of handles of index 2. Condition (h) guarantees that the imbedding $\overline{\overline{W}}_2 \hookrightarrow \overline{W}_2$ is a homotopy equivalence. Consequently, $\overline{\overline{W}_2 \backslash \overline{\overline{W}}_2}$ is an h-cobordism, in which, as is well known, it can be arranged that there are handles of index greater than 3. Shift the handles from h-cobordism to the manifold W_2^N. Thus, we have constructed on W_2^N a decomposition with $\mu(\pi)$ handles of index 1 and the number of handles of index 3 remaining unchanged. We show now that the homotopy system constructed for this decomposition of W_2^N is minimal in dimension 2. Consider the chain complexes associated with the decompositions of the manifolds W_1^N and W_2^N:

$$C(W_1^N): \quad 0 \leftarrow \mathbf{Z} \overset{\varepsilon}{\leftarrow} \mathbf{Z}[\pi] \overset{\partial_1}{\leftarrow} C_1 \overset{\partial_2}{\leftarrow} C_2 \overset{\partial_3}{\leftarrow} C_3 \leftarrow \cdots,$$

$$C(W_2^N): \quad 0 \leftarrow \mathbf{Z} \overset{\varepsilon}{\leftarrow} \mathbf{Z}[\pi] \overset{d_1}{\leftarrow} D_1 \overset{d_2}{\leftarrow} D_2 \overset{d_3}{\leftarrow} D_3 \leftarrow \cdots.$$

Without loss of generality we can suppose that $f\text{-rank}(\partial_4(C_4), C_3) = 0$ and is additive, since elementary operations in this dimension are realized geometrically. The condition that $f\text{-rank}(d_4(D_4), D_3) = 0$ and be additive is satisfied automatically, since the homotopy system $(\hat{\rho}_i, \hat{d}_i)$ is minimal in dimension 3. By Lemma 4.10 of [3],

$$1 - \mu(C_1) + \mu(C_2) - \mu(C_3) = 1 - \mu(D_1) + \mu(D_2) - \mu(D_3).$$

But by construction, $\mu(C_1) \geq \mu(D_1)$ and $\mu(C_3) \geq \mu(D_3)$; hence $\mu(C_2) = \mu(D_2)$. Thus, the homotopy system constructed for the handlebody decomposition of W_2^N is minimal in dimensions 1, 2, and 3. Using stabilization and contraction of the d_i, we can arrange that $f\text{-rank}(d_i(D_i), D_{i-1}) = 0$ and is additive for $i \geq 5$. The homotopy system so obtained is minimal. Indeed, consider the associated chain complex. Performing contractions and stabilizations on the boundary homomorphisms d_2 and d_3, we can arrange that $f\text{-rank}(d_i(D_i), D_{i-1}) = 0$ and is additive for $i = 2, 3$. The chain complex so obtained is minimal, by Theorem 4.7 of [3]; therefore the chain complex associated with the homotopy system was minimal in dimensions greater than 3.

This completes the proof of the theorem.

THEOREM 3.3. *Let $\rho(L^n)$ be a homotopy system associated with a CW-complex L^n. If $\pi = \pi_1(L^n, x)$ satisfies the condition $(h, \chi(k^3(L^n)))$ and is an s-group, then there exists a minimal homotopy system $(\rho_i, d_i) \sim \rho(L^n)$.*

PROOF. Let K be a CW-complex such that the associated homotopy system $\rho(K)$ is homotopy equivalent to $\rho(L^n)$ and minimal in dimension 3. In addition, we can suppose without loss of generality that the chain complex

$$0 \leftarrow \mathbf{Z} \overset{\varepsilon}{\longleftarrow} \mathbf{Z}[\pi] \overset{\partial_1}{\longleftarrow} C_1 \overset{\partial_2}{\longleftarrow} C_2 \overset{\partial_3}{\longleftarrow} C_3 \leftarrow \cdots$$

associated with $\rho(K)$ has the property that $f\text{-rank}(\partial_i(C_i), C_{i-1}) = 0$ and is additive for $i \geq 4$. Write $\pi_2(L^n) \approx \pi_2(K) \approx \pi_2$. We show first that the homotopy system $\rho(K)$ determines a crossed sequence

$$\rho \leftarrow \pi \leftarrow \pi_1(K^1) \leftarrow \pi_2(K, K^1) \leftarrow \pi_2 \leftarrow 0$$

with Euler characteristic equal to $\chi(k^3(L^n)) = \chi_0$. Let

$$e \leftarrow \pi \leftarrow \rho_1 \overset{\bar{\partial}_2}{\longleftarrow} \overline{D}_2 \leftarrow \pi_2 \leftarrow 0$$

be a crossed sequence representing the element $k^3(L^n) \in H^3(\pi, \pi_2)$ and having Euler characteristic χ_0. Pick a minimal set of generators $\bar{d}_1, \dots, \bar{d}_k$ for the crossed module $\overline{D}_2$. Let $(D_2, \rho_1, \partial_2)$ be a free crossed module with basis $d_1, \dots, d_k$ such that $\bar{\partial}_2(\bar{d}_i) = \partial_2(d_i)$. There exists an epimorphism

$$\begin{array}{ccccccccc}
 & & 0 & \longleftarrow & R & \longleftarrow & R & \longleftarrow & 0 \\
 & & & & \downarrow & & \downarrow & & \\
 & \overset{\partial_1}{\nearrow} & \rho_1 & \overset{\partial_2}{\longleftarrow} & D_2 & \longleftarrow & Z_2 & \longleftarrow & 0 \\
e \leftarrow \pi & & \downarrow \text{id} & & \downarrow f_1 & & \downarrow f_2 & & \\
 & \underset{\bar{\partial}_1}{\nwarrow} & \rho_1 & \overset{\bar{\partial}_2}{\longleftarrow} & \overline{D}_2 & \longleftarrow & \pi_2 & \longleftarrow & 0.
\end{array}$$

Assume that R is a finitely generated $\mathbf{Z}[\pi]$-module. Let $\partial_3 \colon D_3 \to R$ be an epimorphism of a finitely generated free module onto R, and consider the homotopy system

$$e \leftarrow \pi \xleftarrow{\partial_1} \rho_1 \xleftarrow{\partial_2} D_2 \xleftarrow{\partial_3} D_3 \leftarrow 0 .$$

By construction, its crossed sequence has Euler characteristic χ_0. By a theorem of Whitehead (Theorem 2 of [28]), this homotopy system can be realized by a CW-complex of dimension 3, which we denote by N^3. Let K^3 be the 3-skeleton of the complex K. By construction, N^3 and K^3 have the same algebraic 2-type, and therefore the same homotopy 2-type. Consequently, by another theorem of Whitehead (Theorem 14 of [29]), there exist m and n such that the CW-complexes

$$N^2 \cup \left(\bigvee_{i=1}^{m} S_i^3\right) \sim K^2 \cup \left(\bigvee_{j=1}^{n} S_j^3\right)$$

are simple-homotopy equivalent. If we seal up the wedge of spheres $\bigvee_{j=1}^{n} S_j^3$ by 4-dimensional disks in the complex $K^3 \cup (\bigvee_{j=1}^{n} S_j^3)$ and attach the remaining cells of K via their previous characteristic mappings, we obtain a complex $\overline{K}$ simple-homotopy equivalent to the complex K. Using $\overline{K}$, build up the complex

$$N^3 \cup \left(\bigvee_{i=1}^{m} S_i^3\right)$$

to a CW-complex N simple-homotopy equivalent to K; for this we can apply Lemma 3.7 of [1]. Now let $\rho(N)$ be the homotopy system associated with N. Realize N by a manifold, and consider the handle decomposition corresponding to the cellular structure. Without loss of generality, we can assume that the chain complex

$$0 \leftarrow \mathbf{Z} \leftarrow \mathbf{Z}[\pi] \xleftarrow{\partial_1} D_1 \xleftarrow{\partial_2} D_2 \leftarrow \cdots$$

constructed from this decomposition has the property f-rank$(\partial_i(D_i), D_{i-1}) = 0$ and is additive for $i \geq 4$. Compare it with the chain complex of the CW-complex K. By Lemma 4.10 of [3],

$$-\mu(C_1) + \mu(C_2) - \mu(C_3) = \chi_0 = -\mu(D_1) + \mu(D_2) - \mu(D_3),$$

and therefore

$$-\mu(C_1) + \mu(C_2) = \chi_0 + \mu(C_3), \quad -\mu(D_1) + \mu(D_2) = \chi_0 + \mu(D_3).$$

Since $\mu(C_3) = \mu(D_3)$, we have $-\mu(C_1) + \mu(C_2) = -\mu(D_1) + \mu(D_2)$. Thus, the crossed sequence determined by the homotopy system $\rho(K)$ has Euler characteristic equal to χ_0.

We show now how to use the system $\rho(K)$ to construct a minimal system $(\rho_i, d_i) \sim \rho(L^n)$. The argument is very similar to that in the proof of sufficiency in the preceding theorem. Let $\rho(T)$ be a homotopy system, associated with a CW-complex T, that is homotopy equivalent to $\rho(K)$ and minimal in dimension 2. Imbed the complexes T and K in a Euclidean space of high enough dimension, and consider their tubular neighborhoods, obtaining manifolds W_2 and W_1 respectively. The cellular structures on T and K induce handle decompositions on W_1 and W_2, so that to each cell of dimension i corresponds a handle of index i. In addition, we can suppose that

$$i_{1*}: \pi_1(\partial W_1) \to \pi_1(W_1) \approx \pi_1$$

and

$$i_{2*}: \pi_1(\partial W_2) \to \pi_1(W_2) \approx \pi_1$$

are isomorphisms. Let W_2^2 be the submanifold in W_2 formed by the handles of index 0, 1, and 2. The decomposition of W_2 gives a presentation of the group $\pi_1(\partial W_2)$ with defect equal to χ_0. Since π_1 satisfies the condition (h, χ_0), there exists a minimal presentation of π_1 with defect χ_0 and a 2-dimensional CW-complex $\overline{K}^2 \sim W_2^2$ constructed from this presentation. Let $j: \overline{K}^2 \to W_2^2$ be an imbedding that is a homotopy equivalence. Consider a tubular neighborhood of $j(\overline{K}^2)$ in W_2^2, and denote it by $\overline{W}_2^2$. Obviously $\Gamma_2 = \overline{W_2^2 \backslash \overline{W}_2^2}$ is an h-cobordism. We can suppose that on Γ_2 there are handles of index greater than 3. Shift them toward the boundary of the manifold W_2. Now consider the chain complexes associated with the decomposition of W_1 and the new decomposition of W_2:

$$0 \leftarrow \mathbf{Z} \leftarrow \mathbf{Z}[\pi_1] \overset{\partial_1}{\longleftarrow} \overline{C}_1 \overset{\partial_2}{\longleftarrow} \overline{C}_2 \overset{\partial_3}{\longleftarrow} \overline{C}_3 \leftarrow \cdots,$$

$$0 \leftarrow \mathbf{Z} \leftarrow \mathbf{Z}[\pi_1] \overset{d_1}{\longleftarrow} E_1 \overset{d_2}{\longleftarrow} E_2 \overset{d_3}{\longleftarrow} E_3 \leftarrow \cdots.$$

As before, we can suppose that $f\text{-rank}(\partial_i(\overline{C}_i), \overline{C}_{i-1}) = 0$ and is additive for $i \geq 4$, and similarly for $f\text{-rank}(d_i(E_i), E_{i-1})$, since the elementary operations in these dimensions can be realized geometrically. The operation of cancelling handles leaves unchanged the homotopy type of the manifold W_2. By Lemma 4.10 of [3],

$$1 - \mu(\overline{C}_1) + \mu(\overline{C}_2) - \mu(\overline{C}_3) = \chi_0 = 1 - \mu(E_1) + \mu(E_2) - \mu(E_3).$$

By construction, $\mu(\overline{C}_1) \leq \mu(E_1)$ and $\mu(\overline{C}_3) \leq \mu(E_3)$; and therefore $\mu(\overline{C}_2) = \mu(E_2)$. Thus, the homotopy system associated with the decomposition of W_2 is minimal in dimensions 1, 2, and 3. We claim it is minimal also in the remaining dimensions. Indeed, by stabilization and contraction of ∂_2 and ∂_3, we can arrange that $f\text{-rank}(\partial_i(\overline{C}_i), \overline{C}_{i-1}) = 0$ and is additive for $i = 2$ and 3; and for the remaining dimensions this condition holds automatically. By Theorem 4.7 of [3], the chain complex

$$0 \leftarrow \mathbf{Z} \leftarrow \mathbf{Z}[\pi_1] \overset{\partial_1}{\longleftarrow} \overline{C}_1 \overset{\partial_2}{\longleftarrow} \overline{C}_2 \leftarrow \cdots$$

is minimal; and therefore so is the homotopy system $\rho(W_2)$.

Finally, we prove that the module R above is finitely generated. If A and B are two stably isomorphic finitely generated modules, and $f\colon F \to A$ is an epimorphism of a finitely generated free module F onto A with $\operatorname{Ker} f$ finitely generated, then for any epimorphism $g\colon G \to B$ of a finitely generated free module G onto B, $\operatorname{Ker} g$ is also finitely generated. This follows from Schanuel's lemma, applied to the epimorphisms in the diagram

$$\begin{gathered} 0 \leftarrow A \oplus m\Lambda \xleftarrow{f\oplus \mathrm{id}} F \oplus m\Lambda \leftarrow \operatorname{Ker} f \leftarrow 0, \\ 0 \leftarrow B \oplus n\Lambda \xleftarrow{g\oplus \mathrm{id}} G \oplus n\Lambda \leftarrow \operatorname{Ker} g \leftarrow 0, \end{gathered}$$

and the composite mapping

$$\operatorname{Ker} f \oplus G \oplus n\Lambda \approx \operatorname{Ker} g \oplus F \oplus m\Lambda \to \operatorname{Ker} g \to 0.$$

Consider the crossed sequences

$$\begin{gathered} e \leftarrow \pi_1 \xleftarrow{\bar\partial_1} \rho_1 \xleftarrow{\bar\partial_2} \overline{D}_2 \leftarrow \pi_2 \leftarrow 0, \\ e \leftarrow \pi_1 \xleftarrow{\partial_1} \pi_1(K^1) \xleftarrow{\partial_2} \pi_2(K, K^1) \leftarrow \pi_2 \leftarrow 0, \end{gathered}$$

and mappings onto them of free crossed modules:

$$\begin{array}{ccccccccccc} & & & & 0 & \longleftarrow & R & \longleftarrow & R & \longleftarrow & 0 \\ & & & & & & \downarrow & & \downarrow & & \\ e & \longleftarrow & \pi_1 & \xleftarrow{\hat\partial_1} & \rho_1 & \xleftarrow{\hat\partial_2} & D_2 & \longleftarrow & Z_2 & \longleftarrow & 0 \\ \bar\partial_1 & & & \nwarrow & \downarrow{\scriptstyle \mathrm{id}} & & \downarrow{\scriptstyle f_1} & & \downarrow{\scriptstyle f_2} & & \\ & & & & \rho_1 & \longleftarrow & \overline{D}_2 & \longleftarrow & \pi_2 & \longleftarrow & 0, \end{array}$$

$$\begin{array}{ccccccccccc} & & & & 0 & \longleftarrow & S & \longleftarrow & S & \longleftarrow & 0 \\ & & & & & & \downarrow & & \downarrow & & \\ e & \longleftarrow & \pi_1 & \xleftarrow{\partial_1} & \pi_1(K^1) & \longleftarrow & \pi_2(K^2, K^1) & \longleftarrow & \pi_2(K^2) & \longleftarrow & 0 \\ & & {\scriptstyle \mathrm{id}}\downarrow & & \downarrow{\scriptstyle \mathrm{id}} & & \downarrow{\scriptstyle i_1} & & \downarrow{\scriptstyle i_2} & & \\ e & \longleftarrow & \pi_1 & \longleftarrow & \pi_1(K^1) & \longleftarrow & \pi_2(K, K^1) & \longleftarrow & \pi_2 & \longleftarrow & 0. \end{array}$$

Obviously the module $S = \partial_3(C_3)$ is finitely generated. Making the segments commutative, we have

$$\begin{array}{ccccccccc}
 & & 0 & \longleftarrow & R & \longleftarrow & R & \longleftarrow & 0 \\
 & & & & \downarrow & & \downarrow & & \\
0 & \longleftarrow & \operatorname{Ker}\widehat{\partial}_1^{\,\mathrm{ab}} & \xleftarrow{\widehat{\partial}_2^{\,\mathrm{ab}}} & D_2^{\mathrm{ab}} & \longleftarrow & Z_2 & \longleftarrow & 0 \\
 & & & & \downarrow{\scriptstyle f_1^{\mathrm{ab}}} & & \downarrow{\scriptstyle f_2^{\mathrm{ab}}} & & \\
0 & \longleftarrow & \operatorname{Ker}\overline{\partial}_1^{\,\mathrm{ab}} & \xleftarrow{\overline{\partial}_2^{\,\mathrm{ab}}} & \overline{D}_2^{\mathrm{ab}} & \longleftarrow & \pi_2 & \longleftarrow & 0,
\end{array}$$

$$\begin{array}{ccccccccc}
 & & 0 & \longleftarrow & S & \longleftarrow & S & \longleftarrow & 0 \\
 & & & & \downarrow & & \downarrow & & \\
0 & \longleftarrow & \operatorname{Ker}\partial_1^{\mathrm{ab}} & \longleftarrow & \pi_2(K^2, K^1)^{\mathrm{ab}} = C_2 & \longleftarrow & \pi_2(K^2) & \longleftarrow & 0 \\
 & & \downarrow{\scriptstyle \mathrm{id}} & & \downarrow{\scriptstyle i_1^{\mathrm{ab}}} & & \downarrow{\scriptstyle i_2} & & \\
0 & \longleftarrow & \operatorname{Ker}\partial_1^{\mathrm{ab}} & \longleftarrow & \pi_2(K, K^1)^{\mathrm{ab}} & \longleftarrow & \pi_2(K) & \longleftarrow & 0.
\end{array}$$

Here C_2 and D_2^{ab} are finitely generated free modules. As noted in the preceding section, the crossed sequences determine exact sequences of $\mathbf{Z}[\pi_1]$-modules

$$0 \leftarrow I[\pi_1] \leftarrow \mathbf{Z}[\rho_1] \otimes \mathbf{Z}[\pi_1] \leftarrow \overline{D}_2^{\mathrm{ab}} \leftarrow \pi_2 \leftarrow 0,$$
$$0 \leftarrow I[\pi_1] \leftarrow \mathbf{Z}[\pi_1(K^1)] \otimes \mathbf{Z}[\pi_1] \leftarrow \pi_2(K, K^1)^{\mathrm{ab}} \leftarrow \pi_2 \leftarrow 0,$$

which are congruent. By Corollary 5.3 of [3], the modules $\overline{D}_2^{\mathrm{ab}}$ and $\pi_2(K, K^1)^{\mathrm{ab}}$ are stably isomorphic. Therefore R is finitely generated. This completes the proof of the theorem.

Observe that we have incidentally proved the following fact: if a crossed sequence $e \leftarrow \pi_1 \leftarrow \rho_1 \leftarrow D_2 \leftarrow \pi_2 \leftarrow 0$ is realized by a finite CW-complex, then every congruent crossed sequence is also realized by a finite CW-complex. Also proved incidentally is the following.

PROPOSITION 3.4. *Suppose $a \in H^3(\pi_1, \pi_2)$, $\chi(a) = \chi_0$, a is realized by a finite CW-complex, and π_1 satisfies condition (h, χ_0) and is an s-group. Then there exists a crossed sequence $s\colon e \leftarrow \pi_1 \leftarrow \mathscr{F} \leftarrow D_2 \leftarrow \pi_2 \leftarrow 0$ representing the element a, with $\chi(s) = \chi_0$ and $\mu(\mathscr{F}) = \mu(\pi_1)$; it will be called a minimal crossed sequence.*

(Obviously, for any two minimal crossed sequences the numbers $\mu(D_2)$ are the same.)

PROPOSITION 3.5. *Suppose the fundamental group of a homotopy system* $\rho(L)$ *is isomorphic to one of the following*:

1) *a free group,*
2) $\mathbf{Z} \oplus \mathbf{Z}$,
3) *a finite abelian group,*
4) *the dihedral group* D_n *(for* n *odd).*

This its homotopy type contains a minimal homotopy system.

Similarly, from Theorem 3.2 and Proposition 2.5 we have

PROPOSITION 3.6. *If a homotopy system has as its fundamental group a finite s-group and the Euler characteristic* χ *of the first k-invariant is* $\geq d^{\mu}(\pi_1) + 1$, *then its homotopy type contains a minimal homotopy system. In particular, there is a minimal homotopy system if* π_1 *is a finite nilpotent s-group and* $\chi \geq d(\pi_1) + 1$.

Putting together the results on existence of a minimal homotopy system and those on r-dim π_1, we can also make inferences concerning existence of a minimal homotopy system in the simple-homotopy type.

We turn now to the question of computing the number of generators in a minimal homotopy system.

Let

$$e \leftarrow \pi_1 \xleftarrow{\partial_1} \rho_1 \xleftarrow{\partial_2} \rho_2 \leftarrow \cdots$$

be a homotopy system, and

$$0 \leftarrow \mathbf{Z} \xleftarrow{\varepsilon} \mathbf{Z}[\pi_1] \xleftarrow{\partial_1} C_1 \xleftarrow{\partial_2} C_2 \leftarrow \cdots$$

the associated chain complex. The homotopy system determines sequences

$$e \leftarrow \pi_1 \leftarrow \rho_1 \leftarrow \rho_2/\partial_3(\rho_3) \leftarrow \pi_2 \leftarrow 0,$$
$$0 \leftarrow [\rho_2/\partial_3(\rho_3)]^{\mathrm{ab}} \leftarrow \rho_2^{\mathrm{ab}} \leftarrow \rho_3/\partial_4(\rho_4) \leftarrow H_3 \leftarrow 0.$$

If π_1 is an s-group and satisfies condition $(h, \chi(k^3))$, then by Proposition 3.4 the congruence class of the crossed sequence contains a minimal crossed sequence. ($\chi(k^3)$ is the Euler characteristic of the element $k^3 \in H^3(\pi_1, \pi_2)$ represented by the crossed sequence.) In general, the mapping $[\rho_2/\partial_3(\rho_3)]^{\mathrm{ab}} \leftarrow \rho_2^{\mathrm{ab}}$ may not be a minimal epimorphism, but if we make two choices of ρ_2^{ab} of the same rank, then by Corollary 1.3 and Lemma 5.2 of [3], the $\mu(\rho_3/\partial_4(\rho_4))$ will also be the same. Hence by analogy with chain complexes we have

PROPOSITION 3.7. *Let* $\rho(L)$ *be a homotopy system with fundamental group* π_1. *Suppose* $\chi(k^3(L)) = \chi_0$ *and* π_1 *is an s-group and satisfies* (h, χ_0). *Then there exists an algorithm for computing the number of generators in a minimal homotopy system* $(\overline{\rho}_i, \overline{d}_i)$.

§4. Minimal Morse functions on nonsimply-connected manifolds with boundary

The need to make use of homotopy systems in order to study Morse functions on nonsimply-connected manifolds with one boundary component arises from the failure of the chain complexes constructed from Morse functions and gradient-like vector fields to capture completely the geometric aspect of the problem. This relates to application of the Whitney lemma to the reduction of points of intersection of manifolds of complementary dimensions. For example, let $f\colon M^n \to I$ be a regular Morse function, ξ a regular gradient-like vector field, and $C(f, \xi, M^n)$ the chain complex of $\mathbf{Z}[\pi_1(M^n)]$-modules associated with f and ξ. Suppose the complex is of the form

$$\begin{array}{c} 0 \leftarrow \mathbf{Z} \leftarrow \mathbf{Z}[\pi_1(M^n)] \xleftarrow{\partial_1} C_1 \xleftarrow{\partial_2} C_2 \xleftarrow{\partial_3} C_3 \leftarrow \cdots \\ \oplus \qquad \oplus \\ 0 \leftarrow F_n \xleftarrow{E} F_n \leftarrow 0, \end{array}$$

where E is the identity matrix. It does not at all follow that the critical points of index 2 and 3 that correspond to the direct summands F_n can be eliminated. A similar observation applies to critical points of index 1 and 2. A suitable algebraic model that permits the determination of whether or not such pairs of critical points can be eliminated is given by the homotopy system constructed from f and ξ. Below, we give conditions, in terms of the homotopy system, for a pair of critical points of index 2 and 3 to be capable of cancellation. Except for free groups, there are no examples known to the author of fundamental groups for which the condition for cancellation of critical points of index 2 and 3 can be given at the level of chain complexes. It is conjectured that if the fundamental group of the manifold is an s-group and its Whitehead group is 0, then a pair of critical points of index 2 and 3 can be eliminated if their (algebraic) intersection index is equal to 1.

Let

$$e \leftarrow \pi_1 \xleftarrow{d_1} \rho_1 \xleftarrow{d_2} \rho_2 \leftarrow \cdots$$

be the homotopy system constructed from a Morse function f and vector field ξ. All Morse functions that we mention will be assumed to be regular and to have one critical point of index 0. As distinguished basis elements in each C_i we choose the median disks of the critical points. Denote by $[h_i^3]$ the element of ρ_3 corresponding to the median disk of a critical point of index 3. Then $d_3[h_i^3]$ can be written in the form

$$d_3[h_i^3] = \varepsilon_1 x_1 [h_{j_1}^2] + \cdots + \varepsilon_s x_s [h_{j_s}^2],$$

where $\varepsilon_k = \pm 1$, $x_k \in \rho_1$ $(1 \leq k \leq s)$, and the $[h_i^2]$ are the generators corresponding to the median disks of critical points of index 2.

LEMMA 4.1. *Suppose* $d_2[h_i^3] = w \pm x_i[h_i^2] \pm v$, *where* $x_i \in \rho_1$ *and* $[h_i^2]$ *does not enter into the words* w *and* v. *Then the critical points of index* 2 *and* 3 *that correspond to* $[h_i^2]$ *and* $[h_i^3]$ *can be cancelled.*

PROOF. The meaning of the condition in the lemma is clear enough: the median sphere of the critical point of index 3 intersects the co-median sphere of the critical point of index 2 in just one point. The median sphere ∂h_i^3 of the critical point of index 3 corresponding to $[h_i^3]$ represents the element $a = w \pm x_i[h_i^2] \pm v$, for which $d_2(a) = 0$. We assume that the corresponding handles are distinguished ones. Realizing the element a by means of a connected sum, consisting of a linear combination of median disks of critical points of index 2, we obtain a disk D_1^2 imbedded in M_2, with $\partial D_1^2 \subset M_1$. Here M_i is the submanifold of M^n containing the critical points with indices $\leq i$ $(i = 1, 2)$. The condition $d_2(a) = 0$ implies that ∂D_1^2 bounds a 2-dimensional disk D_2^2 in M_1. Consider the sphere S^2 formed by identifying D_1^2 and D_2^2 along their boundaries. By construction, S^2 is homotopic to the median sphere ∂h^3. By Corollary 5.9 of [2], ∂h^3 is ambient isotopic to S^2. By construction, S^2 intersects the co-median sphere of the critical point of index 2 corresponding to $[h_i^2]$ in just one point. Carrying out the isotopy from ∂h^3 to S^2, adjust the function f and the field ξ on M_2. There results a Morse function for which the pair of critical points of index 2 and 3 has been eliminated. This proves the lemma.

We note that the lemma appears in [18], but without proof. We note also that since in the crossed module ρ_2 we have the relation $a_1 + a_2 - a_1 = d_2(a_1)\cdot a_2$, the representation $d_3(h_3^i) = \varepsilon_1 x_1[h_{j_1}^2] + \cdots + \varepsilon_s x_s[h_{j_s}^2]$ is nonunique; it can be changed by means of this relation.

Let W^n be a manifold with boundary $\partial W^n = V$, $\pi_1(V) \to \pi_1(W^n)$ an isomorphism, and $n \geq 6$. In this section we study the question of when a manifold of this type has a minimal Morse function. With every regular Morse function f and gradient-like vector field ξ is associated a homotopy system $\rho(f, \xi)$. Similarly to the case of cobordisms in [3] (Proposition 6.1), we have the following.

LEMMA 4.2. *Let* $\rho(f, \xi)$ *and* $\rho(g, \eta)$ *be homotopy systems associated with Morse functions* f *and* g *and vector fields* ξ *and* η, *defined on a manifold* W^n *with boundary* $\partial W^n = V$, *where* $\pi_1(V) \to \pi_1(W^n)$ *is an isomorphism and* $n \geq 6$. *Then* $\rho(f, \xi)$ *and* $\rho(g, \eta)$ *have the same simple-homotopy type. Furthermore, if a homotopy system*

$$\{\rho_i, d_i\} \colon e \leftarrow \pi_1(W^n) \xleftarrow{d_1} \rho_1 \leftarrow \cdots \xleftarrow{d_{n-2}} \rho_{n-2}$$

of length $n-2$ *is simple-homotopy equivalent to* $\rho(f, \xi)$, *then* W^n *has a regular Morse function* $\overline{f}$ *and gradient-like vector field* $\overline{\xi}$ *such that* $\rho(\overline{f}, \overline{\xi})$ *and* $\{\rho_i, d_i\}$ *are simple isomorphic.*

PROOF. Retract the manifold W^n to the median disks of the critical points of the functions f and g, obtaining CW-complexes $K(f)$ and $K(g)$, which are simple-homotopy equivalent. Then the chain complexes constructed on the universal coverings $\widehat{K}(f)$ and $\widehat{K}(g)$ have the same simple-homotopy type. It follows by Whitehead's theorem that also $\rho(f, \xi)$ and $\rho(g, \eta)$ have the same homotopy type.

Now consider the second part of the lemma. The condition on the boundary of the manifold ensures the existence of a Morse function on W^n without critical points of index $n-1$ or n. We can therefore suppose that the homotopy system $\rho(f, \xi)$ has length $n-2$. A theorem of [26] ensures the existence of a sequence of elementary operations taking $\rho(f, \xi)$ into (ρ_i, d_i) without increasing the length of any intermediate homotopy system. Since every elementary operation can be realized, we can suitably alter the Morse function and the vector field to obtain the desired conclusion. This completes the proof.

For a given manifold W^n, different homotopy systems can be constructed either by means of a Morse function and gradient-like vector field or by using cellular decomposition of W^n; but they all have the same simple-homotopy type. It makes sense, therefore, to speak of the simple-homotopy type of the homotopy system of W^n.

COROLLARY 4.3. *Let W^n, $n \geq 6$, be a manifold with boundary $\partial W^n = V$, and $\pi_1(V) \to \pi_1(W^n)$ an isomorphism. In order that W^n have a minimal Morse function, it is necessary and sufficient that its homotopy system be simple-homotopy equivalent to a minimal homotopy system.*

PROOF. Necessity is obvious; and sufficiency follows from the preceding lemma.

THEOREM 4.4. *Let W^n be a manifold with boundary $\partial W^n = V$, $n \geq 6$, $\pi_1(V) \to \pi_1(W^n)$ an isomorphism, and $\chi(k^3(W^n)) = \chi_0$. Suppose $\pi_1(W)$ is an s-group satisfying condition (h, χ_0), and r-$\dim \pi_1 \leq k$. Suppose also that the Morse number $\mu_\lambda(W^n)$ is $\geq k$ $(\lambda \geq 3$; for $n = 6$, $\mu_4(W^n) \geq k)$. Then W^n has a minimal Morse function.*

PROOF. Theorem 3.3 ensures the existence of a minimal homotopy system (ρ_i, d_i) in the homotopy type of the homotopy system of W^n. Realize (ρ_i, d_i) by a CW-complex K. Let K^3 be the 3-skeleton of K. Since K and W^n have the same homotopy type, there exists a mapping $j: K^3 \to \operatorname{Int} W^n$ inducing an isomorphism $j_*: \pi_2(K^3) \to \pi_2(W^n)$. By reason of the dimensional restrictions, we can suppose that j is an imbedding on the 2-skeleton K^2. Consider a tubular neighborhood of $j(K^2) \subset W^n$; call it W_2. Using the composite mapping $K^2 \hookrightarrow K^3 \hookrightarrow W^n$, it is easily seen that $\pi_2(W_2) \to \pi_2(W^n)$ is an epimorphism, since $\pi_2(K^2) \to \pi_2(K^3)$ is an epimorphism. Consider the manifold $W^n \backslash W_2 = \Gamma$. Obviously $\pi_1(\Gamma) \approx$

$\pi_1(\partial W_2) \approx \pi_1(W^n)$. Furthermore, from the exact homology sequence for the pair (W^n, W_2) and the excision theorem for homology of a universal covering, it follows that $\pi_2(\Gamma, \partial W_2) = \pi_1(\Gamma, \partial W_2) = 0$. From the results of [3], or from Wall's theorem [27], it follows that on Γ there exists a regular Morse function $f_2\colon (\Gamma, \partial W_2, \partial W^n) \to [I, 0, 1]$ without critical points of index 0, 1, $n-1$, or n. Using the cellular structure of $j(K^2)$, construct on W_2 a regular Morse function $f_1\colon (W_2, \partial W_2) \to [-1, 0]$ having only critical points of index 0, 1, or 2 (corresponding to the cells of K^2 of dimension 0, 1, or 2). Consider the Morse function $f = f_1 \cup f_2$ on W^2 and a gradient-like vector field ξ; and from them construct the chain complex

$$C(f, \xi)\colon \mathbf{Z} \leftarrow \mathbf{Z}[\pi_1(W^n)] \overset{\partial_1}{\longleftarrow} C_1 \overset{\partial_2}{\longleftarrow} C_2 \leftarrow \cdots.$$

Performing stabilization and contraction on the ∂_i $(i \geq 4)$, and using Theorem 4.3 of [3], we can arrange that f-rank$(\partial_i(C_i), C_{i-1}) = 0$ and is additive. Since in these dimensions all the elementary operations can be realized, we can assume, by suitably altering f and ξ, that the chain complex actually associated with f and ξ has this property; i.e., that f-rank$(\partial_i(C_i), C_{i-1}) = 0$ for $i \geq 4$ and is additive. We assert that f is now a minimal Morse function. Let us show that f has the minimal number of critical points of index 3. Indeed, applying Lemma 4.10 of [3] to the chain complex associated with (ρ_i, d_i) and the chain complex associated with f and ξ, we obtain

$$-\mu(C_1) + \mu(C_2) - \mu(C_3) = -\mu(\rho_1) + \mu(\rho_2^{\text{ab}}) - \mu(\rho_3) = \chi_0.$$

By construction, $-\mu(C_1) + \mu(C_2) = -\mu(\rho_1) + \mu(\rho_2^{\text{ab}})$. Therefore $\mu(C_3) = \mu(\rho_3)$. If now by stabilization and contraction of the ∂_i $(i = 1, 2, 3)$ in the chain complex $C(f, \xi)$ it can be arranged that f-rank$(\partial_i(C_i), C_{i-1}) = 0$ and is additive, it follows from Theorem 4.7 of [3] that the chain complex so obtained is minimal. And since the modules C_i for $i \geq 4$ have been left undisturbed, their ranks must already have been minimal. This completes the proof of the theorem.

Corollary 4.5. *Let W^n be a manifold with boundary $\partial W^n = V$, $n \geq 6$, and $\pi_1(V) \to \pi_1(W^n) = \pi$ an isomorphism, where π is a finite abelian group either isomorphic to $(\mathbf{Z}_2)^n$ or else having every p-Sylow subgroup of the form $\mathbf{Z}_{p^n}$ or $\mathbf{Z}_p \oplus \mathbf{Z}_{p^n}$. Suppose $H_i(\widetilde{W}^n, \mathbf{Z}) \neq 0$, $i \leq 3$ (for $n = 6$, $H_4(\widetilde{W}^n, \mathbf{Z}) \neq 0$). Then W^n has a minimal Morse function.*

Proof. r-dim $\pi \leq 1$ [19], finite abelian groups satisfy condition (h), and the requirement $H_i(\widetilde{W}^n, \mathbf{Z}) \neq 0$ ensures that $\mu_i(W^n) \neq 0$.

Corollary 4.6. *Let π be a finite s-group that has no epimorphic mapping onto any of the groups $\mathbf{Z}_{p^2} \oplus \mathbf{Z}_{p^2}$, $\mathbf{Z}_p \oplus \mathbf{Z}_p \oplus \mathbf{Z}_p$, $\mathbf{Z}_4 \oplus \mathbf{Z}_4$, $\mathbf{Z}_4 \oplus \mathbf{Z}_2 \oplus \mathbf{Z}_2$, or $\mathbf{Z}_p \oplus \mathbf{Z}_2 \oplus \mathbf{Z}_2 \oplus \mathbf{Z}_2$. Suppose also that the Euler characteristic $\chi(k^3(W^n))$ of the* *first k-invariant is* $\geq d^{\mu}(\pi)+1$ *(for nilpotent groups,* $\chi(k^3(W^n)) \geq d(\pi)+1$*) and* $H_i(\widetilde{W}^n, \mathbf{Z}) \neq 0$, $i \geq 3$ *(for* $n=6$, $H_4(W^n, \mathbf{Z}) \neq 0$*). Then* W^n *has a minimal Morse function.*

PROOF. In this case, $r\text{-}\dim \pi \leq 1$, and the condition $(h, \chi(k^3(W^n)))$ is satisfied. For nilpotent groups, $d(G) = d^{\mu}(G)$ [23].

COROLLARY 4.7. *Let* π *be a finite abelian group, and suppose*

$$\mu(H_i(W^n, \mathbf{Z})) + \mu(\operatorname{Tors} H_{i-1}(W^n, \mathbf{Z})) \geq 2$$

(for $n=6$, $\mu(H_4(W^n, \mathbf{Z})) \geq 2$*). Then* W^n *has a minimal Morse function.*

PROOF. π satisfies condition (h), and $r\text{-}\dim \pi \leq 2$.

COROLLARY 4.8. *Let* π *be a finite s-group. Suppose that the Euler characteristic* $\chi(k^3(W^n)) = \chi_0$ *of the first k-invariant of* W^n *is* $\geq d^{\mu}(\pi)+1$ *(for nilpotent groups,* $\chi(k^3(W^n)) \geq d(\pi)+1$*), and* $\mu(H_i(W^n, \mathbf{Z})) + \mu(\operatorname{Tors} H_{i-1}(W^n, \mathbf{Z})) \geq 2$ *(for* $n=6$, $\mu(H_4(W^n, \mathbf{Z})) \geq 2$*). Then* W^n *has a minimal Morse function.*

COROLLARY 4.9. *Let* π *be an s-group and satisfy* (h, χ_0)*, where* $\chi_0 = \chi(k^3(W^n))$*, and* $\operatorname{Wh} \pi = 0$*. Then* W^n *has a minimal Morse function.*

COROLLARY 4.10. *Suppose* π *is one of the following:*

1) *a free group,*
2) $\mathbf{Z} \oplus \mathbf{Z}$,
3) $\mathbf{Z}_2, \mathbf{Z}_3, \mathbf{Z}_4, \mathbf{Z}_6$.

Then W^n *has a minimal Morse function.*

THEOREM 4.11. *Let* W^n *be a manifold with boundary* $\partial W^n = V$, $n \geq 6$, *and* $\pi_1(V) \to \pi_1(W^n) = \pi$ *an isomorphism. Suppose*

1) π *is an s-group, with* $r\text{-}\dim \pi \leq c$;
2) *the essential height of* π *is equal to* a;
3) π *satisfies* (h, b) *for some* b, $d(\pi) \leq b \leq \chi_0 - c$, *where* $\chi_0 = \chi(k^3(W^n)) \geq a-1$.

Then W^n *has a minimal Morse function.*

REMARK. In the preceding theorem, the relation between $r\text{-}\dim \pi$ and the ith Morse number of the manifold W^n allowed the exclusion of critical points of adjacent indices defining a "torsion block." In the present theorem the "torsion block" can be removed by using the structure of the 2-skeleton of W^n, which has the homotopy type $K^2 \bigvee_{i=1}^{l} S_i^2$, where K^2 is a 2-dimensional CW-complex and $\bigvee_{i=1}^{l} S_i^2$ is a wedge of 2-spheres.

PROOF. The hypothesis of the theorem ensures the existence of a minimal homotopy system (ρ_i, d_i) in the homotopy type of the homotopy system of the manifold W^n. As shown during the proof of Theorem 3.3, the crossed sequence of the homotopy system (ρ_i, d_i) has Euler characteristic equal to χ_0. Realize the system (ρ_i, d_i) by a CW-complex K, and consider the 2-skeleton K^2 of K. By assumption, π satisfies the condition (h, b); and therefore K^2 is homotopy equivalent to a 2-dimensional CW-complex of the form

$$L^2 = \overline{L}^2 \bigvee_{i=1}^{\chi_0 - b} S_i^2,$$

where $\overline{L}^2$ is constructed from a presentation $\pi = \{a_1, \dots, a_{\mu(\pi)}, r_1, \dots, r_t\}$. Using the 3-skeleton $K^3 \subset K$ and Lemma 3.7 of [1], build up L^2 to a 3-dimensional complex L^3 homotopy equivalent to K^3. Since L^3 and W^n have equivalent algebraic 2-types, there exists a mapping $j\colon L^3 \to W^n$ inducing an isomorphism $j_*\colon \pi_2(L^3) \to \pi_2(W^n)$. Without loss of generality we can suppose that j is an imbedding on L^2. Obviously $j_*\colon \pi_2(L^2) \to \pi_2(W^n)$ is an epimorphism. Let W_2 be a tubular neighborhood of $j(L^2)$ in W^n. Then W_2 admits a decomposition into handles of index 0, 1, and 2, with one handle of index 0, $\mu(\pi)$ handles of index 1, and $\chi_0 - b$ of index 2, attached by a trivial imbedding of the median spheres. Using this handle decomposition, construct a regular Morse function f_1. Consider the manifold $\overline{W^n \backslash W_2} = \Gamma$. Since Γ obviously satisfies the hypothesis of Proposition 6.4 of [3], there exists on Γ a regular Morse function f_2 and a gradient-like vector field ξ such that the associated chain complex has the form

$$\begin{gathered} 0 \leftarrow H_3(\widetilde{W}^n) \leftarrow C_3 \overset{\partial_4}{\longleftarrow} C_4 \leftarrow \cdots, \\ \oplus \\ 0 \leftarrow R_2 \leftarrow R_3 \leftarrow 0, \end{gathered}$$

where $\mu(R_i) \le c$ and $f\text{-rank}(\partial_{i+1}(C_{i+1}), C_i) = 0$ and is additive. Consider the Morse function $\overline{f} = f_1 \cup f_2$ on W^n. Using the $\chi_0 - b$ critical points of index 2 of the function f_1, whose median spheres realize the trivial element of the fundamental group of the corresponding level surface of f_1, we can apply Lemma 4.10 of [3] to get rid of the critical points of index 2 and 3 that correspond to the elementary complex $0 \leftarrow R_2 \leftarrow R_3 \leftarrow 0$. Let f be the Morse function so obtained. Then it is easily shown that f is minimal; it suffices to use the argument at the end of the proof of the preceding theorem. This completes the present proof.

THEOREM 4.12. *Let W^n be a manifold with boundary $\partial W^n = V$, $n \ge 6$, $\pi_1(V) \to \pi_1(W^n) = \pi$ an isomorphism, and π isomorphic to one of the following groups*:

1) $\mathbf{Z}_k$;
2) $\bigoplus \sum_j \mathbf{Z}_{p_j} \oplus \mathbf{Z}_{p_j^{n_j}}$ (p_j *a prime*);
3) *the dihedral group* D_n (n *odd*).

Then W^n has a minimal Morse function.

PROOF. Let K^2 be a 2-dimensional CW-complex with one 0-dimensional vertex x and $\pi_1(K^2) \approx \pi$. Denote by $\xi(K^2, x)$ the group of classes of homotopy self-equivalences $(K^2, x) \to (K^2, x)$. There is a natural mapping $h: \xi(K^2, x) \to \mathrm{Wh}\,\pi$. It is known that in the above three cases h is an epimorphism; this was proved for $\pi = \mathbf{Z}_k$ in [7], for $\bigoplus\sum \mathbf{Z}_{p_j} \oplus \mathbf{Z}_{p_j^{n_j}}$ in [9], and for D_n in [15]. Furthermore, it is known that $r\text{-}\dim \pi \leq 1$ and π satisfies condition (h); see [4], [9], [19]. Let $\chi_0 = \chi(k^3(W^n))$ be the Euler characteristic of the first k-invariant. Since π is an s-group and satisfies condition (h), the homotopy type of the homotopy system of W^n contains a minimal homotopy system (ρ_i, d_i). Realize this system by a CW-complex. There exists an imbedding of the 2-skeleton $K^2 \subset K$ of this complex into W^n, inducing an isomorphism of fundamental groups and an epimorphism of 2-dimensional homotopy groups. Let W_2 be a tubular neighborhood of the image $j(K^2)$ in W^n. Using the cellular structure, construct on W_2 a regular Morse function and extend it in the obvious fashion to $\overline{W^n \backslash W_2} = \Gamma$. If Γ is not an h-cobordism, there exists on Γ a minimal Morse function, since $\pi_1(\Gamma, \partial W_2) = 0$ and $\pi_2(\Gamma, \partial W_2) = 0$, as is easily proved, and r-$\dim \pi \leq 1$; we can use Proposition 6.6 of [3]. Observe that this function will have no critical points of index 0, 1, 2, $n-1$, or n. Pasting together the Morse functions on W_2 and on Γ, we obtain a Morse function f which is minimal on W^n; the proof of this fact is similar to the proof of Theorem 4.4.

Now suppose Γ is an h-cobordism, with torsion $\tau_0 = \tau(\Gamma, \partial W_2)$. Using the fact that $\xi(K^2, x) \to \mathrm{Wh}\,\pi$ is an epimorphism, find an imbedding $l: K^2 \to \mathrm{Int}\, W^2$ with torsion $-\tau_0$. Consider a tubular neighborhood of $l(K^2)$ in W_2; call it $\overline{W}_2$. By construction, $\overline{W_2 \backslash \overline{W}_2} = P$ is an h-cobordism with torsion $-\tau_0$. Obviously $\Gamma \cup P$ is an h-cobordism. By the addition theorem for h-cobordisms $\tau(\Gamma \cup P, \partial \overline{W}_2) = -\tau_0 + \tau_0 = 0$; i.e., $\overline{W_n \backslash \overline{W}_2} \simeq \partial \overline{W}_2 \times I$. This completes the proof of the theorem.

PROPOSITION 4.13. *Let W^n be a manifold with boundary $\partial W^n = V$, $n \geq 6$, and $\pi_1(V) \to \pi_1(W^n) \approx \pi$ an isomorphism. Suppose π is an s-group and satisfies condition (h, χ_0), where $\chi_0 = \chi(k^3(W^n))$ is the Euler characteristic of the first k-invariant. Then there exists a manifold $\overline{W}^n$ that has a minimal Morse function and is homotopy equivalent to W^n.*

PROOF. Just as in the proof of Theorem 4.4, we find, using the minimal homotopy system (ρ_i, d_i) that by Theorem 3.3 is contained in the homotopy type of the homotopy system of W^n, a submanifold $W_2 \subset W^n$ on which there exists a Morse function with one critical point of index 0, $\mu(\rho_1)$ critical points of index 1, and $\mu(\rho_2)$ of index 2. In addition, the imbedding

$i: W_2 \to W^n$ induces an isomorphism $\pi_1(W_2) \to \pi_1(W^n)$ and an epimorphism $\pi_2(W_2) \to \pi_2(W^n)$. Since the manifold $\Gamma = \overline{W^n \backslash W_2}$ satisfies the hypothesis of Proposition 6.4 of [3], there exists in Γ a submanifold D such that $\overline{\Gamma \backslash D}$ is an h-cobordism, $\partial D = \partial W_2 \cup \partial W^n$, and on D there exists a minimal Morse function. Obviously the manifold $\overline{W}^n = W_2 \cup D$ is homotopy equivalent to W^n (they differ by an h-cobordism). Using the same argument as in the proof of Theorem 4.4, we conclude that also on $\overline{W}^n$ there exists a minimal Morse function. This proves the proposition.

For homotopy systems (ρ_i, d_i) associated with regular Morse functions and gradient-like fields on a manifold W^n, the $\omega_i(\rho_i, d_i)$ are well defined; we denote them by $\omega_i(\widetilde{W}^n)$.

PROPOSITION 4.14. *Let W^n be a manifold with boundary $\partial W^n = V$, $n \geq 6$, $\pi_1(V) \to \pi_1(W) = \pi$ an isomorphism, $\chi(k^3(W^n)) = \chi_0$, and let π satisfy condition (h, χ_0) and be an s-group. Suppose the $\omega_i(\widetilde{W}^n)$ are given, and the manifold W^n has a minimal Morse function. Then there exists an algorithm for computing the number of critical points of a minimal Morse function, using $H_i(\widetilde{W}^n, \mathbf{Z})$, $\omega_i(\widetilde{W}^n)$, and $\Gamma_i(\widetilde{W}^n)$.*

PROOF. This follows from Proposition 3.6.

If the imbedding of the boundary of a manifold does not induce an isomorphism of fundamental groups, then in order to construct minimal Morse functions we must make use of one more homotopy system, associated with a Morse function $f: W^n \to [0, 1]$, $f^{-1}(0) = \partial W^n$. But if $\pi_1(W^n, \partial W^n) \neq 0$, the resultant homotopy systems do not easily lend themselves to study.

The following assertion seems very plausible:

Let W^n be a manifold with one boundary component $\partial W^n = V$, $n \geq 6$, and $\pi_1(V) \to \pi_1(W^n)$ an isomorphism. Suppose π is a finite s-group. Then W^n has a minimal Morse function.

§5. Minimal Morse functions on closed manifolds

Let M^n be a compact manifold without boundary. In this section we examine the question of existence of a minimal Morse function on M^n. The homotopy systems constructed for regular Morse functions f and $-f$, and corresponding regular gradient-like vector fields ξ and $-\xi$, are "dual" to each other; we give a precise definition below. For this reason, closed manifolds provide homotopy systems with a richer structure than those considered above. We shall call them homotopy systems with Poincaré duality. They can be characterized as an adequate algebraic model for describing the Morse functions on closed manifolds.

Let $f: M^n \to I$ be a regular Morse function with one critical point of index 0 and one of index n, and ξ a corresponding gradient-like vector field. From f and ξ (resp. $-f$ and $-\xi$) construct homotopy systems

$$\rho(f, \xi)\colon e \leftarrow \pi \leftarrow \pi_1(M^n, m_0) \leftarrow \rho_1/d_2(\rho_2) \overset{d_2}{\longleftarrow} \rho \leftarrow \cdots,$$

$$\rho(-f, -\xi)\colon e \leftarrow \pi \leftarrow \pi_1(M^n, m_n) \leftarrow \overline{\rho}_1/\overline{d}_2(\overline{\rho}_2) \overset{\overline{d}_2}{\longleftarrow} \overline{\rho}_2 \leftarrow \cdots,$$

where m_0 (resp. m_n) is the critical point of index 0 (resp. n). On M^n choose a base point m and a fixed isomorphism

$$\varphi\colon \pi \to \pi_1(M^n, m).$$

Let $\gamma\colon I \to M^n$ (resp. $\overline{\gamma}\colon I \to M^n$) be a path joining m_0 (resp. m_n) to m. Denote by $\tilde{p}\colon \widetilde{M}^n \to M^n$ the universal covering space, regarded as the set of equivalence classes of paths issuing from the point m. Lifting f and ξ (resp. $-f$ and $-\xi$) to $\widetilde{M}^n$, construct a chain complex $C(f, \xi)$ (resp. $C(-f, -\xi)$) of left modules over the ring $\mathbf{Z}[\pi]$:

$$0 \leftarrow \mathbf{Z} \overset{\varepsilon}{\longleftarrow} \mathbf{Z}[\pi] \overset{\partial_1}{\longleftarrow} C_1 \overset{\partial_2}{\longleftarrow} C_2 \leftarrow \cdots,$$

$$0 \leftarrow \mathbf{Z} \overset{\varepsilon}{\longleftarrow} \mathbf{Z}[\pi] \overset{d_1}{\longleftarrow} D_1 \overset{d_2}{\longleftarrow} D_2 \leftarrow \cdots.$$

We can regard $C(f, \xi)$ (resp. $C(-f, -\xi)$) as associated with the homotopy system $\rho(f, \xi)$ (resp. $\rho(-f, -\xi)$).

LEMMA 5.1. *The homotopy systems $\rho(f, \xi)$ and $\rho(-f, -\xi)$ have the same simple-homotopy type.*

PROOF. Denote by $K(f)$ and $K(-f)$ the CW-complexes obtained by retracting the manifold M^n to the median disks of the critical points of f and $-f$, respectively. They are obviously of the same simple-homotopy type. Consequently, $\rho(f, \xi)$ and $\rho(-f, -\xi)$ are also simple-homotopy equivalent.

Let $\overline{}\colon \mathbf{Z}[\pi] \to \mathbf{Z}[\pi]$ be the involution defined by $\overline{g} = \omega(g)g^{-1}$, $g \in \pi$, where $\omega\colon \pi \to \mathbf{Z}_2 = \{-1, 1\}$ is the orientation homomorphism. If A is a left $\mathbf{Z}[\pi]$-module, then, using the involution, we can turn a right $\mathbf{Z}[\pi]$-module into a left one by the formula $\lambda \cdot b = b \cdot \overline{\lambda}$, where $b \in A^* = \operatorname{Hom}(A, \mathbf{Z}[\pi])$. Applying the functor Hom to the chain complex $C(f, \xi)$, we obtain a chain complex

$$C^*(f, \xi)\colon 0 \leftarrow \mathbf{Z} \leftarrow \mathbf{Z}[\pi] \leftarrow C_n^* \overset{\partial_n^*}{\longleftarrow} C_{n-1}^* \leftarrow \cdots.$$

It is known [2] that the matrix of the homomorphism ∂_i^* coincides up to sign with the matrix of the homomorphism ∂_{n-i}. We assume that for bases in the modules C_i and D_i are taken the median and co-median disks, respectively, and in the modules C_i^* the conjugate bases. Thus, $C^*(f, \xi)$ and $C(-f, -\xi)$ are simple isomorphic.

With any homotopy system (ρ_i, d_i) is always associated a chain complex (C_i, ∂_i). But the converse is false. It is not the case that for every chain complex of the form

$$(C_i, \partial_i): 0 \leftarrow \mathbf{Z} \overset{\varepsilon}{\longleftarrow} \mathbf{Z}[\pi] \overset{\partial_1}{\longleftarrow} C_1 \overset{\partial_2}{\longleftarrow} C_2 \leftarrow \cdots ,$$

in which $H_1(C_i, \partial_i) = 0$ and $\partial(c_j) = g_j - 1$, where the g_j are the generators of π and $\{c_j\}$ is a basis of C_1, there exists a homotopy system $e \leftarrow \pi \leftarrow \rho_1 \leftarrow \rho_2 \leftarrow \cdots$ with which the chain complex (C_i, ∂_i) would be associated. A counterexample is given in [10].

DEFINITION 5.2. A chain complex (C_i, ∂_i) over the ring $\mathbf{Z}[\pi]$ is said to be *realizable* if there exists a homotopy system (ρ_i, d_i) for which (C_i, ∂_i) is the associated chain complex.

We note that, as shown by examples, the property of realizability is not an invariant of simple-homotopy type [6].

Following Maller [18], we give now the definition of homotopy system with Poincaré duality.

DEFINITION 5.3. Let (ρ_i, d_i) be a homotopy system. We say that (ρ_i, d_i) has *Poincaré duality* if the cochain complex (C_i^*, ∂_i^*) is realizable. Here

$$(C_i, \partial_i): 0 \leftarrow \mathbf{Z} \overset{\varepsilon}{\longleftarrow} \mathbf{Z}[\pi] \overset{\partial_1}{\longleftarrow} C_1 \overset{\partial_2}{\longleftarrow} C_2 \leftarrow \cdots$$

is the chain complex associated with (ρ_i, d_i).

The homotopy system (ρ_i^*, d_i^*) with which the chain complex (C_i^*, d_i^*) is associated is called *dual* to the homotopy system (ρ_i, d_i). Obviously it also has Poincaré duality, and its dual is the homotopy system (ρ_i, d_i). Thus, regular Morse functions on closed manifolds give homotopy systems with Poincaré duality. If we perform an elementary operation on the homotopy system associated with the function f, there corresponds an elementary operation on the homotopy system associated with the function $-f$. This fact underlies the following definition.

DEFINITION 5.4. Homotopy systems with Poincaré duality (ρ_i, d_i) and (ρ_i^*, d_i^*) are *simple-homotopy equivalent* if there exist two sequences of homotopy systems (ρ_{ij}, d_{ij}) and (ρ_{ij}^*, d_{ij}^*) such that

1) $(\rho_{i1}, d_{i1}) = (\rho_i, d_i)$ and $(\rho_{i1}^*, d_{i1}^*) = (\rho_i^*, d_i^*)$,
2) (ρ_{ij}, d_{ij}) is dual to (ρ_{ij}^*, d_{ij}^*),
3) $(\rho_{i\,j+1}, d_{i\,j+1})$, $(\rho_{i\,j+1}^*, d_{i\,j+1}^*)$ is obtained from (ρ_{ij}, d_{ij}), (ρ_{ij}^*, d_{ij}^*) by an elementary operation.

ASSERTION 5.5. *Let M^n be a closed manifold. All homotopy systems with Poincaré duality that are associated with regular Morse functions and gradient-like vector fields on M^n are simple-homotopy equivalent. Furthermore, if a homotopy system with Poincaré duality is simple-homotopy equivalent to a homotopy system associated with a regular Morse function f and gradient-like vector field ξ, then there exist a Morse function $\overline{f}$ and a field $\overline{\xi}$ such that $\rho(\overline{f}, \overline{\xi})$ coincides with the given homotopy system.*

PROOF. This was proved by Maller in [18], where instead of Morse functions on M^n he considered handle decompositions.

DEFINITION 5.6. Denote by $(\rho_{ij}, d_{ij})_{j\in\mathscr{J}}$ homotopy systems with Poincaré duality that are simple-homotopy equivalent. We say that $(\overline{\rho}_i, \overline{d}_i) \in (\rho_{ij}, d_{ij})_{j\in\mathscr{J}}$ is *minimal in dimension* k if $\mu(\overline{\rho}_k) \leq \mu(\rho'_k)$ for all (ρ'_i, d'_i) in $(\rho_{ij}, d_{ij})_{j\in\mathscr{J}}$. We say that $(\overline{\rho}_i, \overline{d}_i)$ is *minimal* if it is minimal in all dimensions.

Note that minimal homotopy systems with Poincaré duality need not be minimal as homotopy systems.

The theorems below allow us to determine when there exist minimal homotopy systems with Poincaré duality that are associated with regular Morse functions on closed manifolds, or, in other words, to determine when on closed manifolds there exist minimal Morse functions.

THEOREM 5.7. *Let M^n be a closed manifold, $n \geq 6$, and $\chi(k^3(M^n)) = \chi_0$. Suppose $\pi_1(M^n) = \pi$ is an s-group satisfying (h, χ_0), with r-$\dim \pi \leq k$. Suppose also that the ith Morse number $\mu_i(M^n)$ is $\geq k$, $3 \leq i \leq n-3$ (for $n = 6$, $\mu_3(M^n) \geq 2k$). Then M^n has a minimal Morse function.*

PROOF. Let f be an arbitrary regular Morse function on M^n, ξ a gradient-like vector field, and $\rho(f, \xi)$ the homotopy system associated with f and ξ. Theorem 3.3 ensures the existence of a minimal homotopy system (ρ_i, d_i) homotopy equivalent to $\rho(f, \xi)$. Realize the system (ρ_i, d_i) by a CW-complex K. Since K and M^n have equivalent 2-types, there exists a mapping $j: K^3 \to M^n$ of the 3-skeleton $K^3 \subset K$ into M^n, inducing isomorphisms

$$j_*: \pi_1(K^3) \to \pi_1(M^n),$$
$$j_*: \pi_2(K^3) \to \pi_2(M^n).$$

Without loss of generality we can suppose that j is an imbedding on the 2-skeleton K^2. Since the natural imbedding $i: K^2 \to K^3$ induces an epimorphism $i_*: \pi_2(K^2) \to \pi_2(K^3)$, the mapping $j_*: \pi_2(K^2) \to \pi_2(M^n)$ is an epimorphism. In view of the dimensional restrictions, there exists another imbedding $\overline{j}: K^2 \to M^n$, homotopic to j, such that the images $j(K^2)$ and $\overline{j}(K^2)$ do not intersect. Let M_2 and $\overline{M}_2$ be nonintersecting tubular neighborhoods of $j(K^2)$ and $\overline{j}(K^2)$, respectively. It is easily shown that $\pi_1(\partial M_2) \hookrightarrow \pi_1(\Gamma) \hookleftarrow \pi_1(\partial \overline{M}_2)$ are isomorphisms and $\pi_2(\Gamma, \partial M_2) = \pi_2(\Gamma, \partial \overline{M}_2) = 0$, where $\Gamma = \overline{M^n \backslash (M_2 \cup \overline{M}_2)}$. By Wall's theorem [27], there exists on Γ a regular Morse function

$$f_2: (\Gamma, \partial M_2, \partial \overline{M}_2) \to ([1/3, 2/3], 1/3, 2/3)$$

with no critical points of index $0, 1, n-1$, or $n-2$, if $n \geq 7$. When $n = 6$, it can be arranged that either there are no critical points of index $0, 1, 2,$

$n-1$, n or none of index 0, 1, n, $n-1$, $n-2$. Using the cellular structures of $j(K^2)$ and $\overline{j}(K^2)$, construct regular Morse functions $f_1\colon (M_2, \partial M_2) \to ([0, 1/3], 1/3)$ and $f_3\colon (\overline{M}_2, \partial \overline{M}_2) \to ([2/3, 1], 1)$. Consider the Morse function $g = f_1 \cup f_2 \cup f_3$ and a corresponding gradient-like vector field $\overline{\xi}$. From g and $\overline{\xi}$ construct a chain complex of $\mathbf{Z}[\pi]$-modules:

$$0 \leftarrow \mathbf{Z} \overset{\varepsilon}{\longleftarrow} \mathbf{Z}[\pi] \overset{\partial_1}{\longleftarrow} C_1 \overset{\partial_2}{\longleftarrow} C_2 \leftarrow \cdots \leftarrow \mathbf{Z}[\pi].$$

Suppose $n \geq 7$. Stabilizing and contracting the ∂_i, $4 \leq i \leq n-3$, and using Proposition 4.9 of [3], we can arrange that $f\text{-rank}(\partial_i(C_i), C_{i-1}) = 0$ and is additive. Since in these dimensions all elementary operations can be realized, by altering the function g and the field $\overline{\xi}$ we can suppose that g had been chosen in such a way that the segment

$$C_3 \overset{\partial_4}{\longleftarrow} C_4 \leftarrow \cdots \overset{\partial_{n-3}}{\longleftarrow} C_{n-3}$$

already has this property. By Lemma 4.10 of [3] applied to the chain complex associated with $(g, \overline{\xi})$ and the one associated with (ρ_i, d_i), we have

$$-\mu(C_1) + \mu(C_2^{\varsigma}) - \mu(C_3) = -\mu(\rho_1) + \mu(\rho_2^{\mathrm{ab}}) - \mu(\rho_3).$$

By construction, $-\mu(C_1) + \mu(C_2) = -\mu(\rho_1) + \mu(\rho_2^{\mathrm{ab}})$. Therefore $\mu(\rho_3) = \mu(C_3)$. Now consider the function $-g$ and field $-\overline{\xi}$. Obviously the chain complex associated with $(-g, -\overline{\xi})$ also has the property that $f\text{-rank}(\cdot, \cdot) = 0$ and is additive, since the functor Hom preserves decomposition into direct sum. By a similar argument, using the homotopy system (ρ_i, d_i) and the one associated with $(-g, -\overline{\xi})$, and the fact that they have the same homotopy type, we show that for $-g$ the number of critical points of index 3 is minimal, and therefore for g the number of index $n-3$ is minimal. Consider the chain complex

$$0 \leftarrow \mathbf{Z} \overset{\varepsilon}{\longleftarrow} \mathbf{Z}[\pi] \overset{\partial_1}{\longleftarrow} C_1 \leftarrow \cdots \overset{\partial_{n-1}}{\longleftarrow} C_{n-1} \leftarrow \mathbf{Z}[\pi] \leftarrow 0.$$

Stabilizing and contracting the ∂_i, $1 \leq i \leq 3$, $n-2 \leq i \leq n-1$, we can arrange that $f\text{-rank}(\partial_i(C_i), C_{i-1}) = 0$ and is additive. By Proposition 4.9 of [3], the chain complex so obtained is minimal. Since the modules C_j, $4 \leq j \leq n-4$, have been left undisturbed, their ranks were minimal also.

Consider now the case $n = 6$. Let f be an arbitrary regular Morse function on M^n with one critical point of index 0 and one of index 6. It is known that

$$1 - \nu_1 + \nu_2 - \nu_3 + \nu_4 - \nu_5 + 1 = \chi(M^6),$$

where ν_i is the number of critical points of index i for f. Hence

$$\nu_3 = (-\nu_1 + \nu_2) + (-\nu_5 + \nu_4) + 2 - \chi(M^n).$$

Consequently, ν_3 reaches its minimum when $(-\nu_1 + \nu_2)$ and $(-\nu_5 + \nu_4)$ are least possible. Consider the cobordism $(\Gamma, \partial M_2, \partial \overline{M}_2)$; and suppose the

Morse function on it has critical points of index 2 and 3. By construction, the critical points of $g = f_1 \cup f_2 \cup f_3$ of index 3 are concentrated on Γ. By hypothesis, $\mu_3(M^n) \geq 2k$ and $r\text{-}\dim \pi \leq k$; consequently, using Lemma 4.10 of [3], we can get rid of critical points of index 2. Thus, we find on Γ a Morse function $\hat{f}_2$ having critical points of index 3 only. The functions f_1 and f_3 have been chosen so that $(-\nu_1 + \nu_2)$ and $(-\nu_5 + \nu_4)$ are least possible. Therefore $f_1 \cup \hat{f}_2 \cup f_3$ is a minimal Morse function. This completes the proof of the theorem.

We gather here all the consequences of this theorem into a single assertion, whose proof consists in comparing $r\text{-}\dim \pi$ and the ith Morse number, which is evaluated by using the homology of the manifold.

COROLLARY 5.8. *Let M^n, $n \geq 6$, be a closed manifold and $\pi_1(M^n) = \pi$, where*

1) $\pi = (\mathbf{Z}_2)^n$, *or every p-Sylow subgroup is of the form* $\mathbf{Z}_{p^n}$ *or* $\mathbf{Z}_p \oplus \mathbf{Z}_{p^n}$, *and* $H_i(\widetilde{M}^n, \mathbf{Z}) \neq 0$, $3 \leq i \leq n-3$ *(for* $n = 6$, $\mu(\operatorname{Tors} H_2(M^n, \mathbf{Z})) + \mu(H_3(M^n, \mathbf{Z})) \geq 2$*); or*
2) π *is finite abelian and* $\mu(H_i(M^n, \mathbf{Z})) + \mu(\operatorname{Tors} H_{i-1}(M^n, \mathbf{Z})) \geq 2$, $2 \leq i \leq n-3$ *(for* $n = 6$, $\mu(H_3(M^n, \mathbf{Z})) + \mu(\operatorname{Tors} H_2(M^n, \mathbf{Z})) \geq 4$*); or*
3) π *is a finite s-group, with no epimorphic mapping onto any of the groups* $\mathbf{Z}_{p^2} \oplus \mathbf{Z}_{p^2}$, $\mathbf{Z}_p \oplus \mathbf{Z}_p \oplus \mathbf{Z}_p$, $\mathbf{Z}_4 \oplus \mathbf{Z}_4$, $\mathbf{Z}_4 \oplus \mathbf{Z}_2 \oplus \mathbf{Z}_2$, *or* $\mathbf{Z}_p \oplus \mathbf{Z}_2 \oplus \mathbf{Z}_2 \oplus \mathbf{Z}_2$, $\chi(k^3(M^n)) \geq d^\mu(\pi) + 1$ *(for nilpotent groups,* $\chi(k^3(M^n)) \geq d(\pi) + 1$*), and* $H_i(\widetilde{M}^n, \mathbf{Z}) \neq 0$, $3 \leq i \leq n-3$ *(for* $n = 6$, $\mu(H_3(M^n, \mathbf{Z})) + \mu(\operatorname{Tors} H_2(M^n, \mathbf{Z})) \geq 2$*); or*
4) π *is a finite s-group,* $\chi(k^3(M^n)) \geq d^\mu(\pi) + 1$ *(for nilpotent groups,* $\chi(k^3(M^n)) \geq d(\pi) + 1$*), and for* $n = 6$,

$$\mu(H_3(M^n, \mathbf{Z})) + \mu(\operatorname{Tors} H_2(M^n, \mathbf{Z})) \geq 4$$

(for $n > 6$, $\mu(H_i(M^n, \mathbf{Z})) + \mu(\operatorname{Tors} H_{i-1}(M^n, \mathbf{Z})) \geq 2$*); or*
5) π *is an s-group satisfying the conditions* $\operatorname{Wh} \pi = 0$ *and* (h, χ_0), *where* $\chi_0 = \chi(k^3(M^n))$, *and in particular a free group,* $\mathbf{Z} \oplus \mathbf{Z}$, $\mathbf{Z}_2$, $\mathbf{Z}_3$, $\mathbf{Z}_4$, *or* $\mathbf{Z}_6$.

Then μ has a minimal Morse function.

THEOREM 5.9. *Let M^n be a closed manifold, $n \geq 6$, and $\pi_1(M^n) = \pi$ satisfy the following conditions:*

1) π *is an s-group;*
2) *the essential height of* π *is equal to* a;
3) $r\text{-}\dim \pi \leq b$;
4) (h, c), *where* $d(\pi) \leq c \leq \chi_0 - b$ *and* $\chi_0 = \chi(k^3(M^n)) \geq a - 1$.

Then M^n has a Morse function.

PROOF. We use the argument in the proof of Theorem 4.11. Consider a minimal homotopy system (ρ_i, d_i), which always exists in the homotopy type of the homotopy system associated with an arbitrary regular Morse function of M^n. Realize this minimal system by a CW-complex K. As shown in the proof of Theorem 4.11, K can be chosen so that its 2-skeleton K^2 is of the form $L^2 = \overline{L}^2 \bigvee_i S_i^2$, where $\overline{L}^2$ is constructed from a presentation $\pi = \{a_1, \dots, a_{\mu(\pi)}, r_1, \dots, r_s\}$. There exists an imbedding $j: K^2 \to M^n$, inducing an isomorphism of fundamental groups and an epimorphism $\pi_2(K^2) \to \pi_2(M^n)$. Let $\overline{j}: K^2 \to M^n$ be another imbedding, homotopic to j, such that $j(K^2) \cap \overline{j}(K^2) = \varnothing$. Consider disjoint tubular neighborhoods of $j(K^2)$ and $\overline{j}(K^2)$; denote them by M_1 and M_2, respectively. Using the cellular structure, construct on M_1 and M_2 handle decompositions such that there are $\chi_0 - c$ handles of index 2, attached by trivial imbeddings of the median spheres. From these handle decompositions construct minimal Morse functions

$$f_1: M_1 \to [0, 1/3], \qquad f_1^{-1}(1/3) = \partial M_1,$$
$$f_2: M_2 \to [2/3, 1], \qquad f_2^{-1}(2/3) = \partial M_2.$$

Consider the manifold $\Gamma = \overline{M^n \backslash (M_1 \cup M_2)}$. It is easily shown that $\pi(\Gamma, \partial M_i) = \pi_2(\Gamma, \partial M_i) = 0$. Hence there exist on Γ a regular Morse function f_2 and a gradient-like vector field ξ such that the associated chain complex is of the form

$$\begin{array}{c} C_3 \overset{\partial_4}{\longleftarrow} C_4 \leftarrow \cdots \overset{\partial_{n-3}}{\longleftarrow} C_{n-3} \\ \oplus \\ 0 \leftarrow R_2 \overset{\partial_3}{\longleftarrow} R_3 \leftarrow 0 \end{array}$$

and f-rank$(\partial_i(C_i), C_{i-1}) = 0$ and is additive. Consider the function $\overline{f} = f_1 \cup f_2 \cup f_3$. Using the $\chi_0 - c$ critical points of index 2 of the function f_1, whose median disks are attached by trivial imbeddings of a circle, we can apply Lemma 4.10 of [3] to get rid of the critical points of index 2 and 3 that correspond to the elementary complex $0 \leftarrow R_2 \overset{\partial_3}{\longleftarrow} R_3 \leftarrow 0$. Let f be the Morse function so obtained. Then it is easily shown that f is minimal; it suffices to use the argument in the proof of the preceding theorem. This completes the proof of the present theorem.

THEOREM 5.10. *Let M^n, $n \geq 6$, be a closed manifold, with $\pi_1(M^n) \approx \pi$ isomorphic to one of the following*:

1) $\mathbf{Z}_k$;
2) *an abelian group in which every p-Sylow subgroup is of the form* $\mathbf{Z}_p$ *or* $\mathbf{Z}_p \oplus \mathbf{Z}_{p^i}$ *(p a prime)*;
3) *the dihedral group D_n (n odd)*.

Then M^n has a minimal Morse function.

PROOF. In this case, $r\text{-}\dim\pi \leq 1$, and two 2-dimensional cellular complexes have the same simple-homotopy type if they have the same Euler characteristic $\chi(K) = \chi(L)$, since $\xi(K, x) \to \mathrm{Wh}\,\pi$ is an epimorphism. We use the argument of Theorem 4.12. Let K be a CW-complex associated with the minimal homotopy system of the manifold M^n. Consider imbeddings $j_1: K^2 \to M^n$ and $j_2: K^2 \to M^n$ of the 2-skeleton $K^2 \subset K$ into M^n, inducing an isomorphism of fundamental groups and an epimorphism of 2-dimensional homotopy groups. We can assume $j_1(K^2) \cap j_2(K^2) = \varnothing$. Let M_1 and M_2 be nonintersecting tubular neighborhoods of $j_1(K^2)$ and $j_2(K^2)$, and consider the manifold $\Gamma = \overline{M^n \backslash (M_1 \cup M_2)}$. If Γ is not an h-cobordism, then M^n has a minimal Morse function. Indeed, $r\text{-}\dim\pi \leq 1$, and we can apply Proposition 6.10 of [3]. Starting from the cellular structure, construct regular Morse functions f_1 and f_2 on M_1 and M_2, and extend them to Γ; this gives a minimal Morse function on M^n. If Γ is an h-cobordism, with torsion $\tau_0 = \tau(\Gamma, \partial M_1)$, construct a new imbedding $\bar{j}_1: K^2 \to M_1$ with torsion $-\tau_0$. Then $\overline{M_1 \backslash \overline{M}_1} = P$ is an h-cobordism, and $\tau(P, \partial\overline{M}_1) = -\tau_0$, where $\overline{M}_1$ is a tubular neighborhood of $\bar{j}_1(K^2)$. Obviously $\Gamma \cup P$ is diffeomorphic to $\partial\overline{M}_1 \times I$. This proves the theorem.

As in the case of manifolds with one boundary component, we can construct an algorithm for finding the number of critical points of a minimal Morse function on a closed manifold in terms of $H_i(\widetilde{M}^n, \mathbf{Z})$, $\Gamma_i(\widetilde{M}^n)$, and $\omega_i(\widetilde{M}^n)$.

It seems very plausible that on a closed manifold of dimension greater than 5, for which the fundamental group is a finite s-group, there always exists a minimal Morse function.

BIBLIOGRAPHY

1. J. W. Milnor, *Morse theory*, Ann. of Math. Stud., No. 51, Princeton Univ. Press, Princeton, N. J., 1963.
2. C. P. Rourke and B. J. Sanderson, *Introduction to piecewise-linear topology*, Springer-Verlag, New York, 1972.
3. V. V. Sharko, *K-theory and Morse theory*. I, Akad. Nauk Ukrain. SSR Inst. Mat., Preprint, 1986, no. 39; English transl.
4. W. Browning, *Finite CW-complexes of cohomological dimension* 2 *with finite abelian* π_1, Preprint, Forschungsinstitut für Mathematik, Zurich, 1979.
5. ——, *The homotopy classification of non-minimal* 2*-complexes with given finite fundamental group*, Preprint, Forschungsinstitut für Mathematik, Zurich, 1979.
6. W. H. Cockcroft and R. M. F. Moss, *On the two-dimensional realisability of chain complexes*, J. London Math. Soc. (2) **11** (1975), no. 3, 257–262.
7. M. N. Dyer and A. J. Sieradski, *Trees of homotopy types of two-dimensional CW-complexes*. I, Comment. Math. Helv. **48** (1973), 31–44.
8. M. N. Dyer, *On the essential height of homotopy trees with finite fundamental group*, Compositio Math. **36** (1978), no. 2, 209–224.
9. ——, *Simple homotopy types for* (G, m)*-complexes*, Proc. Amer. Math. Soc. **81** (1981), no. 1, 111–115.

10. M. J. Dunwoody, *Relation modules*, Bull. London Math. Soc. **4** (1972), 151–155.

11. P. J. Hilton and Urs Stammbach, *A course in homological algebra*, Springer-Verlag, New York–Berlin, 1971.

12. D. F. Holt, *An interpretation of the cohomology groups* $H^n(G, M)$, J. Algebra **60** (1979), no. 2, 307–318.

13. Johannes Huebschmann, *Crossed n-fold extensions of groups and cohomology*, Comment. Math. Helv. **55** (1980), no. 2, 302–313.

14. Sushil Jajodia, *Spaces dominated by two-complexes*, Proc. Amer. Math. Soc. **82** (1981), no. 2, 288–290.

15. Sushil Jajodia and B. A. Magurn, *Surjective stability of units and simple homotopy type*, J. Pure Appl. Algebra **18** (1980), no. 1, 45–58.

16. J.-L. Loday, *Cohomologie et groupe de Steinberg relatifs*, J. Algebra **54** (1978), no. 1, 178–202.

17. S. Mac Lane and J. H. C. Whitehead, *On the* 3*-type of a complex*, Proc. Nat. Acad. Sci. U. S. A. **36** (1950), 41–48.

18. Michael Maller, *Fitted diffeomorphisms of nonsimply connected manifolds*, Topology **19** (1980), no. 4, 395–410.

19. B. A. Magurn, R. M. Oliver, and Leonid Vaserstein, *Units in Whitehead groups of finite groups*, J. Algebra **84** (1983), no. 2, 324–360.

20. Wolfgang Metzler, *Über den Homotopietyp zweidimensionaler CW-Komplexe und Elementartransformationen bei Darstellungen von Gruppen durch Erzeugende und definierende Relationen*, J. Reine Angew. Math. **285** (1976), 7–23.

21. Renée Peiffer, *Über Identitäten zwischen Relationen*, Math. Ann. **1212** (1949), 67–99.

22. J. G. Ratcliffe, *Crossed extensions*, Trans. Amer. Math. Soc. **257** (1980), no. 1, 73–89.

23. E. S. Rapaport, *Finitely presented groups: The deficiency*, J. Algebra **24** (1973), 531–543.

24. Michael Shub and D. P. Sullivan, *Homology theory and dynamical systems*, Topology **14** (1975), 109–132.

25. A. J. Sieradski, *A semigroup of simple homotopy types*, Math. Z. **153** (1977), no. 2, 135–148.

26. C. T. C. Wall, *Formal deformations*, Proc. London Math. Soc. (3) **16** (1966), 342–352.

27. ——, *Geometrical connectivity.* I, J. London Math. Soc. (2) **3** (1971), 597–604.

28. J. H. C. Whitehead, *Combinatorial homotopy.* II, Bull. Amer. Math. Soc. **55** (1949), 453–496.

29. J. H. C. Whitehead, *Simple homotopy types*, Amer. J. Math. **72** (1950), 1–57.

30. Y.-C. Wu, $H^3(G, A)$ *and obstructions of group extensions*, J. Pure Appl. Algebra **12** (1978), no. 1, 93–100.

Translated by J. A. ZILBER

Amer. Math. Soc. Transl.
(2) Vol. **149**, 1991

Diffeomorphisms of Two-Dimensional Manifolds with Transitive Foliations

V. Z. GRINES

It is known that many topological and metric properties of Anosov diffeomorphisms follow from the fact that they have pairs of invariant transversal foliations without singularities—one contractive and the other expansive. Recall that a foliation $\{W^s\}$ $(\{W^u\})$ that is invariant with respect to f is said to be contractive (expansive) if for any points $x, y \in W^s \in \{W^s\}$ $(x, y \in W^u \in \{W^u\})$ we have $\rho(f^n(x), f^n(y)) \to 0$ $(\rho(f^{-n}(x), f^{-n}(y)) \to 0$ respectively) as $n \to +\infty$, where ρ is the intrinsic metric on W^s (W^u).

The torus is the only two-dimensional manifold that admits the existence of an Anosov diffeomorphism, and Anosov diffeomorphisms of the torus are topologically conjugate if and only if they are π_1-conjugate [1], [2]. In this paper it is established that on any two-dimensional orientable manifold M_p of genus $p > 1$ there exists a C^∞-diffeomorphism that has both a contractive and an expansive invariant orientable transitive foliation, each of which has a finite number of singular points of saddle type and which are everywhere transversal on M_p, except at singular points.

Diffeomorphisms of this kind will be called T-diffeomorphisms. We recall that a foliation is said to be transitive if there exists a leaf of this foliation that is everywhere dense on M_p. A singular point z of the orientable foliation $\{W\}$ will be called a saddle point if, in a neighbourhood of this point, there is a vector field v such that z is an equilibrium point for the field v and this has at least four sectors all of which are saddle sectors.

It is proved that T-diffeomorphisms given on a manifold M_p of genus $p > 1$ are topologically conjugate if and only if they are π_1-conjugate.

1991 *Mathematics Subject Classification*. Primary 57R50, 57R30, 58F15.

Translation of Methods of the Qualitative Theory of Differential Equations, Gor′kov. Gos. Univ., Gorki, 1982, pp. 62–72; MR 87k:58150.

§1. The structure of T-diffeomorphisms

Let M_2 be a smooth closed orientable two-dimensional manifold of genus $p=2$. According to [3] (Section 5) we can represent M_2 as a C^∞-manifold imbedded in $\mathbf{R}^3 \equiv \{(x_1, x_2, x_3)\}$ and having the form of the standard sphere $x_1^2+x_2^2+x_3^2=1$ with two handles that are situated so that the map $\psi\colon \mathbf{R}^3 \to \mathbf{R}^3$, acting according to the formula $\psi(x_1, x_2, x_3) = (x_1^2-x_2^2, 2x_1\cdot x_2, x_3)$, takes M_2 into a manifold M_1 that is diffeomorphic to the torus $\mathbf{T}^2=\mathbf{R}^2/\mathbf{Z}^2$, where $\mathbf{R}^2$ is the plane and $\mathbf{Z}^2$ is the integral lattice. We let $\phi\colon M_2 \to M_1$ denote the restriction $\psi|M_2$ and let $S(0,0,1)$ and $N(0,0,-1)$ denote the points where M_2 intersects the x_3-axis. Then ϕ is a two-point covering $M_2\backslash(\{S\}\cup\{N\}) \to M_1\backslash(\{S\}\cup\{N\})$ and it is a local diffeomorphism everywhere on M_2 except at the points S and N.

We introduce local coordinates in neighbourhoods $V(S)$ and $V(N)$ of the points S and N on M_2 by assigning to a point $w \in V(S)$ $(w\in V(N))$ with coordinates x_1, x_2, x_3 in $\mathbf{R}^3$ the coordinates x_1, x_2 in the plane $x_3=0$ and we let u_1, u_2 denote the local coordinates in the neighbourhoods $V_1 = \phi(V(S))$, $V_2=\phi(V(N))$ of the points $\phi(S)=S$ and $\phi(N)=N$ connected with x_1 and x_2 by the formulae

$$u_1 = x_1^2 - x_2^2, \qquad u_2 = 2x_1x_2. \tag{1.1}$$

Consider the algebraic automorphism g_0 of the torus $\mathbf{U}^2$ induced by the matrix $A=\begin{pmatrix}5&8\\8&13\end{pmatrix}$. We fix the stationary points z_1, z_2 of the automorphism g_0 that are images of the points $(\frac12, \frac12)$, $(\frac12, 0)$ under the natural projection $\pi\colon \mathbf{R}^2 \to \mathbf{T}^2$. Let h be the diffeomorphism mapping $\mathbf{T}^2$ into M_1 so that $h(z_1)=S$, $h(z_2)=N$, and the diffeomorphism $g=hg_0h^{-1}\colon M_1\to M_1$ in the local coordinates u_1, u_2 has the form $g(u_1, u_2) = (\lambda u_1, \lambda^{-1}u_2)$, where $\lambda>1$ is an eigenvalue of the matrix A.

Following [5] we represent g in the neighbourhoods V_1 and V_2 of the points S and N as the shift by a unit of time along the trajectories of the flow defined by the system

$$\begin{aligned} \dot u_1 &= \ln\lambda\cdot u_1, \\ \dot u_2 &= -\ln\lambda\cdot u_2. \end{aligned} \tag{1.2}$$

We fix the domain $\mathscr{D}_2\{(u_1, u_2)\mid u_1^2+u_2^2<r^2\}$, where the number r is chosen so that the image and the preimage of $\mathscr{D}_2$ under the action of the diffeomorphism g belong to the neighbourhood V_1 (V_2) of the point S (N).

We make a change of time in (1.2):

$$\begin{aligned} \dot u_1 &= \ln\lambda\cdot u_1\psi(u_1, u_2), \\ \dot u_2 &= -\ln\lambda\cdot u_2\psi(u_1, u_2), \end{aligned} \tag{1.3}$$

where

$$\psi(u_1, u_2) = \begin{cases} \sqrt{u_1^2 + u_2^2}, & u_1^2 + u_2^2 < r_1^2 < r^2, \\ 1, & u_1^2 + u_2^2 > r^2, \end{cases}$$

$\psi(u_1, u_2) \in C^\infty$ if $u_1^2 + u_2^2 \neq 0$ and $r_1 < r$ is chosen so that the domain $\mathscr{D}_{r_1}$ belongs to the image of the domain $\mathscr{D}_r$ under the action of g.

We let $\{T^t\}$ denote the flow generated by (1.3) and we define a homeomorphism $\tilde{g}$ on the torus M_1 that coincides with g outside $V_1 \cup V_2$ and acts as a shift along the trajectories of the flow $\{T^t\}$ on $V_1 \cup V_2$. By [4] $\tilde{g}\colon M_1 \to M_1$ lifts to a homeomorphism $\tilde{f}\colon M_2 \to M_2$ such that $\phi\tilde{f} = \tilde{g}\phi$.

LEMMA 1.1. *The homeomorphism* $\tilde{f}\colon M_2 \to M_2$ *is smooth everywhere on* M_2.

PROOF. Obviously $\tilde{f}$ is smooth on $M_2 \backslash (\{S\} \cup \{N\})$. If we make the substitution (1.1) in the system (1.3) we see that in a sufficiently small neighbourhood of the point S (the point N) the homeomorphism $\tilde{f}$ is a shift by a unit of time along the flow

$$\dot{x}_1 = \tfrac{1}{2} \ln \lambda \cdot (x_1^3 - 3x_1 x_2^2),$$
$$\dot{x}_2 = -\tfrac{1}{2} \ln \lambda \cdot (x_2^3 - 3x_1^2 x_2),$$

and consequently is smooth in $V(S)$ $(V(N))$. The lemma is proved.

Similarly to [5] (Section 3) it can be established that the homeomorphism $\tilde{g}$ is topologically conjugate to the diffeomorphism g.* Consequently $\tilde{g}$ has two invariant foliations $\{\widetilde{W}^s\}$ and $\{\widetilde{W}^u\}$ without singularities that are images of the contractive and expansive foliations $\{W^s\}$ and $\{W^u\}$ of the Anosov diffeomorphism g. In [6] it was proved that the foliations $\{\widetilde{W}^s\}$ and $\{\widetilde{W}^u\}$ consist of smooth curves that are continuous and transversal everywhere on M_1. Since $\tilde{g}$ and g are topologically conjugate, for any points $x, y \in \widetilde{W}^s \in \{\widetilde{W}^s\}$ $(x, y \in \widetilde{W}^u \in \{\widetilde{W}^u\})$ the condition $\rho(\tilde{g}^n(x), \tilde{g}^n(y)) \to 0$ $(\rho(\tilde{g}^{-n}(x), \tilde{g}^{-n}(y)) \to 0)$ as $n \to +\infty$ is satisfied. Therefore, although $\tilde{g}$ is not an Anosov diffeomorphism, it has contractive and expansive foliations that are transversal everywhere on M_1.

LEMMA 1.2. *The diffeomorphism* $\tilde{f}$ *has a contractive and an expansive invariant foliation* $\{\Gamma^s\}$ *and* $\{\Gamma^u\}$ *with singularities of saddle type at the points* S *and* N. *The foliations* $\{\Gamma^s\}$ *and* $\{\Gamma^u\}$ *are transversal everywhere except at* S *and* N *and are transitive on* M_2.

PROOF. The construction of the diffeomorphism $\tilde{f}$ implies that on M_2 there are invariant continuous foliations $\{\Gamma^s\}$ and $\{\Gamma^u\}$ that are preimages

* In [5] it was assumed that the function $\psi(u)$, where $u = u_1^2 + u_2^2$, is smooth at $u = 0$ but this condition was used only so that the homeomorphism of the torus constructed there be smooth; the proof of topological conjugacy itself uses only those properties of the function $\psi(u)$ that are satisfied in our case.

under the map $\phi\colon M_2 \to M_1$ of the foliations $\{\widetilde{W}^s\}$ and $\{\widetilde{W}^u\}$. Each foliation $\{\Gamma^s\}$ and $\{\Gamma^u\}$ has exactly two singular points, S and N, of saddle type. Since $\phi\tilde{f} = \tilde{g}\phi$, S and N are fixed points of the diffeomorphism $\tilde{f}$, and the contractive and expansive conditions are satisfied for the foliations $\{\Gamma^s\}$ and $\{\Gamma^u\}$. Since the periodic points of the homeomorphism $\tilde{g}$ are dense on M_1, the periodic points of $\tilde{f}$ are dense on M_2. The map ϕ is a local diffeomorphism $M_2 \to M_1$ everywhere except at the points S and N; therefore the foliations $\{\Gamma^s\}$ and $\{\Gamma^u\}$ are transversal everywhere on M_2 except at these points. This and the density of the set of periodic points of the diffeomorphism $\tilde{f}$ on M_2 imply that each leaf of the foliations $\{\Gamma^s\}$ and $\{\Gamma^u\}$ is everywhere dense on M_2. The lemma is proved.

It is known that for every $p > 2$ there exists a smooth covering map $\phi_p\colon M_p \to M_2$. By [7] there exists a diffeomorphism $f_p\colon M_p \to M_p$ covering $\tilde{f}\colon M_2 \to M_2$, that is, such that $\phi_p f_p = \tilde{f}_2\phi_p$. Since ϕ_p is a local diffeomorphism everywhere on M_p, the following theorem can be proved exactly as Lemma 1.2.

THEOREM 1.1. *On any two-dimensional orientable manifold M_p of genus $p > 1$ there exists a T-diffeomorphism.*

§2. Necessary and sufficient conditions for T-diffeomorphisms to be topologically equivalent

Let us introduce an analytic structure on M_p making M_p into a Riemann surface. We consider a conformal map π of the universal covering $\overline{M}_p$ onto M_p, where $\overline{M}_p$ is the Lobachevskiĭ plane using the Poincaré realization as the interior of the disk $|z| < 1$ in the complex z-plane. It is known [8] that to M_p corresponds a unique discrete group Γ of non-Euclidean translations such that M_p is conformally equivalent to $\overline{M}_p/\Gamma$ and Γ is isomorphic to the fundamental group $\pi_1(M_p)$. Each element $\gamma \in \Gamma$ has exactly two fixed points, a stable point γ^+ and an unstable point γ^-, and these points lie on the absolute $E = \{|z| = 1\}$. As in [9] the set of fixed points of all the elements of the group Γ, apart from the identity, is called the set of rational points. This set is countable and is everywhere dense on the absolute [8]; the points of its complement on the absolute are called irrational points.

We let $\overline{f}\colon \overline{M}_p \to \overline{M}_p$ denote the homeomorphism covering the homeomorphism $f\colon M_p \to M_p$ $(\pi\overline{f} = f\pi)$. The transformation $f_*\colon \gamma' = \overline{f}\gamma\overline{f}^{-1}$, $\gamma, \gamma' \in \Gamma$, is an automorphism of the group Γ. Following [10] we call two homeomorphisms f and f' π_1-conjugate if there exist homeomorphisms $\overline{f}$ and $\overline{f}'$ and an automorphism τ of the group Γ such that $\overline{f}'_* = \tau\overline{f}_*\tau^{-1}$. We recall that any automorphism τ of the group Γ induces in a unique way a homeomorphism τ^* of the absolute and the homeomorphism $\overline{f}^*$ induced by the automorphism $\overline{f}_*$ is a continuous and single-valued extension of the homeomorphism $\overline{f}$ to the absolute.

Let $f: M_p \to M_p$ be a T-diffeomorphism. Since the foliations $\{\widetilde{W}^s\}$ and $\{\widetilde{W}^u\}$ are transversal everywhere on M_p except at the singular points $z_1, z_2, \dots, z_n$, each z_i is a periodic point of the diffeomorphism f with $2k_i$ $(k_i \geq 4)$ saddle sectors. We let $W_1^{is}, \dots, W_{2k_i}^{is}$ $(W_1^{iu}, \dots, W_{2k_i}^{iu})$ denote the curves from $\{\widetilde{W}^s\}$ $(\{\widetilde{W}^u\})$ such that $f^{q_i n}(y) \to z_i$ as $n \to +\infty$ $(n \to -\infty)$ for any point $y \in W_j^{is}$ $(y \in W_j^{iu})$, where q_i is the period of the point z_i, $i = 1, \dots, n$, $j = 1, \dots, 2k_i$.

Let $x \in M_p \setminus \bigcup_{i=1}^n z_i$. We denote by W_x^s, W_x^u the curves from $\{\widetilde{W}^s\}$ and $\{\widetilde{W}^u\}$, respectively, passing through the point x, and by $w_{\overline{x}}^s$, $w_{\overline{x}}^u$ the preimages of these curves, under the map $\pi: \overline{M}_p \to M_p$, passing through the point $\overline{x}$, where $\pi(\overline{x}) = x$.

The fact that the foliations $\{\widetilde{W}^s\}$ and $\{\widetilde{W}^u\}$ are transitive and orientable implies, similarly to [9], [10], the following lemma.

LEMMA 2.1. 1) *If* $\pi(w_{\overline{x}}^s)$ $(\pi(w_{\overline{x}}^u))$ *does not coincide with* W_j^{is} (W_j^{iu}) *for any* i, j, *then* $w_{\overline{x}}^s$ $(w_{\overline{x}}^u)$ *is a smooth curve on* $\overline{M}_p$ *that has exactly two boundary points* s_1, s_2 (u_1, u_2), *and moreover they are irrational.*

2) *If* $\pi(w_{\overline{x}}^s)$ $(\pi(w_{\overline{x}}^u))$ *coincides with* w_j^{is} (w_j^{iu}) *for some* i, j, *then* $w_{\overline{x}}^s$ $(w_{\overline{x}}^u)$ *has precisely one boundary point* s_j^i (u_j^i) *on the absolute, and it is irrational, and one boundary point* $\overline{z}_i \in \overline{M}_p$, $\pi(\overline{z}_i) = z_i$.

3) *If the two curves* $w_{\overline{x}}^s, w_{\overline{y}}^s$ $(w_{\overline{x}}^u, w_{\overline{y}}^u)$ *do not coincide on* $\overline{M}_p$, *then they have no common boundary points on the absolute.*

4) *The sets of boundary points of the curves* $w_{\overline{x}}^s$, $w_{\overline{y}}^u$ *do not intersect for any* $\overline{x}$ *and* $\overline{y}$ *from* $\overline{M}_p$.

LEMMA 2.2. *Let the T-diffeomorphisms* f *and* f' *satisfy the condition* $\overline{f}_* = \overline{f}'_*$.

1) *If both boundary points of the curve* w^s $(\pi(w^s) \in \{W^s\})$ *lie on the absolute, then there exists a curve* w^{is} $(\pi(w^{is}) \in \{W^{is}\})$ *that has boundary points in common with* w^s *on the absolute.*

2) *If one of the boundary points* w^s *is the point* $\overline{z}_i \in M_p$, $\pi(\overline{z}_i) = z_i$, *then there is a point* $\overline{z}'_i \in \overline{M}_p$, $\pi(\overline{z}'_i) = z'_i$ *that is a boundary point for the curve* w^{is}, *where* $\pi(w^{is}) \in \{W^{is}\}$ *and* w^{is} *has a boundary point in common with* w^s *on the absolute.*

PROOF. We consider the singular point z_i of the foliations $\{\widetilde{W}^s\}$ and $\{\widetilde{W}^u\}$ of the diffeomorphism f and let m denote its period. For any preimage $\overline{z}_i$ of the point z_i there exists a map $\overline{f}_m: \overline{M}_p \to \overline{M}_p$ covering f^m for which the point $\overline{z}_i$ is fixed. We let w_j^{is} (w_j^{iu}) denote the preimages of the curves W_j^{is} (W_j^{iu}), $j = 1, 2, \dots, 2k_i$, having $\overline{z}$ as one of its boundary points and by s_j^i (u_j^i) the boundary points of the curves W_j^{is} (W_j^{iu}) on the absolute. One can establish in the same way as in [10] (Corollary 4.1) that the homeomorphism $\overline{f}_m^*$ of the absolute has $2k_i$ stable points $u_1^i, \dots, u_{2k_i}^i$ and $2k_i$ unstable points $s_1, \dots, s_{2k_i}$ and it has no other fixed points.

We consider the diffeomorphism $\overline{f}'_m$ covering f'^m such that $\overline{f}'_{m*} = \overline{f}_{m*}$. By [8] (§14) the sum of indices of the fixed points of the diffeomorphism $\overline{f}'_m$ is equal to $1 - 2k_i$. Therefore the fact that the foliations $\{W^{is}\}$ and $\{W^{iu}\}$ are transversal on M_p (everywhere except at the singular points) implies the existence of a unique fixed point $\overline{z}'_i$ of the diffeomorphism $\overline{f}'_m$, the image $\pi(\overline{z}'_i)$ of which is a singular point of the foliations $\{W^{is}\}$ and $\{W^{iu}\}$. From part 3) of Lemma 2.1 and the fact that the index of the point $\overline{z}'_i$ is equal to $1 - 2k_i$ we get that there exist $2k_i$ curves $w_j'^{is}$ $(w_j'^{iu})$, $j = 1, 2, \dots, 2k_i$, such that $\pi(w_j'^{is}) \in \{W'^s\}$ $(\pi(w_j'^{iu}) \in \{W'^u\})$ and one of the boundary points of the curve $w_j'^{is}$ $(w_j'^{iu})$ is $\overline{z}'_i$ and the other is s_j^i (u_j^i). Now all the assertions follow from [9] (Lemma 4).

THEOREM 2.1. *For the T-diffeomorphisms f and f' to be topological conjugate, it is necessary and sufficient that they be π_1-conjugate.*

PROOF. Necessity is obvious. We prove sufficiency. We shall assume that $\overline{f}_* = \overline{f}'_*$ (the general case reduces to this as in [10] (Theorem 1)).

Let $\overline{x}$ be any point on $\overline{M}_p$ other than the preimage of the singular point of the foliations $\{W^s\}$ and $\{W^u\}$. The following cases are logically possible: 1) the curves $w^s_{\overline{x}}$, $w^u_{\overline{x}}$ have all their boundary points on the absolute; 2) the curve $w^2_{\overline{x}}$ $(w^u_{\overline{x}})$ has exactly one boundary point on the absolute and $w^u_{\overline{x}}$ $(w^s_{\overline{x}})$ has two boundary points on the absolute; 3) both curves have only one boundary point each on the absolute.

In case 1), by Lemma 2.2 there exist curves w'^s and w'^u $(\pi(w'^s) \in \{w'^s\}$, $\pi(w'^u) \in \{W^u\})$ that have boundary points in common with w^s and w^u on the absolute and intersect at the unique point $\overline{x}'$. In case 2) we let $\overline{z}_i$ denote the preimage of the singular point of the foliations $\{W^s\}$ and $\{W^u\}$ that is the boundary point of the curve $w^s_{\overline{x}}$. We denote by $w^{iu}_{j_1}$, $w^{iu}_{j_2}$ the curves whose images in M_p bound, with the curve $\pi(w^s_{\overline{x}})$, adjacent saddle sectors of the singular point $\pi(\overline{z}_i)$ of the foliation $\{W^s\}$. By Lemma 2.2 there exist curves $w'^{\,iu}_{j_1}$, $w'^{\,iu}_{j_2}$, $w'^{\,is}_{j_3}$ that have $\overline{z}'_i$ as one of their boundary points and have boundary points on the absolute in common with the curves $w^{iu}_{j_1}$, $w^{iu}_{j_2}$, $w^s_{\overline{x}}$, respectively, and there also exists a curve w'^u that has boundary points in common with w^u on the absolute. Since $w'^u \cap (w'^{\,iu}_{j_1} \cup w^{iu}_{j_2} \cup \overline{z}'_i) = \varnothing$, w'^u must certainly intersect the curve $w'^{is}_{j_3}$. Again we denote this intersection point by $\overline{x}'$. In case 3), by Lemma 2.2 there exist curves

w'^s, w'^u ($\pi(w'^s) \in \{W'^s\}$, $\pi(w'^u) \in \{W'^u\}$) that have boundary points on the absolute in common with $w^s_{\overline{x}}$ and $w^u_{\overline{x}}$, respectively. Hence, as in case 2), we get that the curves w'^s and w'^u intersect at a unique $\overline{x}'$. If $\overline{x}$ coincides with one of the preimages of the singular point of the foliations $\{W^s\}$ and $\{W^u\}$, then we let $\overline{x}'$ denote the preimage of the singular point of the foliations $\{W'^s\}$ and $\{W'^u\}$ found by the method described in Lemma 2.2.

Let $\overline{\varphi}$ denote the map $\overline{x} \to \overline{x}'$. In the same way as in [10] it can be proved that $\overline{\varphi}$ is a homeomorphism satisfying the conditions $\overline{f}' = \overline{\varphi}\overline{f}\overline{\varphi}^{-1}$ and $\gamma\overline{\varphi} = \overline{\varphi}\gamma$. Therefore there exists a homeomorphism $\varphi\colon M_p \to M_p$ such that $f' = \varphi f \varphi^{-1}$.

REMARK. Recently the author and S. Kh. Aranson have established that any T-diffeomorphism $f\colon M_p \to M_p$ has the least topological entropy among the entropies of all the homeomorphisms on M_p homotopic to f.

REFERENCES

1. D. V. Anosov, *Geodesic flows on closed Riemannian manifolds of negative curvature*, Trudy Mat. Inst. Steklov. **90** (1967), 3–209.

2. Ya. G. Sinai, *Markov partitions and diffeomorphisms*, Funktsional Anal. i Prilozhen. **2** (1968), 64–89.

3. E. Hemmingsen and W. L. Reddy, *Lifting and projecting expansive homeomorphisms*, Math. Systems Theory **2** (1968), 7–15.

4. T. O. Brian, *Expansive homeomorphisms on compact manifolds*, Proc. Amer. Math. Soc. **24** (1970), 767–771.

5. A. A. Gura, *On separating diffeomorphisms of the torus*, Mat. Zametki **18** (1975), 41–49.

6. A. B. Katok, *Bernoulli diffeomorphisms on a surface*, Ann. of Math. (2) **110** (1979), 529–547.

7. T. O. Brian and W. Reddy, *Each compact orientable surface of positive genus admits an expansive homeomorphism*, Pacific J. Mathematics **35** (1970), 737–742.

8. I. Nielsen, *Untersuchungen sur Topologie der geschlossenen zweiseitigen Flächen*, Acta Math. **50** (1927), 189–358.

9. S. Kh. Aranson and V. Z. Grines, *Certain invariant dynamic systems on two-dimensional manifolds (necessary and sufficient conditions for topological equivalence of transitive systems)*, Mat. Sb. (N.S.) **90** (1973), 372–402.

10. V. Z. Grines, *On the topological conjugacy of diffeomorphisms of two-dimensional manifolds on one-dimensional orientable basic sets*. I, Trudy Moskov. Mat. Obshch. **32** (1975), 35–60.

Gorky Agricultural Institute

Translated by A. WEST

Amer. Math. Soc. Transl.
(2) Vol. **149**, 1991

Invariant Two-Dimensional Tori, Their Breakdown and Stochasticity

V. S. AFRAĬMOVICH AND L. P. SHIL′NIKOV

§1

In the study of several nonlinear problems of hydrodynamics, radiophysics, biophysics, etc. that can be described using finite-dimensional dynamical systems, one frequently encounters situations when in the phase space of the corresponding dynamical system there exists an invariant domain V homeomorphic to the product of a two-dimensional torus and a segment or a disk. These are, for instance, the well-known problems on the action of an external periodic force on a self-oscillatory system, or on the interaction of self-oscillatory systems. The simplest attractor of V is a two-dimensional torus. If the motion on an exponentially stable torus is quasi-periodic with two frequencies then the torus, as a smooth attracting manifold, is preserved also for all sufficiently close systems. However, the structure of its partition into trajectories changes depending on the Poincaré rotation number: if it is irrational then the limit set is the whole torus; if it is rational then in general it is not. A. G. Maĭer proved that there is a generic structure on the torus only when the rotation number is rational (equal to p/q) and all the periodic motions are generic. In similar situations we say that we have the resonance case (the ratio of the main frequencies ω_1/ω_2 equals p/q). We must observe that, although from a mathematical point of view there must correspond a periodic regime to the resonant case, in a real oscillatory system the periodic regimes are far from appearing for all rotation numbers—the inner synchronization does not occur for all p/q. The resonant case, corresponding to an inner synchronization, will be called *strong resonance*. Observe that the resonance in a quasilinear system is precisely the strong resonance (as a rule,

---

1991 *Mathematics Subject Classification*. Primary 58F14, 58F12, 58F13; Secondary 58F15, 58F22, 58F21.

Translation of Methods of the Qualitative Theory of Differential Equations, Gor′kov. Gos. Univ., Gorki, 1983, pp. 3–26; MR 88a:58128.

for quasilinear systems p and q are not large(1) and there are two periodic motions on the torus). The strong resonances play an important role since in the transition from a quasiperiodic two-frequency regime of a self-oscillatory system to a stochastic regime, in many cases there previously appears an inner synchronization regime with strong resonance. Precisely this phenomenon is observed in the problems on the Couette flow or on the Raleigh–Benard convection, which were studied experimentally in [1]–[6], when the Reynolds or the Raleigh number, respectively, is increased.

The existence of a complicated nonperiodic regime corresponds to the breakdown of an invariant torus and the creation of a nontrivial attractor in the domain V of the phase space of a self-oscillatory system. The existence of a nontrivial hyperbolic subset in the attractor of the domain V was established earlier by the authors [7]–[9].(2)

Thus, the problem arises of studying the ways invariant two-dimensional attracting tori (with a generic structure on them) break down and the creation of nontrivial limit sets with the variation of the parameters of the system. The present paper is devoted to the consideration of this problem.

§2

We shall consider the n-dimensional system

$$\dot{x} = X(x, \mu), \tag{1}$$

where the right-hand sides belong to the class C^k with respect to x and depend continuously on the vector-valued parameter μ in the C^k-topology, $k \geq 3$. Assume that (1) has a smooth attracting torus $T(\mu_0)$ for $\mu = \mu_0$ in a domain V of the phase space, that the system is structurally stable on this torus, and the limit set consists of two periodic motions: a stable one, $L^+(\mu_0)$, and a saddle-type one, $L^-(\mu_0)$, whose unstable manifold $W^u(\mu_0)$ is two-dimensional, i.e., $T(\mu_0) = W^u(\mu_0) \cup L^+(\mu_0)$. It is not hard to show that if the least multiplier of $L^+(\mu_0)$ in absolute value is real and simple then (1) has a smooth invariant torus for all $|\mu - \mu_0|$ sufficiently small.

Let us denote by G_0 an (open) set in the parameter space such that for every $\mu \in G_0$ the periodic motions $L^+(\mu_0)$ and $L^-(\mu_0)$ that are the continuous extensions of $L^+(\mu)$ and $L^-(\mu)$ are generic, let $\widetilde{G} \in G_0$ be such that for $\mu \in \widetilde{G}$ the set $\overline{W^u(\mu)}$ is the whole torus $T(\mu)$, and let G be the set of interior points of $\widetilde{G}$.

The aim of this section is to establish the structure of the boundary ∂G of G and single out a set B in ∂G such that the torus breaks down when

(1)For strongly nonlinear systems this is not the case. For instance, in [1] a strong resonance was discovered with $p/q = 7/3$. Earlier Van-der-Pol observed synchronization regimes with $p/q \gtrsim 200$, in studying the functioning of a generator with a periodic external force.

(2)This fact is implicitly contained in the well-known papers of Cartwright, Littlewood, and Levinson.

passing through B (from inside G). To this end we introduce the set of continuous curves $H = \{\mu(s)\}$, $0 \leq s \leq 1$, in the parameter space such that for $\mu(s) \in H$:

1) For $\mu = \mu(0)$ system (1) has a smooth torus $T(\mu(0))$, and for $\mu = \mu(1)$ it does not have an invariant set in V homeomorphic to the torus. Therefore, there must be a value s^* such that a) for $0 < s < s^*$ we have $\mu(s) \in G$; b) $\mu(s^*) \in \partial G$; c) at least for some values $s > s^*$, arbitrarily close to s^*, system (1) has no continuous torus.
2) For every $0 \leq s \leq s^*$ the attractor of (1) in the domain V coincides with the torus $T(\mu(s))$.
3) $\mathrm{tl}_{s\uparrow s^*} \overline{W^u(\mu(s))}$ does not contain periodic motions different from $\mathrm{tl}_{s\uparrow s^*}(L^+(\mu(s)) \cup L^-(\mu(s)))$. (tl stands for topological limit.)

Let us divide G into two subsets G_1 and G_2 such that for $\mu \in G_1$, $T(\mu)$ is a smooth torus, and for $\mu \in G_2$ the set $T(\mu)$ is homeomorphic, but not diffeomorphic, to a torus (clearly, the smoothness of $T(\mu)$ may be lost only at the points of $L^+(\mu_0)$). The boundary ∂G may contain the bifurcation surface B^+ corresponding to a structurally unstable motion of saddle type, formed as the result of the amalgamation of $L^+(\mu_0)$ with $L^-(\mu_0)$.([3]) As shown in [9], as we leave G through a point of B^+ not lying in the closure of G_2 the torus is preserved and remains smooth; therefore, there must be a value $s < s^*$ such that $\mu(s) \in G_2$, i.e., $T(\mu(s))$ is a nonsmooth torus. Generally speaking, it is not hard to deduce from 1)–3) that before the torus $T(\mu)$ breaks down it must lose its smoothness. More precisely, the following assertion holds.

For every curve $\mu(s) \in H$ there is an $s_1 < s^*$ and an $\varepsilon > 0$ such that for $\mu = \mu(s)$ system (1) has attracting (in V) torus $T(\mu(s))$, which is smooth for $0 \leq s < s_1$ and nonsmooth for $s_1 < s < s_1 + \varepsilon$. For $s = s_1$ in general([4]) either the least multiplier in absolute value $\xi(L^+)$ of the periodic motion has multiplicity 2 and for $s > s_1$ it becomes complex, or $\chi(L^+)$ has multiplicity 1 and the smoothness of the torus is lost because of the nonsmooth approach of $W^u(\mu(s))$ to $L^+(\mu(s))$ for $s > s_1$ (on $W^u(\mu(s))$ there appear points of the second kind, see the definition below).

The main result of this section is the following assertion.

THEOREM 1 (on the breakdown of the torus). *Let assumptions* 1)–3) *hold. Then for* $\mu(s) \in H$ *for* $s = s^*$ *either* (I) $\mu(s^*) \in B^+$ *and* $\mu(s) \in G_2$ *for* $s_1 \leq \tilde{s} < s < s^*$; *or* (II) *a structurally unstable homoclinic curve appears for*

([3])Clearly, the boundary ∂G may contain bifurcation surfaces corresponding to the creation of other periodic motions on the torus (besides $L^+(\mu_0)$ and $L^-(\mu_0)$). However, we shall not consider these, since their study does not contain new significant features.

([4])For an open subset of curves, dense in H (if we regard H as a subset of the space of continuous maps of the segment into the parameter space).

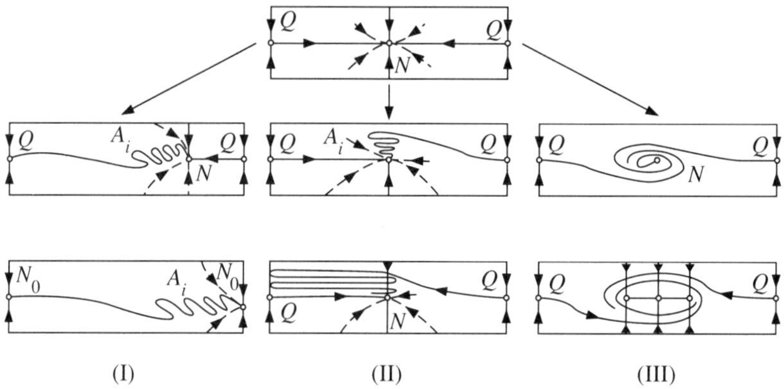

FIGURE 1

the periodic motion L^- , *and the bifurcation surface is attainable*(5) (*in the terminology of* [10]) ; *or* (III) L^+ *loses stability* (*see Figure* 1).

The proof of Theorem 1 follows easily from the assumptions: if none of the options (I)–(III) occur then it is not hard to prove that $\overline{W^u(\mu(s))}$ is a torus up to $s = 1$, which contradicts condition 1).

§3

In this section we consider invariant sets arising in the breakdown of an invariant torus, and their evolution with the subsequent variation of the parameters.

Let us consider first case (I). For $s = s^*$ system (1) has a saddle-node type periodic motion L_0 , and the torus is generated by the closure of its unstable manifold $W^u(\mu_0)$.(6) Let S be a cross-section transversal to L_0 . As it follows from [9], we may introduce on S a system of coordinates $\{(y, z)\}$ such that

1) $0 = L_0 \cap S = (0, 0)$; 2) the Poincaré return map $\tau\colon U \to S$, where U is a neighborhood of zero, is written as

$$\begin{aligned} \bar{y} &= A(\mu)y + f(y, z, \mu)y, \\ \bar{z} &= z + R(z, \mu), \end{aligned} \tag{2}$$

where the eigenvalues of the matrix $A(\mu(s^*))$ are less than 1 in absolute value, $\|y\| < \varepsilon$, and $R(z, 0) = lz^2 + \cdots$, $l > 0$. Let $w = W_0 \cap S$. From

(5)The assumption that there are no bifurcations for $s < s^*$ in case (II) imposes restrictions on the saddle quantity $L^-(\mu(s))$, which hold automatically for the three-dimensional system (1) , if $\operatorname{div} X < 0$; if $L^-(\mu)$ is greater than 1 then, as s varies, limit motions are created independently of the destruction of the torus [10] .

(6)The unstable manifold of L_0 is the set of points of the trajectories tending to L_0 as $t \to -\infty$.

[9] it follows that the points on w may be of the first or of the second kind, according to whether

$$\liminf_{k\to\infty} \frac{|\operatorname{pr}_z d\tau^k \xi|}{\|d\tau^k \xi\|} = 1 \quad (\text{or } = 0),$$

where ξ is a tangent vector to w at a point $P \in U$, pr_z denotes the projection onto the z-axis, and $\tau^k P \in U$.

For a sufficiently small $d > 0$ we consider the "rectangle"

$$D_d = \{(y, z) | \|y\| < \varepsilon, -d \le z \le -d + R(-d, 0)\}$$

in U. It can be proved that any trajectory starting at W_0 meets D_d, i.e., $w^u \cap D_d \neq \varnothing$. Let us state the "big loop" conditions slightly differently than in [9].

Assume that for every sufficiently small fixed $d > 0$ $w \cap D_d$ contains two curves Γ_1 and Γ_2 (see Figure 2), whose equations are written as $y = h_1(z)$ and $y = h_2(z)$ respectively and such that the points of $\Gamma_1 \cup \Gamma_2$ are points of the first kind. Let $0 \in U_1 \subset U$, $U_1^+ = \{(y, z) \in U_1, z \ge 0\}$. Through every point of Γ_j there is a trajectory intersecting U_1^+.
We denote by τ_j the diffeomorphism that associates with a point in Γ_j the first point of intersection with U_1^+ of the trajectory passing through this point, $j = 1, 2$. Let $\tau_j \Gamma_j = \gamma_j$ $(\in \omega)$. Let us set

$$q_j = \min_{p\in\gamma_j} \frac{|\operatorname{pr}_z d\tau_j^{-1}\xi(P)|}{\|\xi(P)\|},$$

where $\xi(P)$ is a tangent vector to γ_j at P. The following theorem holds, which sharpens an earlier result of the authors [9].

Theorem 2. *If the big loop conditions hold for all values μ close to $\mu(s^*)$ for which L_0 vanishes then system* (1) *has a limit set containing countably*

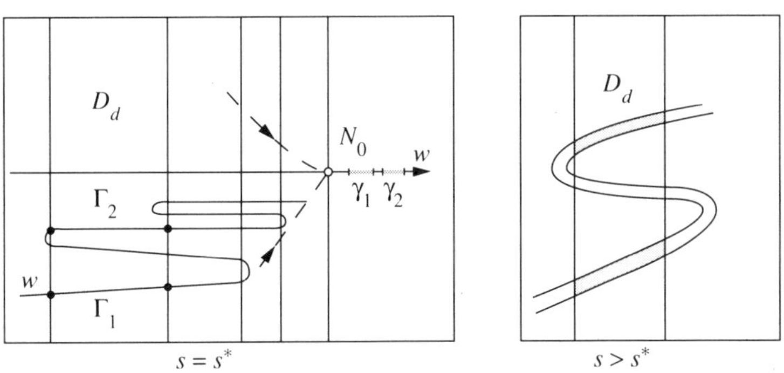

FIGURE 2

many periodic trajectories. Moreover, if $q_j > 1$, $j = 1, 2$, *this limit set contains a nontrivial hyperbolic subset.*

We give the following definition [11], [12]. An attracting set containing a nontrivial hyperbolic subset is called an *ε-quasiattractor* if it does not contain stable periodic motions of period less than ε^{-1}. If ε is sufficiently small an ε-quasiattractor may be regarded as the mathematical model of stochastic oscillations. Theorem 2 implies the following result.

COROLLARY. *If the assumptions of Theorem 2 hold, for* $s^* < s < s^* + \varepsilon_0$ *system* (1) *has an ε-quasiattractor in the domain* V *with* $\varepsilon = \text{const}\cdot|\mu(s) - \mu(s^*)|^{1/2}$.

The proof of Theorem 2 (and its corollary) is carried out the same as in [9] and [13].

If the big loop condition does not hold and the torus is not smooth for $s = s^*$ the problem on the creation of nontrivial limit sets for $s > s^*$ remains open. Sufficient conditions are indicated in [9] for the existence of a sequence $\{s_i\}$, $s_i \downarrow s^*$ as $i \to \infty$, such that system (1) has a nontrivial limit set for $\mu = \mu(s_i)$ ("small loop" conditions).

Now let us proceed to case (II). For $s > s^*$ two structurally stable homoclinic curves appear for the periodic motion $L^-(\mu(s))$, and therefore, a nontrivial hyperbolic set appears; however, the representative point remains in a neighborhood of $L^+(\mu(s))$: the system remains in a "periodic regime". The arising hyperbolic set may belong to the limit set of the trajectories, which generally speaking is not hyperbolic; in particular, it may contain structurally unstable homoclinic curves, countably many stable periodic trajectories, etc. As s increases, after the bifurcation creating the structurally unstable homoclinic trajectory a bifurcation may occur that destroys the periodic motion L_0, i.e., the saddle-node type periodic motion having a structurally stable homoclinic trajectory vanishes. This situation was studied in detail in [14]; in particular, it implies the possibility of the creation of a quasiattractor when L_0 disappears, in this case as well.

In case (III) L^+ may lose its stability as a consequence of the following bifurcations a)–e) (see, for instance, [15] and [20]). We assume here that the periodic motion is stable when the bifurcation occurs.

a) "Period-doubling" bifurcation: $\chi(L^+) = -1$.
b) Generation of a torus: $\chi(L^+) = e^{i\alpha}$, $\alpha \neq 0, \pi, 2\pi/3, \pi/2$.
c) "Period-tripling" bifurcation: $\chi(L^+) = e^{2\pi i/3}$.
d) "Period-quadrupling" bifurcation: $\chi(L^+) = e^{\pi i/2}$.
e) The presence of a multiple multiplier (of modulus 1), etc.

The latter four bifurcations cannot be realized in a system with negative divergence for $n = 3$. Their presence is an indirect indication that the model describing the real physical system with dissipation must be at least four-dimensional.

The periodic motion created as a result of one of the bifurcations a), c), or d) may change again its stability properties and have one of the bifurcations a)–e), etc. In particular, as a result of an infinite sequence of period-doubling bifurcations there may arise a nontrivial limit set and a quasiattractor (see, for instance, [16]).

§4

In this section we give the result of the analytic study of the concrete map

$$\begin{aligned} \overline{x} &= (x + b\mu(1 + a\sin\theta))^{\nu}, \\ \overline{\theta} &= \theta - \ln(x + b\mu(1 + a\sin\theta)), \end{aligned} \tag{3}$$

where $0 < x \leq x_0$, $\theta \in \mathbf{R}^1$, $\nu > 1$, $0 < \mu \leq \mu_0$, $0 < b < 1$, and $0 \leq a < 1$. This map models the problem on the interaction of an external periodic force on a self-oscillatory system under the assumption that the limit cycle passes near a saddle (see [7] and [8] for more details). In spite of the particular form of (3), it models all the main features of the breakdown of tori in three-dimensional systems with a negative divergence in a neighborhood of the torus. The map (3) is the composition of two maps: a local one—along the trajectories passing through a neighborhood of the saddle taking into account only the linear part (for a suitable choice of the variables (x, y) it may be written as $x = y^{\nu}$, $\overline{\theta} = \theta - \ln y$), and a global one—along the trajectories near a "global piece" of a stable periodic motion; in the simplest case it has the form $y = x + b\mu + A\mu\sin\theta$, $\theta_0 = \theta + \text{const}$. The function $\mu b(1+a\sin\theta)$, where $a = A/b$, is Mel′nikov's function. The invariant curve $x = x_*$, where x_* is the solution of the equation $x = (x+b\mu)^{\nu}$, corresponds to a limit cycle of the system in the absence of a periodic perturbation, i.e., for $a = 0$.

It is not hard to show that there are positive b_0 and x_0 such that for every $0 < b < b_0$ the map (3) is a diffeomorphism onto the image of the annulus $\mathscr{K}: 0 < x \leq x_0$, $\theta \in \mathbf{R}/\mathbf{Z}$ into itself. Moreover, all the positive semitrajectories of (3) enter the annulus

$$0 < x \leq (2b\mu)^{\nu} \left(\frac{\nu}{\nu - 1}\right)^{\nu},$$

which contains also the whole nonwandering set. The map (3) is quite suitable for our goals, since for some values of the parameters it has a closed invariant curve, and for other values it has no such curve. In order to establish the existence of an invariant curve we use the contraction mapping principle in the form indicated in [17]. Applying it to the map (3) and taking into account the fact that its determinant is less than 1 in absolute value for small b we obtain the following assertion: there are $b_1 > 0$, $c_1 > 0$, and $b_1 \leq b_0$, such that (3) has a closed invariant curve $x = h(\theta)$ as a unique attracting set in the annulus $\mathscr{K}$ for the values of the parameters belonging to the domain $D_1: 0 < b \leq b_1$, $a < 1/\sqrt{2} - c_1(G\mu)^{(\nu-1)/2}$; h is 2π-periodic and satisfies a

Lipschitz condition. On the other hand, there are values of the parameters for which (3) has a nontrivial hyperbolic set. In order to prove this fact we use the sufficient conditions for the existence of such a set stated as the "annulus principle" in [7] and [8]. Applying this principle to the map (3) we obtain the following assertion: there are $0 < b_2 < b_1$ and $C_1, C_2 > 0$ such that for every $0 < b \le b_2$ and $\tanh 2\pi + C_1(b\mu)^\nu < a < 1 - C_2(b\mu)^{\nu-1}$ the map (3) has an invariant set in $\mathscr{K}$, and on this set it is topologically conjugate to a Bernoulli shift over two symbols. Of course, for these values of the parameters the map (3) can no longer have a closed invariant curve. In order to clarify how an invariant curve ceases to be the homeomorphic image of a circle we need to know the bifurcations of the fixed points. On Figure 3 we show the bifurcation diagram of the map (3).([7]) The equations on B^+_{1k} and B^+_{2k} are

$$-\ln b\mu = \ln(1+a) - \ln(e^{-2\pi k} - e^{-2\pi k\nu}),$$
$$-\ln b\mu = \ln(1-a) - \ln(e^{-2\pi k} - e^{-2\pi k\nu})$$

respectively.

For the values of the parameters on B^+_{2k} at the points of the annulus $x = e^{-2\pi k\nu}$, $\theta = -\pi/2$ a fixed point of (3) arises, which splits into a saddle Q_k and a node N_k as b decreases. For fixed a and decreasing b the saddle moves clockwise and the node, counterclockwise. For $(\mu b, a) \in B^+_{1k}$ it joins the saddle and forms a saddle-node at the point $x = e^{-2\pi k\nu}$, $\theta = \pi/2$, which disappears as b decreases further. From the form of the curves B^+_{1k} and B^+_{2k} it follows directly that as $a \uparrow 1$ the number of fixed points of (3) increases indefinitely. As b decreases there are no bifurcations for the saddle Q_k. But the node N_k can lose its stability by means of a period-doubling bifurcation.

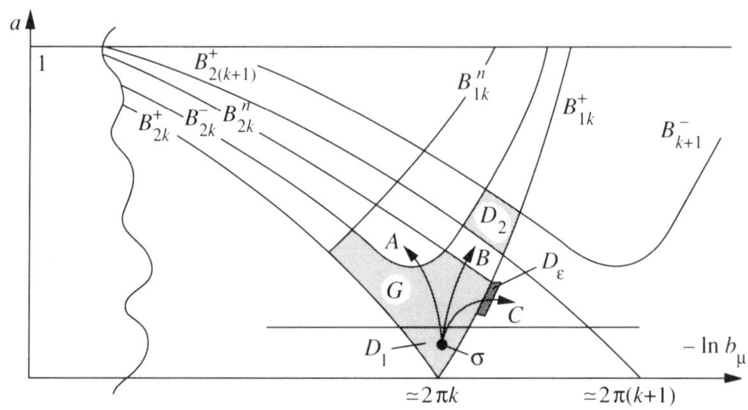

FIGURE 3

([7])It is convenient to illustrate our assertions on the plane $(-\ln b\mu; a)$.

The corresponding bifurcation curve B_k^- has the equation

$$a^2 = \frac{4}{(b\mu)^2}(e^{-2\pi k} + \nu e^{-2\pi k \nu})^2 + \frac{1}{(b\mu)^2}(e^{-2\pi k} - e^{-2\pi k \nu} - b\mu)^2$$

and lies in the "sector" generated by the curves B_{1k}^+ and B_{2k}^+.

On Figure 3 we show two additional bifurcation curves corresponding to the generation of a structurally unstable homoclinic trajectory on each of the two connected components of $w^u \backslash Q_k$, where w^u is the unstable manifold of Q_k. Their equations have the form

$$B_{1k}^n: \ -\ln b\mu = 2\pi k + \ln \frac{1+\sqrt{2a^2-1}}{2} - \frac{5\pi}{4} - \arcsin \frac{1}{a\sqrt{2}} - \arcsin \frac{e^{-2\pi k} - e^{-2\pi k \nu} - b\mu}{ab\mu} + \alpha(b);$$

$$B_{2k}^n: \ -\ln b\mu = 2\pi k + \ln \frac{1-\sqrt{2a^2-1}}{2} + \frac{7\pi}{4} - \arcsin \frac{1}{a\sqrt{2}} - \arcsin \frac{e^{-2\pi k} - e^{2\pi k \nu} - b\mu}{ab\mu} + \beta(b),$$

where $\alpha, \beta \to 0$ as $b \to 0$.

These curves divide the "sector" bounded by the curves B_{1k}^+, B_{2k}^+, and $a = 1$, into several domains. Let us denote by G the pentagonal domain adjacent to $a = 0$, whose boundary is formed by all five bifurcation curves. An invariant curve may break down only when exiting G. Let us fix an arbitrary point O in G, sufficiently close to the x-axis, and three paths A, B, and C, starting at this point and meeting the curves B_k^-, B_{2k}^n, and B_{1k}^+ respectively (see Figure 3).

Case A. As the parameters vary along the path A there is a moment when the multipliers of the node N_k become equal and complex afterwards. The closed invariant curve consisting of the unstable manifold W^u of Q_k and the point N_k becomes nonsmooth at this instant; however, its length is finite and it is still a topological circle. But as the path A meets the bifurcation curve B_k^n the length of $\overline{w^u}$ becomes infinite (see Figure 3).

Case B. On Figure 3 we show the path B that meets B_{2k}^n. The closed invariant curve becomes clearly nonsmooth when w^u meets the nonleading manifold of the node N_k. When B meets B_{2k}^n the unstable manifold of the saddle Q_k is tangent to its stable manifold along a structurally unstable homoclinic trajectory, and we have the attainable case. Clearly, at this moment $\overline{w^u}$ is no longer homeomorphic to S^1.

Case C. As we have already mentioned, if the path C meets B_{1k}^+ at a point lying in $\overline{D}_1$ then the invariant curve not only does not disappear, but also continues to satisfy a Lipschitz condition. We assume therefore that $a > 1/\sqrt{2} + C_2(b\mu)^{\nu-1}$ at the point of transition through B^+_{1k}. If this inequality is satisfied we have the so-called "large loop" or "small loop" cases described earlier. The "large loop" conditions and inequalities $a_j > 1$ are clearly satisfied if the transition point through B^+_{1k} is sufficiently close to $B^+_{1k} \cap B^n_{2k}$. This implies the existence of a nontrivial hyperbolic set in the domain adjacent to B^+_{1k} (outside G, see Figure 3). The fact that the period of any periodic point arising after the saddle-node vanishes is not less than $\text{const} \cdot |\mu(b - b_*)|^{-1/2}$, where b_* is the value of b at the bifurcation instant, implies the following assertion: on the curve B^+_{1k} there is a segment J such that one of its endpoints is $B^n_{2k} \cap B^+_{1k}$ and for every small $\varepsilon > 0$ there is a domain D_ε (outside G), adjacent to J, such that for the parameter values in this domain the map (3) has a quasiattractor.

Summing up, we observe that all three possibilities of the breakdown of the invariant torus stated in Theorem 1 are realized in this example. Moreover, it is not hard to check the assumptions of Theorem 2 for this example.([8])

To conclude this section we mention yet another property of the map (3) related to the breakdown of an invariant curve. From the form of the bifurcation curves B^+_{1k}, B^-_k, and B^n_k it follows directly that on the parameter plane there is a domain bounded by B^+_{1k}, B^-_k, $B^+_{2(k+1)}$, and B^-_{k+1} (see Figure 3) such that for the parameter values inside this domain the map (3) has two stable periodic trajectories with rotation numbers $2\pi k$ and $2\pi(k+1)$ respectively. And this means that the original system, whose model is the map (3), has two stable periodic motions with periods differing by that of the exterior force. We call this phenomenon *Vander-Pol's effect*, since he was the first to observe it in the experimental investigation of the interaction of a periodic external force on a self-generator.

§5

Let us discuss the transitions from a laminar to a turbulent flow that were empirically studied in [1]–[4]. The transition from the resonant periodic flow to an irregular, aperiodic, one, occurs either immediately [1], or after a succession of period-doubling bifurcations [3], or after a period-tripling bifurcation [3], or after a period-quadrupling bifurcation [4], etc.

In our opinion, the set of bifurcations given in this paper is sufficient to explain these rearrangements of the flow in the language of qualitative theory. The study of spectral characteristic flows [1]–[4] suggests the assumption that the transition to an aperiodic flow in [1] corresponds to a bifurcation that annihilates a saddle-node-type periodic motion, followed by the creation of a quasiattractor; it is customary to say that "alternation" occurs. The situation of [3] and [4] corresponds to case (III) of Theorem 1 followed by a sequence

([8])The possibilities of the breakdown of an invariant curve of concrete maps of the plane into the plane were analyzed numerically in [18] and [19].

of period-doubling bifurcations that occurs either immediately or after the bifurcations c) and d). It was established in [2] that as the Reynolds number increases the periodic regime is replaced by a stochastic one; afterwards, it again becomes periodic, and only later on it becomes and remains irregular. We may assume that the first transition corresponds to case (I) of Theorem 1; the second transition corresponds to the creation of a stable periodic motion in the quasiattractor, and the representative point falls into this domain; and the third transition corresponds to a bifurcation that annihilates a saddle-node-type periodic motion having a homoclinic trajectory.

Subsequent experiments and numerical investigations of these problems will make our hypotheses more precise (or will reject them).

In conclusion we mention that the results of this paper were reported in the VI Gor′kiĭ School on Nonlinear Oscillations (1981), the IX International Conference on Nonlinear Oscillations in Kiev (1981), in the School on Nonlinear Problems of Hydrodynamic Stability in Kolyubakino (1982), and in the Seminars in Moscow State University on Synergetics (1982) and on Ergodic Theory and Dynamical Systems (1982).

BIBLIOGRAPHY

1. J. P. Gotlub and S. V. Benson, *Many routes to turbulent convection*, J. Fluid Mech. **100** (1980), 449–470.

2. V. S. L′vov and A. A. Predtechenskiĭ, *Stagewise transition to turbulence in Couette flow*, in: *Nonlinear waves. Stochasticity and turbulence*, IPF Akad. Nauk SSSR, Gor′kiĭ, 1980, 57–77. (Russian)

3. J. Maurer and A. Libhaber, *Rayleigh-Benard experiments in liquid helium: frequence locking and the onset of turbulence*, J. Physique Lett. **40** (1979), 419–423.

4. A. Arncodo, P. Cullet, C. Tresser, A. Libhaber, J. Maurer, and D. d'Humieres, *About the observation of the uncompleted cascade in Rayleigh-Benard experiments*, Preprint, 1981.

5. G. Ahlers and R. P. Beringer, *The Rayleigh-Benard instability and the evolution of turbulence*, Suppl. of the Progress of Theor. Phys. **64** (1978), 186–201.

6. Yu. N. Belyaev and I. M. Yavorskaya, *Transition to a stochastic regime in the flow between rotating spheres*, in: *Nonlinear waves. Stochasticity and turbulence*, IPF Akad. Nauk SSSR, Gor′kiĭ, 1980, 78–88. (Russian)

7. V. S. Afraĭmovich and L. P. Shil′nikov, *On small periodic perturbations of autonomous systems*, Dokl. Akad. Nauk SSSR **24** (1974), 739–742.

8. ——, *The annulus principle and the problem on the interaction of two self-oscillatory systems*, Prikl. Mat. Mekh. **42** (1977), 618–627.

9. ——, *On certain global bifurcations connected with the breakdown of a fixed point of saddle-node type*, Dokl. Akad. Nauk SSSR **219** (1974), 1281–1285.

10. N. K. Gavrilov and L. P. Shil′nikov, *On three-dimensional dynamical systems close to systems with a structurally unstable homoclinic curve*. I. II. Matem. Sb. **88 (13)** (1973), 475–492; **90 (132)** (1973), 139–156.

11. V. S. Afraĭmovich and L. P. Shil′nikov, *On strange attractors and quasi-attractors*, in: *Nonlinear dynamics and turbulence*, Boston-London-Melbourne, 1983, 1–34.

12. V. S. Afraĭmovich, *The annulus principle and quasi-attractors*, Thesis, IX International Conference on Nonlinear Waves, Kiev, 1981, p. 42.

13. ——, *On certain global bifurcations leading to the creation of countably many periodic motions*, Dissertation, Gor′kiĭ, Gor′k. Gos. Univ., 1974. (Russian)

14. V. I. Luk′yanov and L. P. Shil′nikov, *On certain bifurcations of dynamical systems with homoclinic structures*, Dokl. Akad. Nauk SSSR **243** (1978), 26–29.

15. V. I. Arnol′d, *Complementary chapters of the theory of ordinary differential equations*, "Nauka", Moscow, 1978.

16. M. Hénon and J. Pomeau, *Two strange attractors with a simple structure*, Springer-Verlag, Lecture Notes in Math., Vol. 565, 1975, 29–68.

17. V. S. Afraĭmovich, V. V. Bykov, and L. P. Shil′nikov, *On attracting structurally unstable limit sets of the Lorentz-attractor type*, Trudy Moskov. Matem. Obshch. **44** (1982), 150–212.

18. J. Curry and J. Yorke, *A transition from Hopf bifurcation to chaos: Computer experiments with maps of* $\mathbf{R}^2$, Springer-Verlag, Lecture Notes in Math., Vol. 470, 1978, 48–66.

19. D. G. Aronson, M. A. Chory, G. R. Hall, and R. P. McGehee, *A discrete dynamical system with subtly wild behavior*, in: *New approaches to nonlinear problems in dynamics*, SIAM, Philadelphia, 1980, pp. 339–359.

20. J. Marsden, M. McCracken, and G. Oster, *The Hopf bifurcation and its applications*, Applied Mathematical Sciences **19**, Springer-Verlag, New York, 1976, xiii+408 pp.

Research Institute of Mathematics and Cybernetics
Gor′kiĭ State University

Translated by J. J. TOLOSA

Amer. Math. Soc. Transl.
(2) Vol. **149**, 1991

Stable Random Variables and Convexity of Banach Function Spaces

UDC 513.881

I. YA. NOVIKOV AND O. P. SKACHKOVA

Let $1 < p < \infty$. We consider a sequence $\{f_i\}_{i=1}^{\infty}$ of independent random variables identically distributed on some probability space (Ω, Σ, ν) such that for all $\lambda \geq 1$

$$\nu\{\omega \in \Omega: |f_i(\omega)| > \lambda\} = \lambda^{-p}.$$

Maurey and Pisier [1] proved that for all $r > p$ there exists a constant $K = K(p, r) < \infty$ such that

$$K^{-1} \int_{\Omega} \left(\sum_{i=1}^{n} |a_i f_i(\omega)|^2 \right)^{1/2} d\nu \leq \left(\sum_{i=1}^{n} |a_i|^p \right)^{1/p} \leq K \int_{\Omega} \max_{1 \leq i \leq n} |a_i f_i(\omega)| \, d\nu \tag{1}$$

for all finite sequences $\{a_i\}_{i=1}^{n}$ of numbers.

The inequality (1) is central in the proof that satisfying an upper r-estimate implies p-convexity of a Banach lattice for $p < r$. The inequality (1) can be strengthened.

THEOREM 1. *For an arbitrary finite sequence* $\{a_i\}_{i=1}^{n}$

$$\int_{\Omega} \max_{1 \leq i \leq n} [i^{1/p} \cdot (a_i f_i(\omega))^*] \, d\nu \leq \frac{p}{p-1} \left(\sum_{i=1}^{n} |a_i|^p \right)^{1/p}.$$

Here $\{(a_i f_i(\omega))^*\}_{i=1}^{n}$ *is the decreasing rearrangement of the sequence* $\{|a_i f_i(\omega)|\}_{i=1}^{n}$.

PROOF. We prove the inequality for the simplest realization of the random variables $\{f_i\}_{i=1}^{n}$: $\Omega = [0, 1]^n$, ν is the product of Lebesgue measures, and $f_i(\omega) = y_i^{-1/p}$, where $\omega = (y_1, \dots, y_n) \in \Omega$.

1991 *Mathematics Subject Classification*. Primary 46B20; Secondary 46E30, 46B42, 46A45.
Translation of Studies in the Theory of Functions of Several Real Variables, Yaroslav. Gos. Univ., Yaroslavl′, 1986, pp. 84–92; MR 88f:60032.

The change of variables $x_i = a_i^{-p} y_i$ gives us that

$$\begin{aligned}\int_\Omega & \max_{1\le i\le n} [i^{1/p}(a_i y_i^{-1/p})^*]\, dy_1 \cdots dy_n \\ &= \left(\prod_{i=1}^n |a_i|^p\right) \int_0^{a_n^{-p}} \int_0^{a_{n-1}^{-p}} \cdots \int_0^{a_1^{-p}} \max_{1\le i\le n} [i^{1/p}(x_i^{-1/p})^*]\, dx_1 \cdots dx_n \\ &= \left(\prod_{i=1}^n |a_i|^p\right) \int_0^\infty \mu(\{x = (x_1, \dots, x_n) \colon 0 \le x_i \le a_i^{-p}, \\ &\qquad\qquad \max_{1\le i\le n} [i^{1/p}(x_i^{-1/p})^*] > \lambda\})\, d\lambda.\end{aligned}$$

Let $P = \{x = (x_1, \dots, x_n) \colon 0 \le x_i \le a_i^{-p}\}$. Without loss of generality it can be assumed that $a_1 \ge \cdots \ge a_n > 0$. We prove that

$$\mu\left(\left\{x \in P \colon \max_{1\le i\le n} [i^{1/p}(x_i^{-1/p})^*] > \lambda\right\}\right) = \left(\prod_{i=1}^n a_i^{-p}\right) \cdot \left(\sum_{i=1}^n |a_i|^p\right) \cdot \lambda^{-p} \tag{2}$$

for $\lambda \ge n^{1/p} \cdot a_i$. Let Π_n be the set of all permutations of the numbers $1, \dots, n$, $A_\lambda = \{x \in P \colon \max_{1\le i\le n}[i^{1/p} \cdot (x_i^{-1/p})^*] \le \lambda\}$, and $C_k(B)$ the set of distinct samples of k elements of the set B. Since

$$\max_{1\le i\le n} [i^{1/p} \cdot (x_i^{-1/p})^*] = \min_{\pi\in\Pi_n} \max_{1\le i\le n} [i^{1/p} \cdot (x_{\pi(i)}^{-1/p})],$$

it follows that

$$A_\lambda = \bigcup_{\pi\in\Pi_n} \left\{x \in P \colon \max_{1\le i\le n} [i^{1/p} \cdot (x_{\pi(i)}^{-1/p})] \le \lambda\right\}.$$

Using the combinatorial formula for the measure of a union of sets, we get that

$$\mu(A_\lambda) = \sum_{k=1}^{n!} (-1)^{k+1} \sum_{\{\pi_j\}_{j=1}^k \in C_k(\Pi_k)} \mu\left(\left\{x \in P \colon x_i \ge \max_{1\le j\le k} \pi_j(i) \cdot \lambda^{-p}\right\}\right). \tag{3}$$

Since $\lambda \ge n^{1/p} \cdot a_1$, it follows that

$$\mu\left(\left\{x \in P \colon x_i \ge \max_{1\le j\le k} \pi_j(i) \cdot \lambda^{-p}\right\}\right) = \prod_{i=1}^k \left(a_i^{-p} - \max_{1\le j\le k} \pi_j(i) \cdot \lambda^{-p}\right).$$

From the last equality and (3),

$$\mu(A_\lambda) = \sum_{l=0}^n \frac{(-1)^{n-l}}{\lambda^{(n-l)p}} \sum_{\{i_m\}_{m=1}^l \in C_l(\{1, 2, \dots, n\})} b_{i_1, \dots, i_l}^{(n)} \cdot \left(\prod_{m=1}^l a_{i_m}^{-p}\right), \tag{4}$$

where

$$b_{i_1, \dots, i_l} = \sum_{k=1}^{n!} (-1)^{k+1} \sum_{\{\pi_j\}_{j=1}^k \in C_k(\Pi_n)} \prod_{i \notin \{i_m\}_{m=1}^l} \max_{1\le j\le k} \pi_j(i).$$

It is easy to see that the coefficients $b^{(n)}_{i_1,\dots,i_l}$ depend only on l and n. In what follows they will be denoted by $b^{(n)}_l$.

For $a_1 = \cdots = a_n = 1$ we get from (4) that

$$\mu\left(\left\{x \in \Omega\colon \max_{1\le i\le n}[i^{1/p}(x_i^{-1/p})^*] \le \lambda\right\}\right) = \sum_{l=0}^{n} b_l^{(n)} \cdot C_l^n \cdot (-1)^{n-l} \cdot \lambda^{(l-n)p},$$

where $C_l^n = n!/(l! \cdot (n-l)!)$.

On the other hand, for $\lambda > n^{1/p}$

$$\begin{aligned}\mu&\left(\left\{x \in \Omega\colon \max_{1\le i\le n}[i^{1/p} \cdot (x_i^{1/p})^*] \le \lambda\right\}\right)\\ &= n!\mu\left(\left\{x\in\Omega\colon 0 \le x_1 \le \cdots \le x_n \le 1;\ \max_{1\le i\le n}[i^{1/p}\cdot x_i^{-1/p}] \le \lambda\right\}\right)\\ &= n!\int_{n\lambda^{-p}}^{1}\int_{(n-1)\lambda^{-p}}^{x_n}\cdots\int_{\lambda^{-p}}^{x_2} dx_1\cdots dx_n.\end{aligned}$$

It is not hard to prove by induction that the last multiple integral is equal to $1/n! - 1/(\lambda^p \cdot (n-1)!)$. We get finally that for $\lambda > n^{1/p}$

$$\sum_{l=0}^{n} b_l^{(n)} \cdot C_l^n \cdot (-1)^{n-l} \cdot \lambda^{(l-n)p} = 1 - n\cdot\lambda^{-p}.$$

Comparing the coefficients of the same powers of λ, we have that

$$b_n^{(n)} = b_{n-1}^{(n)} = 1; \qquad b_l^{(n)} = 0, \quad 0 \le l \le n-2.$$

Let us substitute the coefficients found into (4). Then

$$\mu(A_\lambda) = \left(\prod_{i=1}^{n} a_i^{-p}\right)\cdot\left(1 - \lambda^{-p}\cdot\sum_{i=1}^{n} a_i^p\right). \tag{5}$$

This implies (2).

We consider values $\lambda < n^{1/p}\cdot a_i$. As usual, $\operatorname{card} B$ denotes the number of elements in a set B. Then

$$P\backslash A_\lambda = \bigcup_{k=1}^{n}\{x\in P\colon \operatorname{card}\{i\colon x_i^{-1/p} > \lambda\cdot k^{-1/p}\} \ge k\}$$

or

$$P\backslash A_\lambda = \bigcup_{k=1}^{n}\{x\in P\colon \operatorname{card}\{i\colon x_i < k\cdot\lambda^{-p}\} \ge k\}. \tag{6}$$

Let $\mathbf{\Phi}$ be the set of all mappings φ_λ defined on $\{1, \dots, n\}$ and satisfying the condition

$$\varphi_\lambda(i) \in \{1, 2, \dots, n, a_i^{-p}\cdot\lambda^p\}, \quad 1 \le i \le n.$$

Let

$$\Phi_\lambda^0 = \left\{ \varphi_\lambda \in \Phi_\lambda : \begin{array}{c} \exists k \in \{1, 2, \dots, n\} \text{ with} \\ \operatorname{card}\{i : \varphi_\lambda(i) < k ; \varphi_\lambda(i) \neq a_i^{-p} \cdot \lambda^p\} \geq k \end{array} \right\},$$

$$\varphi_\lambda'(i) = \begin{cases} \varphi_\lambda(i) - 1 & \text{if } \varphi_\lambda(i) = 1, 2, \dots, n; \\ n & \text{if } \varphi_\lambda(i) = a_i^{-p} \cdot \lambda^p. \end{cases}$$

From (6),

$$P \backslash A_\lambda = \bigcup_{\varphi_\lambda \in \Phi_\lambda^0} \{x \in P : \lambda^{-p} \cdot \varphi_\lambda'(i) \leq x_i < \lambda^{-p} \cdot \varphi_\lambda(i)\}. \tag{7}$$

We remark that to distinct φ_λ there correspond disjoint sets; therefore, for $\lambda \geq n^{1/p} \cdot a_1$

$$\mu(P \backslash A_\lambda) = \sum_{\varphi_\lambda \in \Phi_\lambda^0} \left(\prod_{i=1}^n (\lambda^{-p}(\varphi_\lambda(i) - \varphi_\lambda'(i))) \right).$$

Since $\lambda^{-p}(\varphi_\lambda(i) - \varphi_\lambda'(i))$ is equal either to λ^{-p} or to $(a_i^{-p} - n \cdot \lambda^{-p})$, we get after grouping similar terms in the last expression that

$$\mu(P \backslash A_\lambda) = \sum_{l=0}^n \lambda^{(l-n)p} \cdot \sum_{\{i_k\}_{k=1}^l \in C_l(\{1, 2, \dots, n\})} B_{i_1 \cdots i_l}^{(n)} \left(\prod_{k=1}^l (a_{i_k}^{-p} - n \cdot \lambda^{-p}) \right),$$

where

$$B_{i_1 \cdots i_l}^{(n)} = \operatorname{card} \left\{ \varphi_\lambda \in \Phi_\lambda^0 : \begin{array}{l} \lambda^{-p} \cdot \varphi_\lambda(i_k) = a_{i_k}^{-p}, \ i \leq k \leq l; \\ \lambda^{-p} \cdot \varphi_\lambda(j) \neq a_j^{-p}, \ j \neq i_k. \end{array} \right\}.$$

It is easy to see that the $B_{i_1 \cdots i_l}^{(n)}$ coincide for distinct samples for a fixed l, and we denote these coefficients by $B_l^{(n)}$ in what follows.

Accordingly, for $\lambda \geq a_i \cdot n^{1/p}$

$$\mu(P \backslash A_\lambda) = \sum_{l=0}^n B_l^{(n)} \cdot \lambda^{(l-n)p} \cdot \sum_{\{i_k\}_{k=1}^l \in C_l(\{1, 2, \dots, n\})} \left(\prod_{k=1}^n (a_{i_k}^{-p} - n \cdot \lambda^{-p}) \right). \tag{8}$$

The expression

$$\left(\prod_{i=1}^n a_i^{-p} \right) \cdot \left(\sum_{i=1}^n a_i^p \right) \cdot \lambda^{-p}$$

can be written analogously. Indeed,

$$\left(\prod_{i=1}^n a_i^{-p} \right) \cdot \left(\sum_{i=1}^n a_i^p \right) \cdot \lambda^{-p} = \sum_{i=1}^n \mu(\{x \in P : 0 \leq x_i \leq \lambda^{-p}\}).$$

Using the set Φ_λ, we get that

$$\left(\prod_{i=1}^n a_i^{-p} \right) \cdot \left(\sum_{i=1}^n a_i^p \right) \cdot \lambda^{-p} = \sum_{\varphi_\lambda \in \Phi_\lambda} \mu(\{x \in P : x_i \in [\varphi_\lambda'(i) \cdot \lambda^{-p}, \varphi_\lambda(i) \cdot \lambda^p]\}) \times \operatorname{card}\{i : \varphi_\lambda(i) = 1\}. \tag{9}$$

For $\lambda \geq a_1 \cdot n^{1/p}$

$$\left(\prod_{i=1}^{n} a_i^{-p}\right)\cdot\left(\sum_{i=1}^{n} a_i^{p}\right)\cdot\lambda^{-p} = \sum_{\varphi_\lambda\in\Phi_\lambda}\left(\prod_{i=1}^{n}(\lambda^{-p}(\varphi_\lambda(i)-\varphi'_\lambda(i)))\cdot\operatorname{card}\{j\colon \varphi_\lambda(j)=1\}\right).$$

Grouping similar terms, we get that

$$\begin{aligned}\left(\prod_{i=1}^{n} a_i^{-p}\right)&\cdot\left(\sum_{i=1}^{n} a_i^{p}\right)\cdot\lambda^{-p}\\ &= \sum_{l=0}^{n}\lambda^{(l-n)p}\cdot\mathscr{D}_l^{(n)}\cdot\sum_{\{i_k\}_{k=1}^{l}\in C_l(\{1,2,\dots,n\})}\left(\prod_{k=1}^{l}(a_{i_k}^{-p}-n\cdot\lambda^{-p})\right), \quad (10)\end{aligned}$$

where

$$\mathscr{D}_l^{(n)} = \sum_{m=1}^{l} m\cdot\operatorname{card}\left\{\varphi_\lambda\in\Phi_\lambda\colon \begin{array}{c}\lambda^{-p}\varphi_\lambda(i)=a_i^{-p},\ i=1,2,\dots,l;\\ \lambda^{-p}\varphi_\lambda(j)\neq a_j^{-p},\ j=l+i,l+2,\dots,n;\\ \operatorname{card}\{i\colon\varphi_\lambda(i)=1\}=m.\end{array}\right\}.$$

By (2), the polynomials (8) and (10) coincide, and are independent of the values $\{a_i\}_{i=1}^n$. Therefore, $\mathscr{D}_l^{(n)} = B_l^{(n)}$ for $l = 0, 1, 2, \dots, n$.

The last equality means that there exists a mapping $F\colon \Phi_\lambda^0 \to \Phi_\lambda$ with the following properties:

1) $(F\varphi_\lambda)(i) \in \{1, \varphi_\lambda(i)\}$, and if $\varphi_\lambda(i) = a_i^{-p}\cdot\lambda^p$, then $(F\varphi_\lambda)(i) = a_i^{-p}\cdot\lambda^p$;
2) $\operatorname{card}\{\varphi_\lambda\in\Phi_\lambda^0\colon F\varphi_\lambda=\overline{\varphi}_\lambda\} = \operatorname{card}\{i\colon \overline{\varphi}_\lambda(i)=1\}$ for an arbitrary fixed $\overline{\varphi}_\lambda\in\Phi_\lambda$.

In view of property 1), for all λ

$$\begin{aligned}\mu(\{x\in P\colon [F\varphi_\lambda]'(i)\cdot\lambda^{-p}\leq x_i<[F\varphi_\lambda](i)\cdot\lambda^{-p}\})&\\ \geq \mu(\{x\in P\colon \lambda^{-p}\cdot\varphi'_\lambda(i)\leq x_i<\lambda^{-p}\cdot\varphi_\lambda(i)\}).&\end{aligned}$$

By (7), (9), and property 2), the last inequality gives us that for all λ

$$\mu(P\backslash A_\lambda)\leq\left(\prod_{i=1}^{n} a_i^{-p}\right)\cdot\left(\sum_{i=1}^{n} a_i^{p}\right)\cdot\lambda^{p}.$$

Finally,

$$\begin{aligned}\int_{\Omega}\max_{1\le i\le n}[i^{1/p}\cdot(a_i\cdot y_i^{-1/p})^*]\,dy_i\cdots dy_n &= \left(\prod_{i=1}^n a_i^p\right)\cdot\left(\int_0^{(\sum_{i=1}^n a_i^p)^{1/p}}\mu(P\backslash A_\lambda)\,d\lambda+\int_{(\sum_{i=1}^n a_i^p)^{1/p}}^{\infty}\mu(P\backslash A_\lambda)\,d\lambda\right)\\ &\le\left(\prod_{i=1}^n a_i^p\right)\left[\left(\prod_{i=1}^n a_i^{-p}\right)\cdot\left(\sum_{i=1}^n a_i^p\right)^{1/p}\right.\\ &\qquad\left.+\left(\prod_{i=1}^n a_i^{-p}\right)\cdot\left(\sum_{i=1}^n a_i^p\right)\cdot\int_{(\sum_{i=1}^n a_i^p)^{1/p}}^{\infty}\lambda^{-p}\,d\lambda\right]\\ &=\frac{p}{p-1}\left(\sum_{i=1}^n a_i^p\right)^{1/p}.\end{aligned}$$

Thus, the proof of Theorem 1 is complete. We note that the constant $p/(p-1)$ is sharp, since for $a_1=\cdots=a_n=1$, by the equality

$$\mu\left(\left\{x\in\Omega\colon\max_{1\le i\le n}[i^{1/p}(x_i^{-1/p})^*]>\lambda\right\}\right)=n\cdot\lambda^{-p},\qquad\lambda\ge n^{1/p},$$

we get that

$$\int_{\Omega}\max_{1\le i\le n}[i^{i/p}\cdot(x_i^{-1/p})^*]\,dx_1\cdots dx_n=n^{1/p}+\int_{k^{1/p}}^{\infty}n\cdot\lambda^{-p}\,d\lambda=\frac{p}{p-1}\cdot n^{1/p}.$$

To formulate a geometric consequence of Theorem 1 we introduce the following definition.

DEFINITION 1. Let X be a Banach function space, and let A and B be sequence spaces. We say that X is (A, B)-convex if there exists a constant M such that

$$\|\,\|x_i(t)\|_A\|_X\le M\cdot\|\,\|x_i\|_X\|_B$$

for all finite sequences $\{x_i\}_{i=1}^n$ in X. The inside norm in the expression $\|\,\|x_i(t)\|_A\|_X$ is taken with respect to i for fixed t, and hence $\|x_i(t)\|_A$ is a function in X, and a finite sequence is completed to an infinite sequence in the standard way.

We note that (l_p, l_p)-convexity coincides with the concept of p-convexity, and (l_∞, l_p)-convexity coincides with the concept of satisfying an upper p-estimate.

Denote by $l_{p,\infty}$ the space of sequences $x=\{x_k\}_{k=1}^\infty$ with the norm $\|x\|_{l_{p,\infty}}=\sup_n n^{1/p-1}\sum_{k=1}^n x_k^*$. We note that for $p\in(1,\infty)$ the norm in $l_{p,\infty}$ is equivalent to the functional $G(x)=\sup_k k^{1/p}x_k^*$. Here $\{x_k^*\}_{k=1}^\infty$ is a rearrangement of the sequence $\{x_k\}_{k=1}^\infty$ in decreasing moduli.

COROLLARY 1. *Let $1<p<\infty$. If the Banach function space X is $(l_\infty, l_{p,\infty})$-convex, then X is p-convex.*

The following is an equivalent formulation of Corollary 1.

If for every $x \in X$ and every partition $[0, 1] = \bigcup_k e_k$ of $[0, 1]$ into disjoint subsets the inequality $\|x\|_X \le C_1 \| \, \|x \cdot \chi_{e_k}\|_X \|_{e_{p,\infty}}$ holds with a constant c_1 independent of x and e_k, then the space X is p-convex.

PROOF. Let $\{f_i\}_{i=1}^{\infty}$ be the sequence of random variables on (Ω, Σ, ν) defined in Theorem 1. Since X is $(l_\infty, l_{p,\infty})$-convex, it follows that

$$\left\| \sup_{1\le i\le n} |f_i(\omega)x_i| \right\|_X \le M \cdot \max_{1\le i\le n} i^{1/p} \cdot (\|f_i(\omega)x_i\|_X)^*. \tag{11}$$

From (1),

$$\begin{aligned}\left\| \left(\sum_{i=1}^n |x_i|^p \right)^{1/p} \right\|_X &= K \left\| \int_\Omega \sup_{1\le i\le n} |f_i(\omega)x_i| \, d\nu \right\|_X \\ &\le K \int_\Omega \left\| \sup_{1\le i\le n} |f_i(\omega)\cdot x_i| \right\|_X d\nu.\end{aligned}$$

Using (11) and Theorem 1, we have that

$$\begin{aligned}\int_\Omega \left\| \sup_{1\le i\le n} |f_i(\omega)x_i| \right\|_X d\nu &\le M \int_\Omega \max_{1\le i\le n} [i^{i/p} (\|f_i(\omega)x_i\|_X)^*] \, d\nu \\ &\le M \frac{p}{p-1} \left(\sum_{i=1}^n \|x_i\|_X^p \right)^{1/p}.\end{aligned}$$

Corollary 1 is proved.

Since $l_{p,\infty} \subset l_r$ for $r > p$, it follows that Corollary 1 is an improvement of the result that satisfying an upper r-estimate implies that X is p-convex ([2], p. 85).

COROLLARY 2. *Let* $1 < p < \infty$. *If a Banach lattice* X *satisfies an upper p-estimate, then there exists a constant* C *such that*

$$\left\| \left(\sum_{i=1}^n |x_i|^p \right)^{1/p} \right\|_X \le C(1 + \ln n)^{1/p} \left(\sum_{i=1}^n \|x_i\|_X^p \right)^{1/p}$$

for every finite collection $\{x_i\}_{i=1}^n$ *of elements in* X.

PROOF. Using (1), Theorem 1, and the obvious inequality

$$\left(\sum_{i=1}^n |a_i|^p \right)^{1/p} \le C_1 \cdot (1 + \ln n)^{1/p} \cdot \max_{1\le i\le n} (i^{1/p} \cdot a_i^*),$$

we get that

$$\begin{aligned}\left\| \left(\sum_{i=1}^{n} |x_i|^p \right)^{1/p} \right\|_X &\leq K \int_\Omega \left\| \max_{1\leq i\leq n} |f_i(\omega)x_i| \right\|_X d\nu \\ &\leq KM \int_\Omega \left(\sum_{i=1}^{n} \|f_i(\omega)x_i\|^p \right)^{1/p} d\nu \\ &\leq KMC_1(1+\ln n)^{1/p} \int_\Omega \max_{1\leq i\leq n} i^{1/p} \|f_i(\omega)x_i\|_X^* \, d\nu \\ &\leq KMC_1 \frac{p}{p-1}(1+\ln n)^{1/p} \left(\sum_{i=1}^{n} \|x_i\|_X^p \right)^{1/p}.\end{aligned}$$

Corollary 2 is proved.

We remark that the power $1/p$ cannot be decreased. Indeed, consider the Marcinkiewicz space $L_{p,\infty}$ with $\|f\|_{L_{p,\infty}} = \sup_{0\leq t\leq 1}(\int_0^t f^*(\tau)\,d\tau)\cdot t^{(1/p)-1}$. It is known that $L_{p,\infty}$ satisfies an upper p-estimate. Let

$$x_i = \sum_{j=1}^{n} \frac{1}{i+j \pmod n} \cdot \kappa_{((j-1)/n,\, j/n)}.$$

Then it is easy to see that

$$\left\| \left(\sum_{i=1}^{n} |x_i|^p \right)^{1/p} \right\|_{L_{p,\infty}} \geq C \cdot (\ln n)^{1/p} \cdot \left(\sum_{i=1}^{n} \|x_i\|_{L_{p,\infty}}^p \right)^{1/p}.$$

BIBLIOGRAPHY

1. B. Maurey and G. Pisier, *Séries de variables aléatoires vectorielles indépendantes et propriétés géométriques des espaces de Banach*, Studia Math. **58** (1976), 45–90.

2. J. Lindenstrauss and L. Tzafriri, *Classical Banach Spaces*. II. *Function Spaces*, Ergebnisse der Mathematik und ihrer Grenzgebiete **97**, Springer-Verlag, Berlin-New York, 1979, x+243 pp.

Voronezh University

Translated by H. H. McFADEN